CROP GROWTH AND CULTURE

CROP GROWTH AND CULTURE

ROGER L. MITCHELL

The Iowa State University Press · Ames

❧ ❧ ❧ To Joyce, Laura, Susan, Sarah, and Martha

❧ ROGER L. MITCHELL is Chairman of the Department of Agronomy at the University of Missouri. He received his B.S. and Ph.D. degrees at Iowa State University, where he taught for several years before going to his present position. He is a member of numerous professional and honorary organizations in the field of agronomy and was a Danforth Fellow, 1956–61. He has written articles for several agronomy and education journals.

Composed and printed by The Iowa State University Press. FIRST EDITION, 1970. International Standard Book Number: 0–8138–0377–2. Library of Congress Catalog Card Number: 72–88006. Second Printing, 1972. Third Printing, 1975. Fourth Printing, 1976, Fifth Printing, 1977. Sixth Printing, 1979.

CONTENTS

PREFACE

❧ THE UNIQUE CONTRIBUTION of agronomy as a discipline, represented by the subdivisions of crops, soils, and climatology, is to provide the integration of biological, chemical, and physical knowledge into useful production systems. With the present trend of many biological scientists to move more and more in the direction of molecular biology (the reductionist approach), it becomes increasingly imperative for the agronomist to work toward synthesis and composition. The first goal of this text, then, is to strive for this integration. Secondly, the present treatment moves away from the crop-by-crop approach followed in many crops texts. Instead, the subject matter is developed on the basic concepts and common factors of growth and culture among crops. This approach is intended to point up the key principles that may find useful application with many crops. Basic concepts in agronomy are often an integration or synthesis of still other basic concepts from various disciplines such as biochemistry and plant physiology.

This discussion thus has two major purposes:

1. To develop an understanding of the important principles underlying the practices used in the culture of crop plants.

2. To develop the ability to apply these principles to production situations.

During the past decade there has been a desirable emphasis on broadening agricultural curricula to include more courses in the humanities, social sciences, and business administration. Partly because of this, students come to agronomy without the useful prerequisite in plant physiology and with the likelihood of not getting full courses in weed science, plant pathology, and economic entomology. Therefore, an effort is made to treat these subjects in this text.

This approach to crop physiology and production was developed in outline form with Franklin P. Gardner in 1963. To him I wish to express appreciation for the many hours of planning and developing we spent together on this course outline. Fertile and provocative discussions with Jack Tanner provided expansion and depth, especially for several of the early chapters, and for these I am grateful.

I also thank Robert Loomis, Dale Smeltzer, and A. W. Burger for a most helpful critique of the entire manuscript; and my colleagues I. C. Anderson, C. J. deMooy, E. R. Duncan, F. P. Gardner, Detroy Green, Frank Schaller, D. L. Stamp, David Staniforth, and D. G. Woolley for reading chapters and offering useful comments. My appreciation is extended to Mrs. Ina Couture and Mrs. Ida Morgan for typing the manuscript and to Mrs. Sharon Hendricks for several drafts. The figures were enthusiastically prepared by Tom Kelly and Larry Barr.

CROP GROWTH AND CULTURE

❧ CHAPTER ONE ❧ THE LEAF AND LIGHT ENERGY TRANSFORMATION

❧ AGRICULTURE is basically a system of exploiting photosynthesis. Photosynthesis serves as the primary source of all energy for mankind—from the food he eats and the feed for his livestock to the fuel that powers his heating plants and automobiles. A study of crop growth and culture is built on the fact that the yield of agricultural plants ultimately depends on the size and efficiency of this photosynthetic system. All other crop management practices proceed from this point. Because photosynthesis is the cornerstone of crop production it is important to develop an understanding of the energy available to drive photosynthesis, how this energy is stored and released, and finally to consider how the anatomical features of the leaf and the biochemical processes in the plant interact to capture and store the radiant energy.

ENERGY TRANSFORMATIONS

All the biological energy on the earth comes initially from the nuclear fusion reactions in the sun (Fig. 1.1). The radiant energy released as hydrogen is fused into helium and provides the driving force to form carbohydrate (e.g. glucose) from carbon dioxide and water. This synthesis of carbon dioxide and water into hydrocarbon chains is an energy-requiring (endergonic) reaction in which there is a large increase in free energy and a decrease in disorganization (entropy) when the energy of the sun (radiant energy) is fixed into the chemical bond of the carbon atom. Through this process the high-grade energy of the sun becomes the medium-grade energy of organic molecules. This

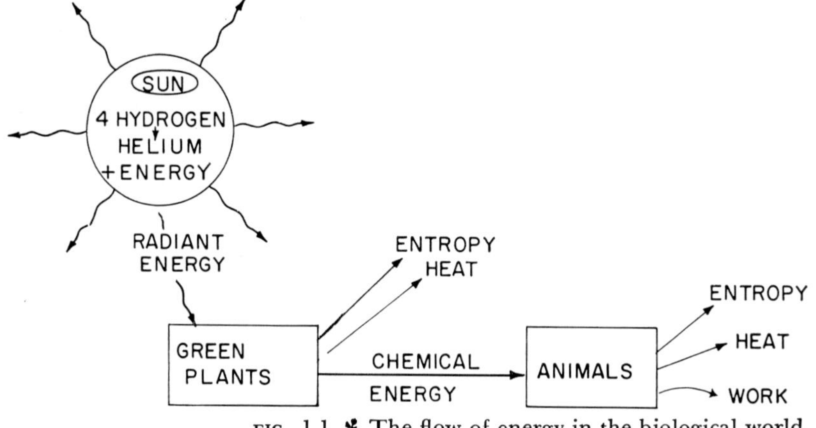

FIG. 1.1 The flow of energy in the biological world.

medium-grade energy is dissipated as low-grade energy in the form of heat which finally becomes useless as entropy (randomness of molecules). Photosynthesis therefore represents a tightly integrated and complex system for capturing radiant energy and reversing the trend toward high entropy.

The radiant energy released from the sun covers a much wider spectrum than that used in photosynthesis. Some of this radiation is filtered out by the layers of carbon dioxide, ozone, and water above the earth's surface; the remainder falling on the earth can provide a source of heat or energy to drive photosynthesis. That portion of the spectrum used in photosynthesis is absorbed primarily by chlorophyll in green plants, although other organisms do have unique pigment systems that function in a manner similar to the chlorophyll system. With the energy captured by chlorophyll, synthesis reactions occur which combine carbon dioxide and water into stored energy forms. An analysis of plant dry matter emphasizes the tremendous contribution carbon dioxide and water make to final yield, with as much as 90–95% of the weight synthesized from these two substances. In other words, only 5–10% comes from the mineral nutrients and nitrogen, which must be purchased as part of the soil resource or in commercial fertilizer, whereas the carbon dioxide and water are available at no direct cost, except in irrigated agriculture.

What is photosynthesis? It can be defined as photoreduction in which water serves as the ultimate donor of hydrogen to reduce the valence of carbon in carbon dioxide as expressed in the following equation:

$$6\ CO_2 + 12\ H_2O^* \xrightarrow[\text{chlorophyll}]{\text{light}} C_6H_{12}O_6 + 6\ O_2^* + 6\ H_2O$$

(* designates the observation that oxygen released in the process comes from the water)

THE LEAF

The leaf serves as the major photosynthetic organ of higher plants, and evolution has provided a structure that will both withstand the rigors of the environment and yet provide effective light absorption for photosynthesis. The leaf has (1) a large external surface, 2) extensive internal surface and interconnected air spaces, (3) an abundance of chloroplasts, particularly in the palisade cells, and (4) a close relation between the vascular and photosynthetic cells (Fig. 1.2). A leaf ideal for gaseous exchange and light interception might be only one cell thick, but the rigors of the natural environment demand several layers of cells for survival.

Each leaf part has evolved to a modification lending efficiency to its function. The epidermis serves as a barrier to water loss, primarily because this single layer of cells is covered by a waxy layer of cuticle. Both the cuticle and the epidermis are nearly transparent and allow light to enter the leaf readily. Palisade cells are rectangular in shape, and their long axis is commonly oriented perpendicular to the leaf surface. Each palisade cell contains a large number of chloroplasts. During photosynthesis, especially under high light intensity, these chloroplasts often congregate along the palisade cell wall.

The spongy mesophyll cells have interconnected spaces between

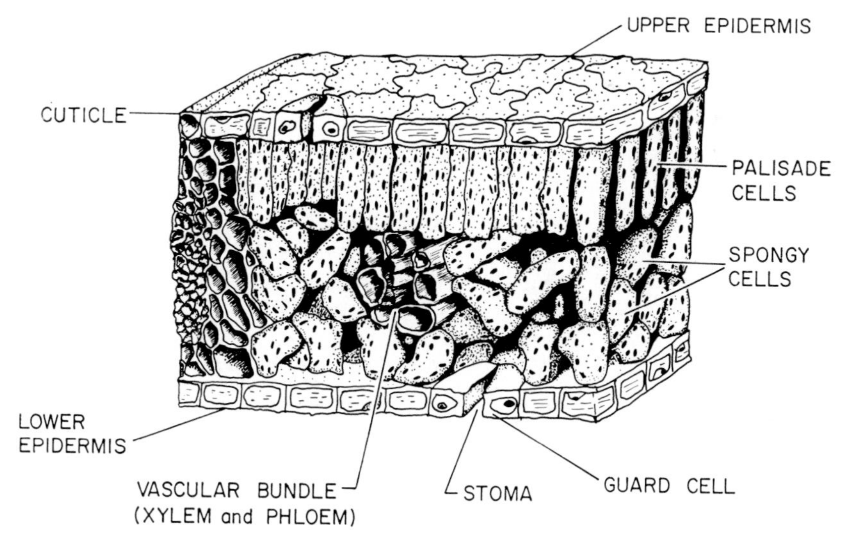

FIG. 1.2 ☙ A perspective view of the leaf. Note shape differences of palisade and spongy mesophyll cells.

them which permit carbon dioxide to move easily to the palisade cells and the area of carbon dioxide use. On the surface of the palisade cells, carbon dioxide is dissolved in water (available from the xylem) and passes through to the chloroplasts. Thus the chloroplasts are well placed for both light interception and gaseous supply.

In leaves with parallel veins stomates are arranged in rows, but in netted vein leaves the stomates are scattered. Guard cells surround the stomatal openings and represent the only epidermal cells which contain chlorophyll. In addition, the stomates create the only intercellular spaces in the epidermis. When the stomates are open, both carbon dioxide and water move through them freely. Relative humidity is commonly 100% in the plant leaf and less than 100% outside. This is conducive to rapid transpiration and allows transpiration to continue even when the plant's water supply is limited. This water loss may be detrimental to final crop yield. Thus extensive research is under way seeking ways to reduce water loss by transpiration (Zelitch, 1963).

Bulliform cells commonly appear in the grass leaf, either concentrated at the midrib or spread across the leaf. These large thin-walled cells collapse under drouth stress, causing the leaf to fold or roll; as a result, water loss is reduced. However, bulliform cells may actually be more important as a leaf unrolling mechanism which keep the leaf flat when the cells are turgid and environmental conditions favor maximum photosynthetic rates.

The composition of a typical leaf palisade cell (Table 1.1) serves to place the numerical frequency of the several cellular parts in perspective. Chloroplasts can be removed from the cell (as can other cell particles) and studied independently. Plant breeders and physiologists have shown that differences are observable in the rate of adenosine triphosphate (ATP) formation of illuminated chloroplasts isolated from different strains and varieties of a species (Kleese, 1966; Miflin and Hageman, 1966). Thus a very basic component of the photosynthetic apparatus is shown to differ from one genetic line to another.

TABLE 1.1. Subcellular particles in typical plant cell (after Bonner and Varner, 1965)

Subcellular Particle	Diameter	Number/Cell
Nucleus	5–20 μ	1
Chloroplasts	5–20 μ	50–200
Mitochondria	1–5 μ	500–2,000
Ribosomes	250 Å	$5\text{–}50 \times 10^5$
Enzyme molecules	20–100 Å	$5\text{–}50 \times 10^8$

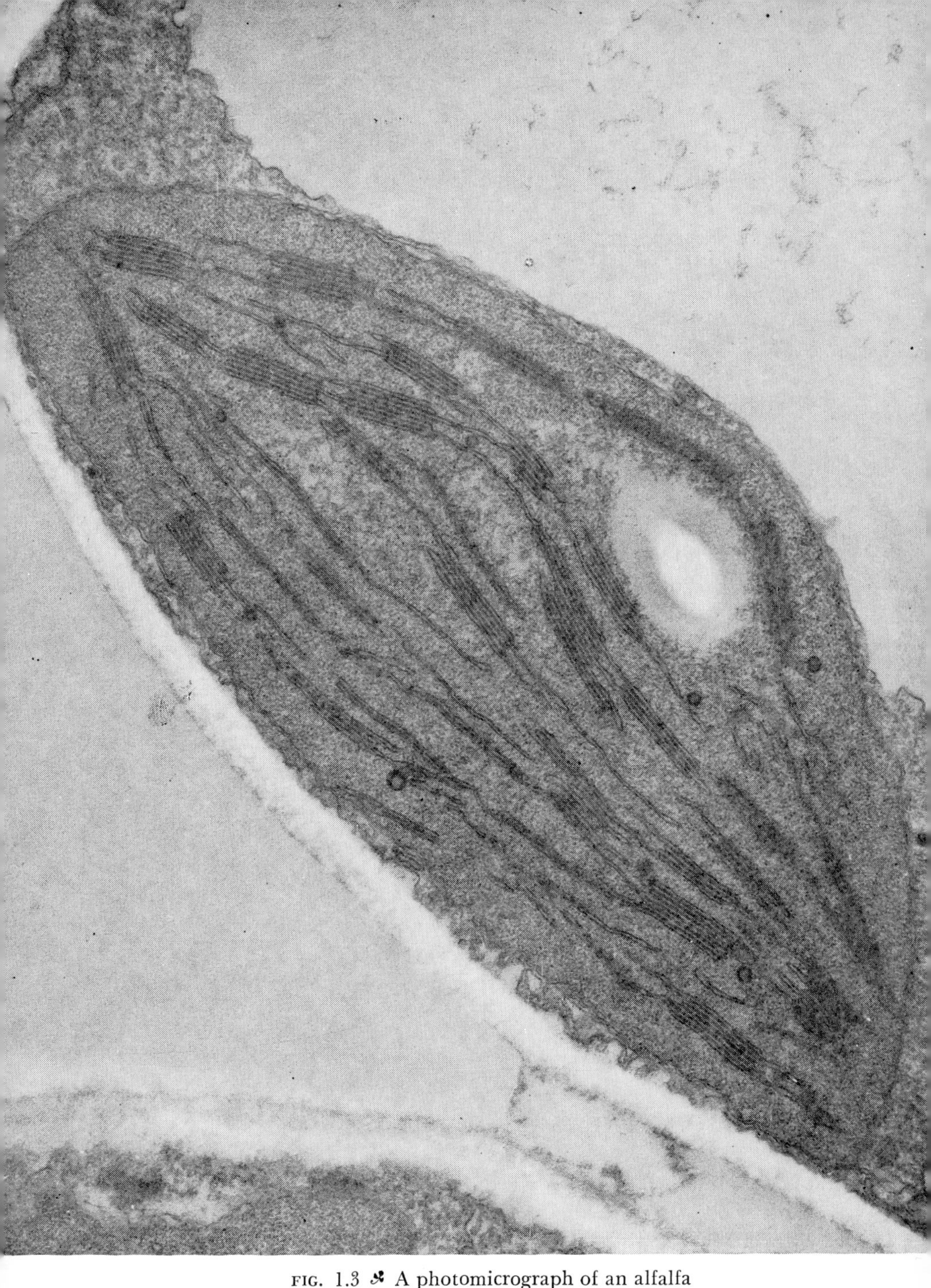

FIG. 1.3 ☙ A photomicrograph of an alfalfa chloroplast enlarged 64,500 times (Stifel et al., 1968).

This observation may provide the breeder a tool by which he can select for higher productivity lines at an early and preliminary level.

PHOTOSYNTHETIC APPARATUS

In recent years electron microscopy has made it possible to look more closely at the photosynthetic apparatus of the plant. The chloroplast of alfalfa (Fig. 1.3) displays two key areas: (1) the grana, concentrated areas of chlorophyll in which the light reaction occurs, and (2) the stroma, less dense light-colored areas where the reduction of carbon dioxide occurs in the dark reaction. The chloroplast is an elliptically shaped body 1–10 microns across and is usually characterized by a series of lamellae piled one upon the other to form grana (Fig. 1.4). A granum consists of layers of protein, chlorophyll, and phospholipids, all tightly integrated and capable of facilitating electron transport readily.

On a dry weight basis, chloroplasts contain:

	%
Protein, structural	29–32
Protein, stroma	16–18
Phospholipids	21–22
RNA	5
Chlorophyll a + b (3:1)	5–10
Carotenoids	1–2

The primary pigment chlorophyll, a magnesium-porphyrin compound, is concentrated in the grana. The chlorophyll molecule consists of four pyrol rings joined together with a magnesium ion at their center and an alcohol tail of phytol (Fig. 1.5). The alternating single and double bonds in the porphyrin ring give chlorophyll its unique light absorption characteristics. This system is effective in capturing light energy and converting it to potential chemical energy.

PHOTOSYNTHESIS

As a first step to understanding the light absorption mechanism, it is important to review the nature of light. Light moves in discrete packets called photons. As each photon strikes the chlorophyll molecule, it excites an electron, raises the electron to a high energy level, and makes it capable of transferring this energy to other compounds in the photosynthetic system. Different photons (which may be broadly

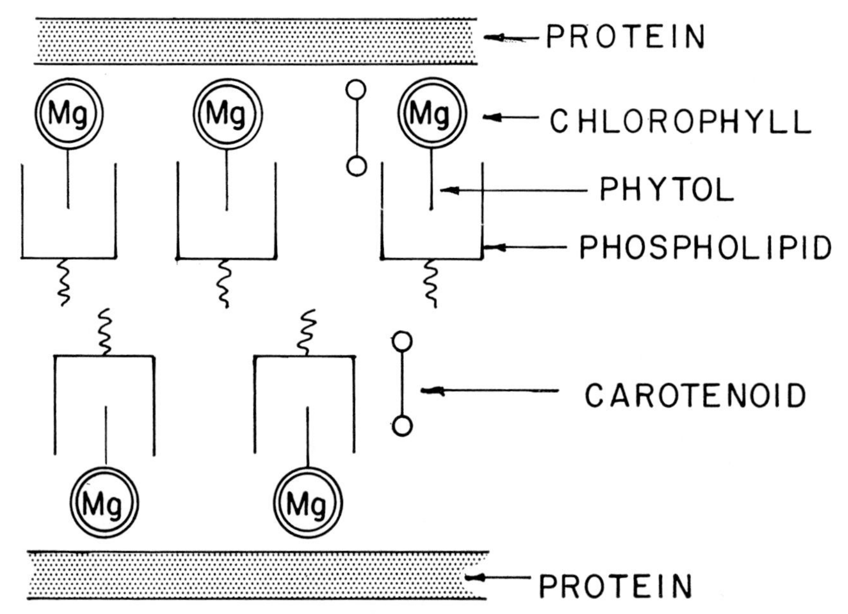

FIG. 1.4 ❧ The cross section of a lamella, describing the close relationship of the several components of the photosynthetic apparatus.

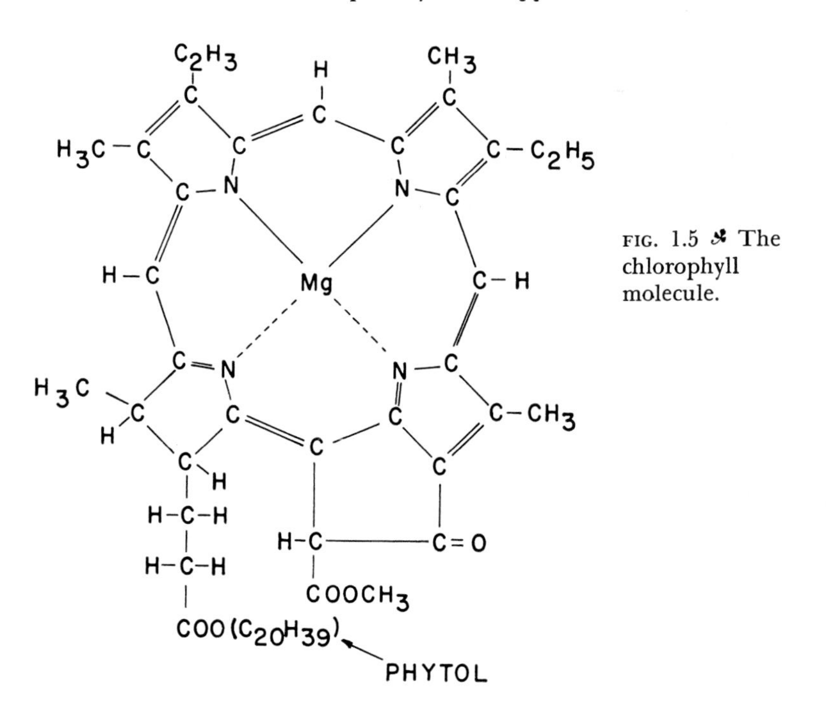

FIG. 1.5 ❧ The chlorophyll molecule.

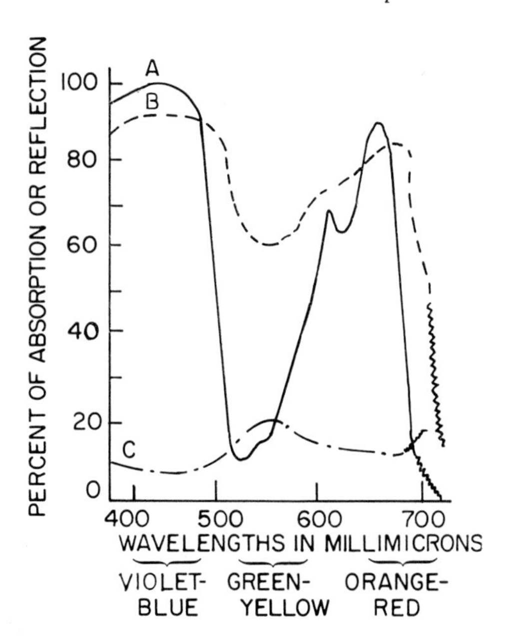

FIG. 1.6 ❧ Absorption and reflection of light by chlorophyll and by an intact leaf. Curve A (—): Absorption of light by chlorophyll in solution. Curve B (– – –): Absorption by a leaf with the same concentration of chlorophyll. Chlorophyll in the leaf absorbs several times more green light than the same chlorophyll in solution. Curve C (–.–): The green color of a leaf is due partially to a slightly greater reflection of green than of other light (Wilson and Loomis, 1967).

considered analogous to wave lengths) have differing effects on the degree of electron excitation they cause. The photons in the red range have just enough energy to excite an electron and this provides a high efficiency ratio. Blue light in contrast contains much more energy—in fact, more energy than needed to excite an electron—and this results in a low energy efficiency ratio because a single photon of blue light still performs the singular step of exciting an electron. The action spectrum of visible light on a green leaf is described in Figure 1.6. The absorption spectrum of chlorophyll *a* in ether differs from the action spectrum by less reduction in apparent activity in the green range, but emphasis on greater action and light absorption in the blue and red ranges is similar.

Carotenoids serve as auxiliary pigments in the light utilization process. These carotenoids play a small role in absorbing light energy and transferring it to chlorophyll. In addition, the carotenoids appear to have the capacity to absorb excess energy absorbed by chlorophyll and thus slow the rate of chlorophyll photodestruction.

An increased light intensity may not increase the plant's photosynthetic rate. F. F. Blackman demonstrated in 1905 that above a certain intensity, photosynthesis could not be accelerated by increasing the intensity of illumination. He suggested this to be evidence for a dark (nonphotochemical) part of the synthesis reaction. Subsequently it was shown that the photosynthetic yield from flashing light of high intensity could be increased to a maximum by lengthening the dark period between flashes to about 0.06 seconds.

More recent studies have shown that the light reaction produces a "chemical potential" which subsequently generates reduced nicotinamide-adenine dinucleotide ($NADPH_2$) and adenosine triphosphate (ATP). $NADPH_2$ serves as a very effective hydrogen transport system, and because of its high oxidation potential, it is one of the most powerful reductants known in biological systems. ATP is synonymous with available energy in the biological system. The ATP molecule has one more phosphate group than ADP (adenosine diphosphate). Solar or food energy is required to combine ADP and bring together the phosphates in ATP. Conversely, when the phosphate group is released from ATP, it can attach to some material such as a 6-carbon (hexose) sugar and raise its energy level, thus allowing it to undergo more chemical reactions than before it was phosphorylated.

Generation of ATP during the light reaction may occur through (1) cyclic photophosphorylation (Fig. 1.7) or (2) noncyclic photophosphorylation. During cyclic photophosphorylation, an electron in chlorophyll is raised to an "excited" state, captured by a series of carriers (possibly cytochromes b_6 and f), and passed through a "circuit" of cytochrome, losing energy along the way. From this process two ATPs are formed. Noncyclic photophosphorylation is part of the light reaction. In this case, both ATP and $NADPH_2$ are formed (Fig. 1.8). $NADPH_2$ generated by these steps is available for the reduction

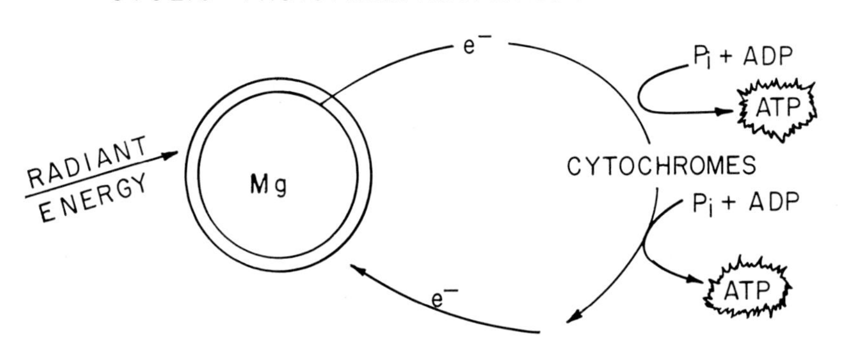

FIG. 1.7 ☙ Cyclic photophosphorylation. Direct generation of ATP following chlorophyll excitation by radiant energy.

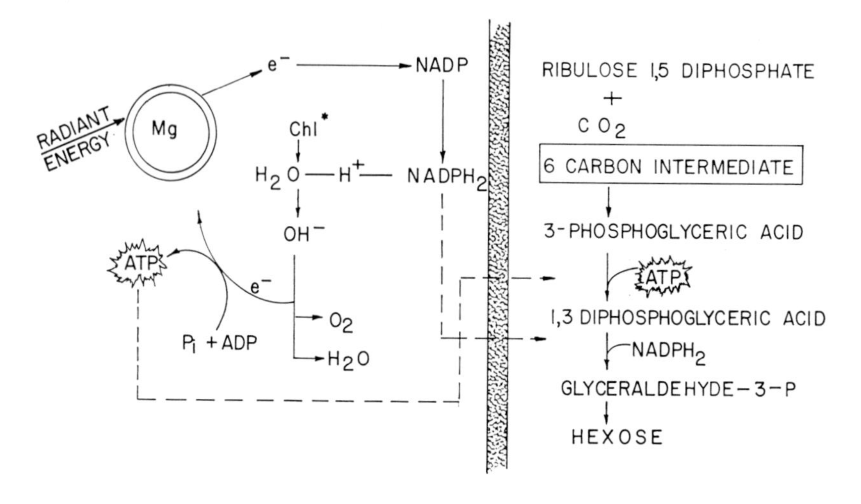

FIG. 1.8 ❧ Complete photosynthesis. The light reaction of photosynthesis results in the generation of ATP and $NADPH_2$, which are used in the dark reaction to reduce CO_2 to form glyceraldehyde-3-P.

of CO_2 to carbohydrates. $NADPH_2$ may also function to reduce $SO_4^{=}$ and NO_3^- to a form usable by the plant. The OH^- ions produced may combine to form peroxide (H_2O_2) which can be split into water and O_2 by the peroxidase enzyme.

The dark reaction consists of the synthesis of glucose from CO_2 and H_2. It does not require light but does require the ATP and $NADPH_2$ produced in the light process. The pathway of CO_2 during fixation is not directly to glucose. It first attaches to ribulose-1,5-diphosphate producing a 6-carbon product which is very unstable. This 6-carbon molecule cleaves into the 3-carbon phosphoglyceric acid to which phosphate and hydrogen are added. The ATP formed in the light reaction is used to energize the phosphoglyceric acid. This highly energized 1,3 diphosphoglyceric acid is then capable of accepting the hydrogen from $NADPH_2$ and reduction occurs. From the 3-carbon glyceraldehyde-3-P glucose is finally formed.

Thus the photosynthetic process consists of three main stages: (1) the removal of hydrogen atoms from water and the production of oxygen molecules; (2) the transfer of the hydrogen atoms to an intermediate compound; (3) the use of the hydrogen atoms to convert carbon dioxide into a carbohydrate.

ENVIRONMENTAL EFFECTS ON PHOTOSYNTHESIS

To integrate these ideas and move from the cellular to the organismal level of activity, it is useful to consider how the organism responds to environmental factors. The photosynthetic process may be viewed as several partial processes: (1) a photochemical process resulting in the conversion of radiant energy to chemical energy—ATP and $NADPH_2$—which are used to reduce CO_2 to carbohydrate; (2) processes involved in transporting CO_2 from the external air toward the chloroplast reaction center; and (3) biochemical processes preceding and following CO_2 reduction. The direct effects of light intensity, CO_2 concentration, and temperature on the rates of the photochemical, transport, and biochemical processes differ. The photochemical process is affected by light intensity only. The transport of CO_2 from external air to the chloroplast is essentially a diffusion process. This diffusion rate is mainly dependent on the difference in CO_2 concentration in the external air and the chloroplast; it is only slightly affected by temperature. The biochemical processes are mainly affected by temperature.

At low light intensities, the photochemical process limits photosynthesis in cultivated plants. With normal CO_2 (300 ppm) and light intensity equivalent to full sunlight, leaf photosynthesis is strongly affected by variation of the external CO_2 concentration. Temperature has only a slight effect over a wide range; therefore, the diffusion process is limiting. At saturating light intensities and high CO_2 (1,300 ppm), photosynthesis is strongly affected by temperature because the biochemical processes are limiting. Under field conditions, light intensity and CO_2 concentration appear to be low enough to limit photosynthesis through the photochemical and diffusion processes over a wide range of temperatures (Hesketh, 1963).

USE OF ENERGY STORED DURING PHOTOSYNTHESIS (RESPIRATION)

The generation of adenosine triphosphate by the photosynthetic process may be likened to a charged storage battery. This energy may be released to other molecules which in turn become energized and more reactive.

STORAGE BATTERY

ATP (A-P-P-P) *"charged"* ADP + P (A-P P + P) *"discharged"*

The rapid combustion of stored energy in an engine gives energy release as heat and converts the stored energy to useful work at approximately 35% efficiency. In contrast, digestion and respiration in biological systems are slow energy releases. Substrates are energized by phosphorylation and the entire process is done in many small steps at isothermal conditions. For example, glucose is converted to CO_2 in approximately 30 steps.

$$C_6H_{12}O_6 + 6\ O_2 \xrightarrow[\text{steps}]{30} 6\ CO_2 + 6\ H_2O + 673\ \text{Kcal}$$

(Potential energy)

These steps include sequences referred to as glycolysis and the Krebs cycle. Together these cycles are known to generate 38 ATP. Since 1 ATP = 12 Kcal, the 38 ATP generate 456 Kcal or

$$\frac{456\ \text{(actual)}}{673\ \text{(potential)}} = 68\%\ \text{efficiency.}$$

Carbohydrate utilization occurs in part in the cytoplasm but is completed in the mitochondrion. Conversion of the carbon structure to the 3-carbon pyruvate is necessary for passage into the mitochondria. There ATP, CO_2, and H_2O are generated. Thus a cycle is completed in which ATP is first generated in photosynthesis, stored in a variety of carbohydrates, and then released via respiration for use by both photosynthetic and nonphotosynthetic organisms. ❧

LITERATURE CITED

Arnon, D. I. 1960. The role of light in photosynthesis. *Sci. Am.* 203:105–18.

Bassham, J. A. 1962. The path of carbon in photosynthesis. *Sci. Am.* 206:88–100.

Bonner, J., and J. E. Varner. 1965. *Plant Biochemistry.* Academic Press, New York.

Emerson, R., and W. Arnold. 1932. The photochemical reaction in photosynthesis. *J. Gen. Physiol.* 16:191–205.

Esau, K. 1965. *Plant Anatomy,* 2nd ed. John Wiley and Sons, Inc., New York.

Galston, A. W. 1964. *The Life of the Green Plant,* 2nd ed. Prentice-Hall, Inc., Englewood Cliffs, N.J.

Hesketh, J. D. 1963. Limitations to photosynthesis responsible for differences among species. *Crop Sci.* 3:493–96.

Kleese, R. A. 1966. Photophosphorylation in barley. *Crop Sci.* 6:524–27.

Lehninger, A. L. 1961. How cells transform energy. *Sci. Am.* 205:63–73.

Miflin, B. J., and R. H. Hageman. 1966. Activity of chloroplasts isolated from maize inbreds and their F_1 hybrids. *Crop Sci.* 6:185–87.

Rabinowitch, E. I., and Govindjee. 1965. The role of chlorophyll in photosynthesis. *Sci. Am.* 213:74–83.

Stifel, F. B., R. L. Vetter, R. S. Allen, and H. T. Horner, Jr. 1968. Chemical and ultrastructural relationships between alfalfa chloroplasts and bloat. *Phytochemistry* 7:355–64.

Wilson, C. L., and W. E. Loomis. 1967. *Botany,* 4th ed. Holt, Rinehart and Winston, Inc., New York.

Zelitch, I. 1963. Stomata and water relations in plants. *Conn. Agr. Exp. Sta., Bull.* 664.

❦ CHAPTER TWO ❦ LIGHT UTILIZATION BY LEAF AND CROP SURFACES

❦ THE UTILIZATION of light by crop plants is a complex phenomenon and reflects the interweaving of many factors. The biochemical aspects of photosynthesis discussed in Chapter 1 suggest a rather definite step-by-step procedure for light utilization. Duncan et al. (1967) suggest "a great deal has been learned about the capabilities of individual leaves to assimilate carbon under controlled conditions, (but) much less is known about the rate and efficiency of this process as it applies to plant communities." To emphasize the complexity of the problem when considered for an entire plant community, Duncan considered nine factors influencing light utilization alone to study the problem of photosynthetic activity in a computer model, with the assumption that minerals, water, and temperature were not limiting. The factors influencing light utilization will be discussed by first separating out factors essential for photosynthesis, then those affecting leaf growth and function, and finally to consider radiation utilization by a crop surface.

Milthorpe (1956) states the problem very succinctly:

> All aspects of agricultural production are intimately associated with the growth of leaves. This is not immediately obvious with such crops as sugarbeets or corn of which the harvested product form but part of the plant. But even with these crops the yield ultimately depends upon the rate of addition of dry matter per unit land area, that is, on the efficiency of the photosynthetic processes and on the extent of the photosynthetic surfaces.

TABLE 2.1. Light intensity terminology

Unit	Type
1. lux (lx) (meter candle)	Illumination
2. footcandle (ft.c.)	Illumination
3. langley (ly)	Irradiance
4. Watt (W)	Power
5. British thermal unit (Btu)	Power
6. Calorie (cal)	Power
7. Photon	Power

1 ft.c. = 10.76 lux
1 lux = 10^4 photons (quanta) = 9.29×10^{-2} ft.c.
1 cal/cm^2 = 1 langley = 6.978×10^{-2} watts/cm^2
1 ly-minute (natural conditions) corresponds to 6,700 ft.c. sunlight, clear
7,000 ft.c. cloudy
1 quantum = 1 photon

FACTORS ESSENTIAL FOR PHOTOSYNTHESIS

Of the several factors essential for photosynthesis, light would obviously be emphasized as the primary input. Carbon dioxide, temperature, water, minerals, leaf composition, and pigment composition and concentration are also to be considered.

LIGHT

Light must be discussed in terms of quality, intensity, and duration.

Table 2.1 outlines some of the common ways of describing light intensity and interrelates them. One term most familiar to the layman, footcandles, is not an accurate measure of light useful to the plant, but footcandles is still used in many reports, and it is useful to have its approximate value in mind for making comparisons and extrapolations. Light quality is defined as the wavelengths available and active in a particular biological phenomenon. As previously discussed (Chapter 1), light visible to the eye is also active in photosynthesis. The quality that arrives at the top surface of the canopy may be changed as it moves through a leaf or is reflected from a leaf. While quality is of interest and a factor that should be understood, it has less relative importance than the next factor, light quantity—a product of irradiance and duration. Light intensity is not readily changed, but much crop management is conditioned to making the best possible use of that available. As noted in Figure 2.1, the quantity of radiation increases sharply as daylength increases. This is an effect not only of longer daylight hours but of the more direct angle of the sun over a particular latitude.

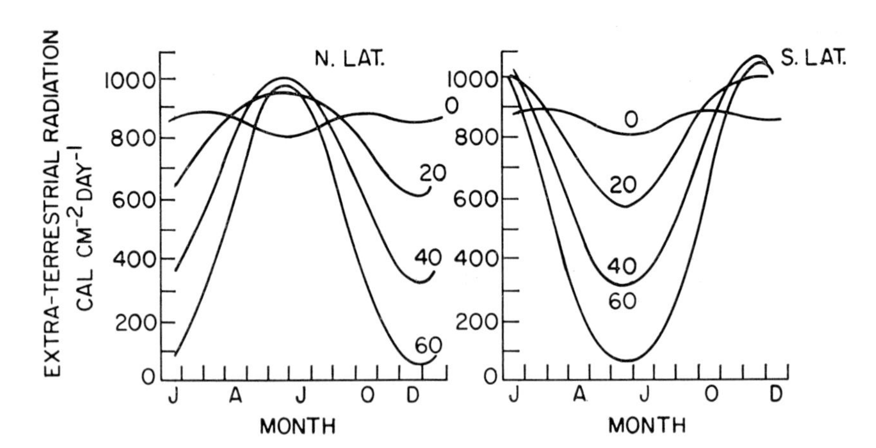

FIG. 2.1 ❧ Variation in daily totals of solar radiation received at the earth's surface at different latitudes and months of the year (after Slayter, 1967).

In the northern temperate region this duration factor suggests that whenever possible a crop plant should be producing its economic yield (the harvested portion of the dry matter) at the peak of this radiation curve. It is obvious that more energy is available to drive photosynthesis each day in June than in August, and the untutored observer might question why the agriculturist does not always have his crop ready to make its peak growth in June and July (for instance, to have the sorghum crop producing its head at this time). However, the opportunity to utilize this peak on the curve is limited by seasonal temperature boundaries and the fact that commonly a superstructure of vegetative material must be built before many crops produce their economic yield. The student of crop physiology must take care not to confound the plant's reproductive response to photoperiod with the total radiation phenomenon described above when he is analyzing a particular experiment or production situation. Nevertheless, the general rule seems clear-cut—long, high intensity radiation June days favor maximum photosynthesis. The challenge then presents itself of developing crop management practices which will place the crop in an appropriate growth cycle to take advantage of this radiation peak.

Loomis and Williams (1963) present a very thoughtful analysis of plant response to the combination of duration and intensity of light available (Table 2.2). They calculate the maximum daily productivity by considering calories per square centimeter of total solar radiation, estimating how much of that would be in the visible range, and convert

TABLE 2.2. Calculation of potential daily productivity by a crop surface (adapted from Loomis and Williams, 1963)

1. Total solar radiation	500 cal/cm²
2. Visible light (400–700 mμ) = 44.4%	222 cal/cm²
3. Total quanta: 400–700 mμ (8.64 μEinsteins/cal)	4320 μE/cm²
This item based on an estimate made from the total in 1.	
a. Albedo (reflection loss) (6–12% over the visible range)	—360 μE/cm²
b. Inactive absorption loss (10%) (cell walls, etc.)	—432 μE/cm²
4. Total quanta usefully absorbed (400–700 mμ) and available for photosynthesis 4320—360—432 = 3528	3528 μE/cm²
5. Amount of carbohydrate (moles CH_2O) produced (10 quanta or 10 μE required to reduce $1CO_2$)	353 μ moles/cm²
6. *Respiration loss (33%)	—116 μ moles/cm²
7. Net production of CH_2O	237 μ moles/cm²

237 μmoles/cm² = .000237 moles = 2.37 moles/m² of CH_2O

$C_6H_{12}O_6$ = 180 g/mole
CH_2O = 30 g/mole

2.37 moles $CH_2O \times$ 30 g/mole = 71 g/m²/day

8. If CH_2O is 92% of the dry weight and inorganic constituents are 8%, then
71 × 100/92 = 77 g dry matter/m²/day

$$77\ \text{g/m}^2\text{/day} = \frac{77\ \text{g}}{454\ \text{g/lb per acre}} \times 4840\ \text{sq yd} \times .8361\ (\text{converting m}^2\text{ to yards}^2)$$

= 687 lb/acre/day (maximum dry weight production)
= 68,700 lb/100 days
= 34.35 tons of dry matter in a 100-day season

*This respiration loss is an estimate. Measured values range widely, often occurring between 25 and 50%.

this amount to μEinsteins. While researchers estimate that values between 2.8 and 10 μE are required to reduce carbon dioxide to carbohydrate, these authors chose the more conservative value of 10 to represent the number of excited chlorophylls that finally result in transferring hydrogen to carbon dioxide and reducing it. The remainder of the table represents manipulations to convert these ideas to a unit of yield.

Species differ in their response to light intensity, as shown in Figure 2.2 for corn and soybeans. With increasing light a unit area of corn leaf shows increasing photosynthetic capacity when compared to the same area of soybean leaf. A soybean leaf is only two-thirds as efficient as a corn leaf in fixing carbon at full sunlight. Waggoner et al. (1963) further describe species differences (Fig. 2.3). It is common to observe that species botanically referred to as shade species (i.e. tobacco, dogwood) are saturated (i.e. cease to respond to increased light intensity) more rapidly than do sun species. Dogwood and tobacco are saturated near one-quarter full sunlight. Corn and sunflowers

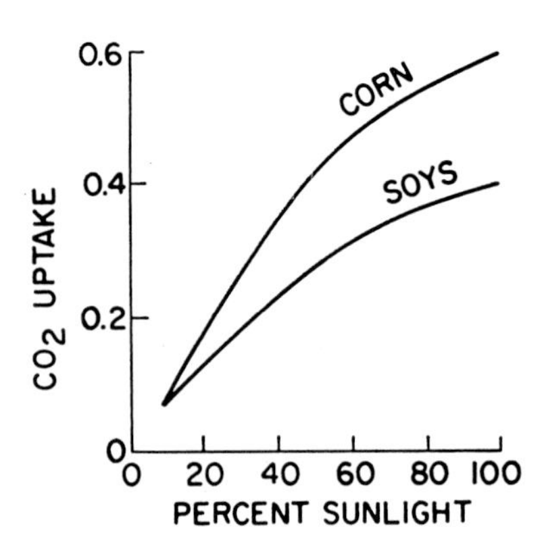

FIG. 2.2 ☙ Photosynthetic rate per cm² of corn and soybean leaves. Rates are given in mg CO_2 taken up per cm² leaf area (Shibles and Weber, 1966).

continue to respond to light intensity changes in the range near full sunlight.

CARBON DIOXIDE

Carbon dioxide supply interacts extensively with light intensity in determining photosynthetic rate (Fig. 2.4). Research work reported in terms of "apparent photosynthesis," represents net CO_2 fixation as measured by intake and exhaust differences in a leaf chamber. Such measurements are made with an infrared gas analyzer that can measure CO_2 concentration in the air passing through it. Light saturation curves advance upward as more CO_2 is made available, and photosynthetic rates between species may vary at apparently saturating light intensities due to varying CO_2 concentrations. The shade species do not respond as much to the greater CO_2 present. They do not appear capable of responding significantly to either more light or more

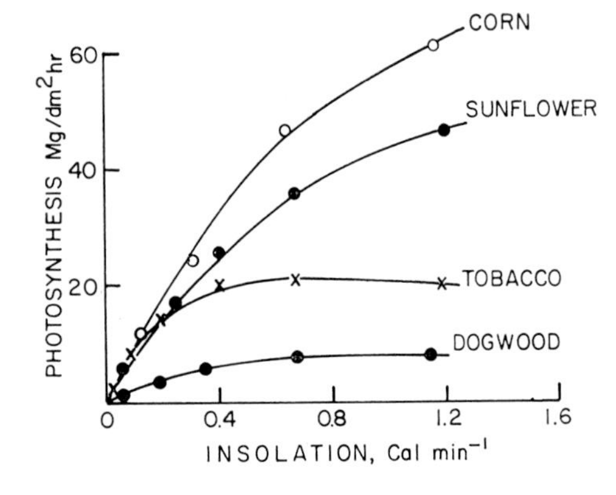

FIG. 2.3 ☙ The response of the photosynthesis of four species to increasing light (Waggoner et al., 1963).

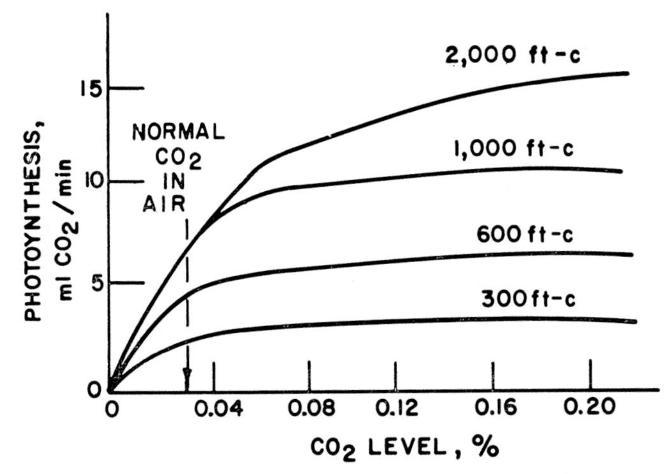

FIG. 2.4 ❧ Carbon dioxide response curves for photosynthesis of wheat seedlings at four light intensities (Hoover et al., 1933).

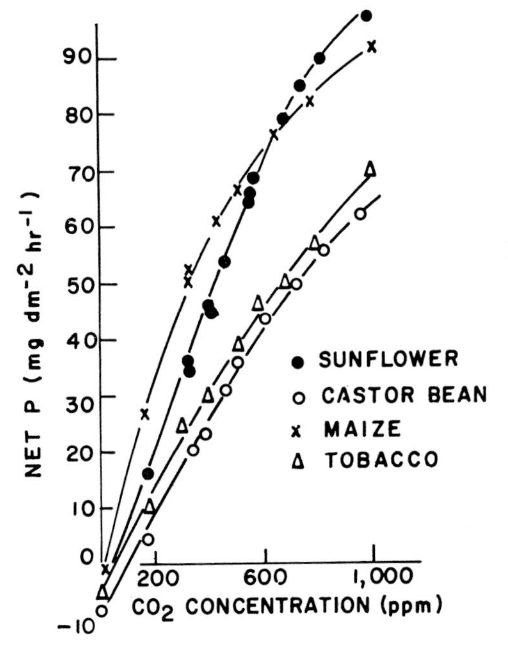

FIG. 2.5 ❧ Relation of CO_2 concentration of the net photosynthesis (net P) of 4 species at 2.4 ly min.$^{-1}$ (Hesketh, 1963).

CO_2. One further point to note here is that species may change positions in their response to increasing levels of light and CO_2. Two species—corn and sunflower—demonstrate this shift in Figure 2.5. Intensive study of photosynthetic rate differences among species and races has been done by Hesketh (1963) in an attempt to explain these position changes. He emphasized that during the transport of CO_2 from the external air toward the reaction center in the chloroplasts, several diffusion resistances are encountered. The most important resistances are as follows: the external air near the leaf surface, the stomata, and the mesophyll cells. Hesketh observed that among nine species studied, the photosynthesis in normal CO_2 and bright light varied manyfold. As a result of his studies, he concluded that these species differences must lie in mesophyll diffusion and the kinetics of the dark reaction.

While it is not readily obvious how to enrich the CO_2 supply under field conditions, the wind does play a key role in stirring the air which enhances photosynthesis (Fig. 2.6) by renewing the supply of air containing a normal CO_2 content next to the leaf. Wind does not allow air which has had its CO_2 content depleted to accumulate near the leaf.

Considerable debate continues as to the relative diffusion resistance caused by the stomates partially closing and the mesophyll cells losing turgidity under water stress. Present evidence suggests that CO_2 movement is more severely affected by mesophyll cell collapse than by partial stomatal closure.

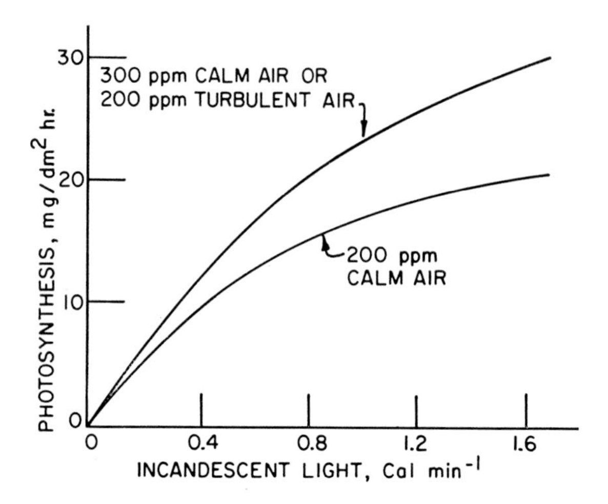

FIG. 2.6 ❧ Response of the photosynthesis of sugarcane to increasing light in calm or turbulent air containing 200 ppm CO_2 and in calm air containing 300 ppm CO_2 (Waggoner et al., 1963).

TEMPERATURE

Temperature effects in the total process referred to as photosynthesis must be carefully separated to be meaningful. The initial photochemical step is independent of temperature in the range of normal plant environment. This means that chlorophyll excitation, the splitting of water, and ATP generation may be considered independent of temperature.

Respiration is known to increase rapidly as temperature rises. Since "apparent photosynthesis" is total photosynthesis minus respiration it often decreases at high temperature (Fig. 2.7). In addition, the dark reaction of photosynthesis is temperature dependent. This is due to the temperature response of enzymes in the dark reaction and the effects of rate of reaction on the accumulation of end products.

WATER

The effects of moisture availability on photosynthesis are challenging to evaluate. Yet there is little question that the total growth of the plant is dependent on an adequate supply of water and a slow enough transpiration rate to keep the plant turgid, and there is no question that a fuller understanding of plant-water relationships will be necesary to maximize crop productivity. The present state of knowledge on this point may be summarized as follows.

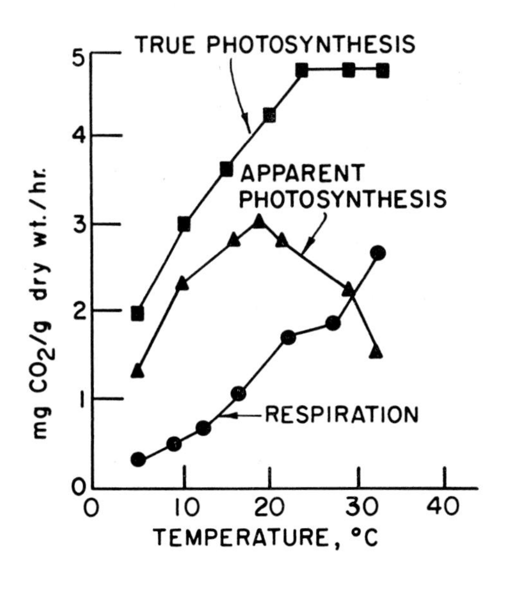

FIG. 2.7 ☙ Temperature effects on photosynthesis in Bryophyllum evaluated by measuring the apparent photosynthesis in the light and the respiration in the dark (Stalfelt, 1937).

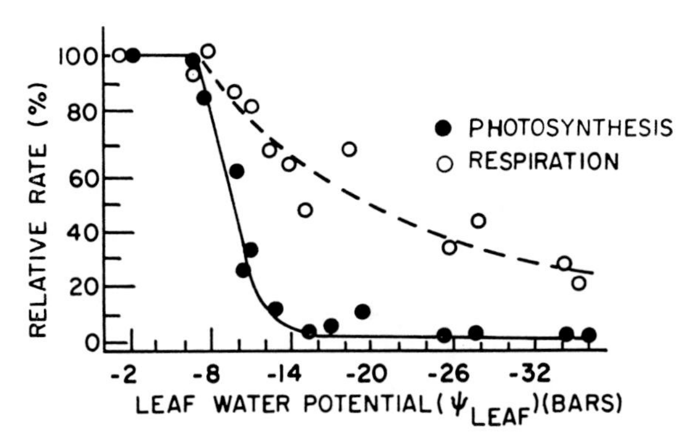

FIG. 2.8 ☙ Effect of decreasing water potential on relative rates of photosynthesis and respiration of tomato plants (Brix, 1962).

Transpiration accounts for nearly 99% of the water used by plants. Approximately 1% is used to hydrate the plant, maintain turgor pressure, and make growth possible. Only 0.1% of the water is chemically bound in the plant. As water becomes limiting, transpiration slows first, then the water of hydration is reduced and causes a slowing of growth. Photosynthesis is most directly involved with generating the hydrogen which will become "chemically bound water" and is the last affected under water stress. Photosynthesis is so basic to plant survival that if the process were very sensitive to environmental flux, plants would not have been as successful in their wide adaptation as we observe them to be. Decreasing water potential has differing effects on the relative rates of photosynthesis and respiration (Fig. 2.8). The plant will therefore lose dry weight long after it has ceased to synthesize new carbohydrate. If the relative rates of photosynthesis and transpiration are graphed over time, a close relationship is observed between the two phenomena. Figure 2.9 shows a related response by comparing net photosynthesis and leaf turdigity. As stated above, these results characterize field observation and summarize the generalized total effect on the plant of a limiting water supply.

Considerable interest has centered on the use of long-chain alcohols (i.e. hexadecanol) to reduce transpiration rates. Water loss per unit surface area changes some when these alcohols are used, but the major change is in total leaf area because such transpiration inhibitors appear to inhibit leaf growth and/or photosynthesis as well. In summary, when the leaf does wilt, photosynthesis is reduced in many species; but evidence is not available to determine whether an increased stomata resistance or an increased mesophyll resistance is more important in slowing CO_2 diffusion.

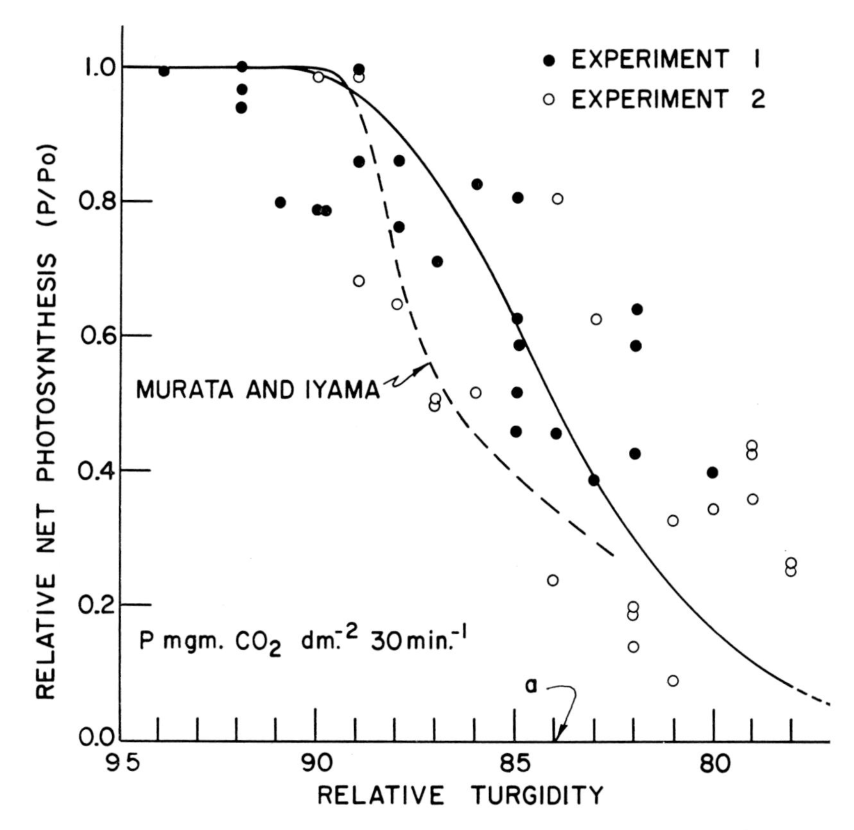

FIG. 2.9 Relative net photosynthesis plotted as a function of mean relative turgidity for each 30-minute measurement period. Experiment 2 carried out using a drying element in the air stream to increase stress levels (Laing, 1966).

LEAF AGE AND MINERAL STATUS

Leaf age has an effect on photosynthetic capacity, and senescence will cause a reduction in that capacity. However, the major factor which influences the rate of senescence is the mineral nutrient status of the leaf. Adequate supplies of the mineral nutrients will allow both older and younger leaves a supply to meet their needs; if the nutrients are limiting, they will be preferentially distributed to the young leaves. In addition to this effect on aging, the mineral nutrients —especially nitrogen and potassium—can be shown to have enhance-

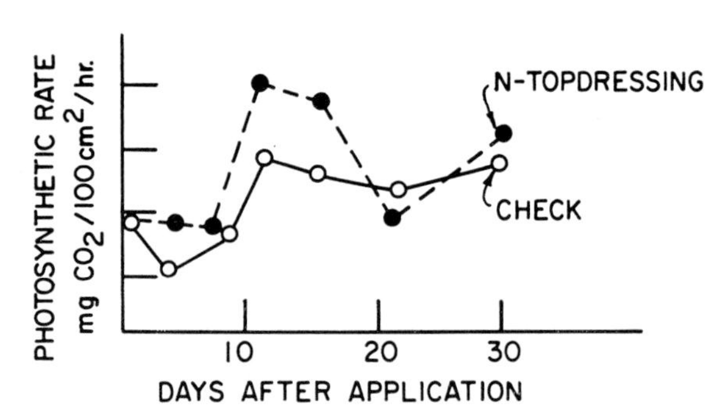

FIG. 2.10 Promotion of photosynthetic activity in the rice plant after application of nitrogenous topdressing (International Rice Research Institute, 1965).

ment effects on the rate of photosynthesis. Chandraratna (1964) describes the promotion of photosynthetic activity in the rice plant after an application of nitrogen topdressing (Fig. 2.10). This suggests a temporary enhancement of photosynthetic activity; whether this can be maintained by continued additions of nitrogen is not clear.

Potassium has a direct effect (Table 2.3) on net photosynthesis (Moss and Peaslee, 1965). A lack of potassium is presumed to affect stomatal closure and thereby decrease the photosynthetic rate. This conclusion was reached because an increased supply of CO_2 did not correct the rate but an increased rate of potassium did (Peaslee and Moss, 1968).

Numerous authors implicate loss of nutrients from a leaf as the cause for reduced photosynthetic capacity. Bottom leaves were observed to have a significantly lower net assimilation rate than top leaves in orchardgrass, often only half as much (Brown et al., 1966), and this appeared to be related to their nutrient status. Moss and Peaslee (1965) considered the contribution to grain yield of upper and lower corn leaves. Lower leaves were less efficient, and such an

TABLE 2.3. Potassium content and photosynthesis of corn leaves—9,000 ft.c., 30° C (Moss and Peaslee, 1965)

Leaf Number (from top)	K Content μg/g fresh wt.	Photosynthesis mg CO_2 dm^{-2} h^{-1}
Well fertilized		
2	6,100	40
4	5,500	38
7	5,000	36
11	4,350	36
K-stressed		
2	2,150	33
6	800	15
7	600	14
11	250	1

inefficiency could be induced by reduced rates of potassium, phosphorus, magnesium, and nitrogen. Apparently if these nutrients are in short supply they are translocated from the older to the younger tissue, thus causing what appears as either more rapid aging or reduced photosynthetic activity in the lower leaves.

CHLOROPHYLL CONTENT

Chlorophyll content is not often a limiting factor in photosynthesis. Evans (1963) reported that fully developed leaves commonly contain more than 4 mg/dm^2 of chlorophyll, an amount that gives maximal photosynthesis. There is the possibility that under low light conditions, the rate of the light reaction will limit the overall process, thus making the chlorophyll content a limiting factor. Chandraratna (1964) thus suggests that under heavy fertilizer applications where a dense foliage results and there is a low quantity of light down in the canopy, the chlorophyll content would be more important for total photosynthesis of the population than under less fertile conditions. However, Gaastra (1962) emphasizes that the properties of the photochemical process of different leaves are similar enough that light intensity is still the most important factor. Hesketh (1963) supports this view after finding rather small variations in chlorophyll content between species.

FACTORS AFFECTING LEAF GROWTH AND FUNCTION

Leaf growth and leaf function are influenced by many of the same factors discussed above. It is useful, however, to consider them separately for appropriate emphasis.

LIGHT

Light intensity and duration both affect the size and longevity of the leaf. In general, leaf expansion is closely linked with light intensity. Because the leaf is a modified stem, the midrib acts like an internode; therefore, more light intensity results in less leaf extension. Above 300–600 footcandles leaf extension is often reduced. Short days accelerate the leaf production of long-day plants and retard leaf production in short-day plants (Krizek et al., 1966). The leaves of both short- and long-day plants age more rapidly on long days.

TEMPERATURE

Leaves differentiated at low temperatures are never as large as those differentiated at high temperatures, even though they both grow at the same medium to high temperatures after differentiation. Friend et al. (1962) found that the area of the fully grown leaf lamina varied according to its position on the stem and that in continuous illumination, raising the temperature by 5° intervals between 10 and 25° resulted in progressively higher rates of leaf initiation, emergence, and expansion. The greatest leaf area (according to Friend's work) came at 20° C and 1,000–1,750 footcandles of light.

MINERAL NUTRIENTS

Leaves lack the capacity for secondary growth or for regeneration of a leaf part. Each leaf is thus limited in both its growth and longevity. Therefore, it is very important to provide the leaf with adequate mineral nutrients, especially nitrogen, phosphorus, and potassium, so that a minimum of primary growth is lost after it is once developed. Langer (1959) found that nitrogen, phosphorus, and potassium all had a pronounced effect in increasing leaf area per plant in timothy. In addition to significant individual nutrient effects, there were positive NP, NK, and NPK interactions (Fig. 2.11); that is, all combinations of these nutrients also increased leaf area per plant.

Nitrogen results in a greater leaf area whenever it is applied to

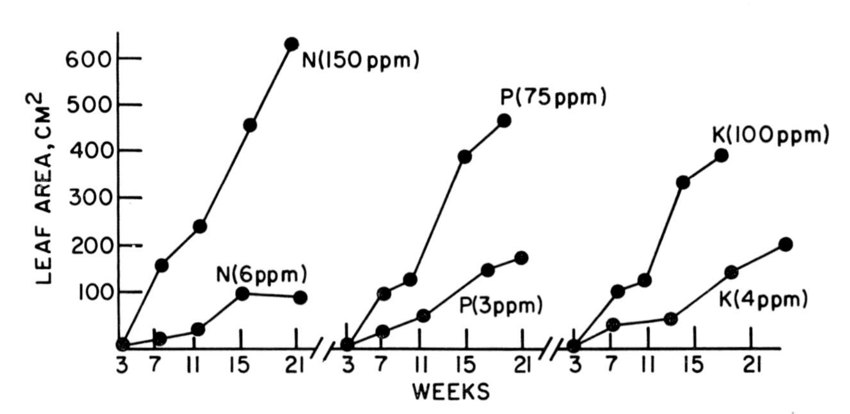

FIG. 2.11 ☙ Effects (means) of nitrogen, phosphorus, and potassium on leaf area per plant in timothy (Langer, 1959).

the actively growing plant in a vegetative condition. However, if applied at the wrong time, it may increase leaf area and total dry weight but not increase grain yield. This emphasizes the need to coordinate certain management practices with the developmental rhythm of the plant for maximum benefit.

Although leaf area throughout the growth period is the main determinant of the total yield of dry matter, it does not necessarily follow that the same is true of the yield of a particular organ or chemical constituent (for example, the grain yield of a cereal). When nitrogen fertilizer was applied to fall wheat in April, it had the effect of nearly doubling the leaf area index in May and June (Thorne and Watson, 1955). When the same amount of nitrogen was applied at the time of head emergence in late May and early June, it caused a smaller increase in the leaf area index. This effect did not appear until mid-June, by which time the leaf area was rapidly decreasing on other wheat plants. The increase in grain yield was nearly equal for the two times of application.

Eik and Hanway (1965) found that a greater supply of nitrogen applied to the corn plant early increased the number of leaves formed per plant and increased leaf size even more consistently than leaf number. Greater nitrogen also increased the rates of leaf emergence and leaf area expansion.

Phosphorus increases leaf area particularly in the early stages of growth. Late in the plant's life, phosphorus hastens the senescence of leaves and may eventually reduce the area of leaves, particularly if it is applied in an imbalance.

Potassium has a twofold effect on leaf development. In addition to delaying senescence it may stimulate greater leaf area either alone or by interacting with nitrogen and/or phosphorus.

RADIATION UTILIZATION BY A CROP SURFACE

The radiation which reaches the earth's surface is some 50–100 times the energy fixed in plant constituents. For the wavelengths 400–700 $m\mu$ (visible light), maximum efficiency conversions of 5–7% are reported (Alberda, 1962; Loomis and Williams, 1963); estimates suggest that as much as 12% of the visible spectrum energy could be converted to plant energy when growing conditions are optimal and a closed green crop surface is present. The rate of the photochemical process is influenced more by a variation in light intensity than by

TABLE 2.4. Some high dry matter yields and daily growth rates for selected species (Loomis and Williams, 1963)

Crop	Location	Tons/Acre	Growing Season Days	Mean Growth Rate $g/m^2/day$	Maximum $g/m^2/day$
Napiergrass	Puerto Rico	43.0	365	26	—
Sugarcane	Hawaii	28.6	365	18	38
Alfalfa	California	14.5	250	13	23
Sudangrass	California	13.3	160	18	51
Corn	Iowa	10.0	120–180	27	27–51

differences in species, although both intensity and species are important. At a given geographic location, this means the maximum potential is definable.

Loomis and Williams (1963) summarized certain notably high dry matter yields and daily growth rates (Table 2.4). It is of interest to note that these large yields and highest mean growth rates are from crops adapted for producing vegetative rather than grain yield.

In consideration of maximum yield potentials, the following questions arise:

How much radiant energy is present?

How much radiant energy is intercepted?

What is the efficiency of conversion to a useful product?

What is the pattern of distribution in the plant?

How much yield may result from all this?

In discussing light interception by a crop surface, these primary factors are to be considered: (1) total area of leaves, (2) efficiency of light interception by the crop canopy, (3) factors which modify the leaf area and efficiency, (4) leaf area duration, and (5) duration of exposure (hours of daylight as determined by daylength and season length).

LEAF AREA

The total area of leaves is described by a leaf area index (LAI, L). This value indicates the number of unit areas of leaf per unit area of ground surface; for example, the total square feet of leaf surface above one square foot of soil. One flat layer of leaves over a unit area would reflect 10% of the incoming radiation, absorb 80%, and transmit only 10% to the next leaf area. Since leaves are seldom oriented in such a flat layer, the L serves as an indicator of the surfaces available for light absorption and also provides a more useful common denominator for discussing the photosynthetic potential of a given crop than to evaluate the leaf area per plant (which may vary drastically as plant population, plant distribution, or variety is changed).

Three workers—Watson (1952) in England, Nichiporovich (1960) in Russia, and Brougham (1956) in New Zealand—did much of the early work to define and embellish these concepts. Nichiporovich suggested that an optimal leaf area is definable for every economic crop. He considered the optimal area range to be between 2.5 and 5.0 and outlines the following points as a useful basis for discussion:

1. The L during growth should be sufficient to intercept as much of the incoming radiation as possible.
2. The L should be of a magnitude which prevents parasitism; that is, a condition of lower leaves using carbohydrates at a greater rate than they photosynthesize.
3. The L must suit the conditions and purposes for which the crop is grown. Maximum L does not always equal maximum grain yield, nor does it always yield maximum dry matter production.

The L is computed at specific stages of growth (i.e. 30 days after emergence, at anthesis, etc.), and comparisons are made between varieties at a given stage of development. A related value, leaf area duration (LAD or D), is used to describe the length of time the leaf area is functional. For example, a field of corn might have an L of 4.5 at the time of pollination, but it would be useful also to know how long this L is maintained. A comparison could be made using the D for two varieties, one of which maintains an L of 4.5 for 40 days and another with an L of 5.3 for 30 days. This would provide additional management information as to how to use these varieties for given environmental conditions of radiation supply, fertility level, or expected periods of dry weather.

There is convincing evidence that the grain yield of cereals is related to leaf area after the ears emerge but not before. The relationship between grain yield and L can be described conveniently by the ratio between them, the grain-to-leaf ratio G, where G is a measure of the photosynthetic efficiency of the leaves in producing dry matter for the grain in the same way as E (see below) measures the photosynthetic activity of the leaves during the vegetative stage.

NET ASSIMILATION RATE

The net assimilation rate (NAR or E) represents the net photosynthetic activity of a unit area of leaf—that is, photosynthesis minus respiration—and it is commonly evaluated over all the leaf areas in a canopy. Current respiration utilizes between 25 and 50% of the total daily photosynthate and may be considered to average 33%. The E per unit leaf area is less as the L increases, but the rate of E loss does not decline directly proportional to the L increase. It declines at a slower rate; as a result increasing L gives a yield increase. The following example suggests how these factors are related:

First L	Max. E = 1.0	
Second L	½ Max. E = 0.5	
Third L	⅓ Max. E = 0.3	
		1.8 ÷ 3 = 0.6 average E.

The first L has an E of 1.0, thus giving a unit of yield. Two Ls give 1.5 units (1.0 + 0.5) and three Ls give 1.8 units of yield. Therefore, while the average E for the three Ls is 0.6, the total production has risen from 100 to 180 units in this generalized example—a total production increase of 80%. Watson (1952) noted that the E declined over the season but concluded that this was often due to reduced radiant energy availability and not just to leaf aging.

When the E is equal to zero, the organ or plant is said to be at the compensation point. If the E is greater than zero, net photosynthesis has occurred. If the E is less than zero, the organ (unit leaf area) or plant is parasitic. Whether or not leaves under field conditions are ever parasitic is a matter of continuing debate. In summary, differ ences in E are relatively small, and most workers report that the variation in dry matter production is far more dependent on the L.

OPTIMUM VERSUS CRITICAL LEAF AREA

Within the range of commonly developed Ls, crops vary as to the best L for maximum productivity. For instance, while kale has a distinct optimal L at approximately 3, sugar beets continue increasing in yield over the range of normally developed L up to 6 (Fig. 2.12).

Brougham (1956) found that both percentage of solar radiation interception and rate of dry matter production increased with leaf area development and approached a maximum at similar leaf area

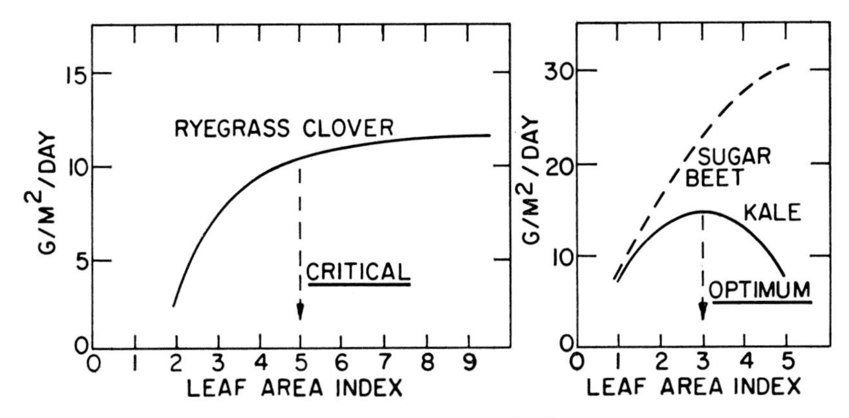

FIG. 2.12 Relationships between crop growth rate (dry matter accumulation per m² day) and leaf area index. Left: The plateau response obtained with ryegrass-clover pasture. The critical leaf area index indicated is for 95% light interception. Right: Data for kale showing an optimum leaf area index and for sugar beet for which neither an optimum nor a critical index has been established (Loomis and Williams, 1963).

indices. However, the nature of photosynthetic response at an L exceeding that required for maximum interception and net productivity is not fully agreed upon. Subterranean clover, kale, and rice develop an optimum L (i.e. the rate of dry matter production is maximum at a particular L and less at L below or exceeding this value). In contrast, the ryegrass-red clover-white clover mixture Brougham studied had a constant rate of dry matter production after the maximum rate was reached. He defined this as the "critical L" (Fig. 2.12) required to give 95% interception at noon.

There is interest in defining whether a crop has an optimal or a critical L response because this would be useful in determining how the critical L or plant population should be modified for a particular crop variety. Soybeans exhibit a critical L response (Shibles and Weber, 1965). Lower, shaded leaves of the soybean canopy are not "parasitic" upon the productive portion of the canopy and do not, therefore, detract from the net production of photosynthate.

Now that the parameters often used to describe a leaf complex have been discussed, it is important to note that several factors influence leaf area efficiency and total L. Population density, population distribution, the angle of leaf exposure, the carbon dioxide diffusion to the site of photosynthesis, the quantity or renewal rate of ribulose 1,5-

diphosphate, and the chlorophyll content have all been suggested as playing a role in leaf efficiency.

LIGHT SATURATION

It is appropriate to review the question of light saturation for an individual leaf before considering canopy effects. Studies over the past 40 years have indicated that light saturation of individual leaves (no more photosynthesis even though light intensity increased) was occurring at approximately 2,500 footcandles (full sunlight = 9,000–10,000 footcandles). This figure is very much under question and, as has been previously discussed, may be a carbon dioxide diffusion rate problem or a leaf angle question and not really light saturation. It appears quite certain from the work of Moss, Hesketh, and others that species like corn and sunflower do not become light saturated in the range of natural light intensity. Nevertheless, even if the individual leaf did become light saturated, the canopy of leaves could presumably benefit from more light if that light were distributed down through the canopy.

LEAF ANGLE

Light saturation of a canopy of leaves is due at least in part to the way in which the several layers of leaves are displayed. Experiments with whole plant communities which used Watson's concept of leaf area index have underlined the importance of foliage configuration as a major determinant of net productivity. Tanner and Stoskopf (1967) lay heavy emphasis on the idea that broad, lax-leafed varieties were selected in earlier years to assist in shading weeds, whereas more erect leaf types are more efficient photosynthetically. If the leaves of the crop are oriented horizontally (as is the case in soybeans), light penetrates into the completed canopy only about twelve inches (Sakamoto and Shaw, 1967). The saturation level under this kind of condition is between 6,000 and 6,400 footcandles, although it shifts through the season, declining to 3,000 in August. Brougham (1960) has shown that ryegrass with vertically disposed leaves transmits 74% of the light per unit leaf area index. Clover, with more nearly horizontal leaves, transmits 50% per unit leaf area index. For sugar beets grown in spaced plant tests, the growth of a prostrate leaf type was superior to that of an erect leaf type, but the erect leaf type gave greater growth

TABLE 2.5. Grain yields and barren plants from leaf angle study (Pendleton et al., 1968)

Comparisons	Yield (kg/ha)	Plants Barren (%)
Genetic isolines of hybrid C103 × Hy		
1. Normal leaf	6,202 a	28 a*
2. Upright leaf	8,769 b	14 b
Mechanical manipulation of leaf angle of Pioneer 3306		
3. Normal (untreated)	10,683 c	4 c
4. All leaves positioned upright	11,386 cd	6 bc
5. Leaves above ear positioned upright	12,202 d	3 c

* Means with the same letter are not significantly different at the 5% level.

under conditions of plant competition (Oshima, 1962). Observations like these have led to the proposal that a nearly vertical orientation of the uppermost leaves of a canopy and with the lower leaves approaching the horizontal might be effective in increasing photosynthetic efficiency. Pendleton et al. (1968) present results to support these ideas (Table 2.5).

Genetic isolines of a single-cross hybrid differed only in their leaf angle. The upright leaf gave a significantly higher grain yield and also had fewer barren stalks. Secondly, a commercial hybrid had its leaves manipulated to (a) all upright or (b) those above the ear upright. Both treatments increased yields over the normal leaf positioning, but those with upper leaves upright and lower leaves untreated gave the greatest yield increase.

The reasoning behind such proposals and observations may be stated as follows. It has been previously noted that some species reach light saturation between 2,000 and 3,000 footcandles. The highest efficiency occurs below this saturation point; this supports the conclusion that the more uniformly light is received by all leaves throughout the canopy, the greater will be the productivity. Calculations by Duncan (1965) suggest that an individual leaf of corn may be most efficient at 800 footcandles.

If one considered a surface one centimeter square on the horizontal with 5,000 footcandles of light shining on it (half the intensity of full sunlight), and then tilted this surface through an arc until there were 800 footcandles on the surface, 6.3 square centimeters of surface could be exposed in the same light beam. Although photosynthetic rate per leaf area would be low, efficiency would be higher and community photosynthesis could be increased as much as two times by this means. By the use of a digital computer and series of related assumptions, Duncan (1965, 1967) has tested this model for corn and con-

FIG. 2.13 ☙ Theoretical values for dry matter accumulation by corn at two leaf angles and three leaf area indices (Duncan et al., 1967).

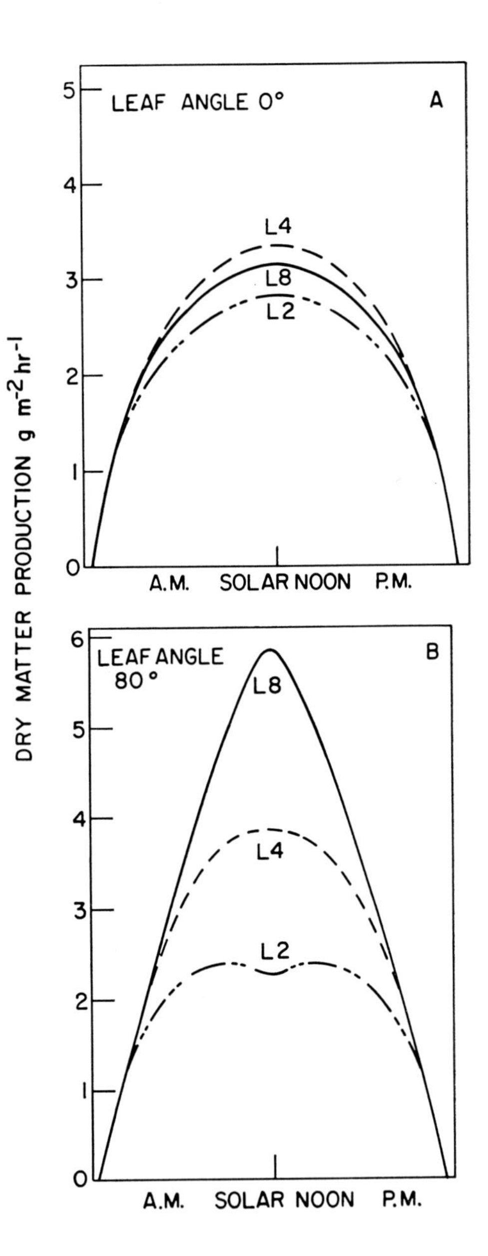

cludes that an 80° angle is better in temperate regions during the summer than smaller angles. The optimum angle must always be stated in reference to solar elevation and L. Figure 2.13 describes two of his models with different L values.

Gardener (1966) provides further experimental evidence to support these concepts. Using barley lines with erect versus lax leaves, he obtained yield increases of 16% from lines with erect leaves on lines with slightly lower Ls. In summary, it is important to recognize that leaf angle is a potent new factor in plant structure. But it is only one factor in the yield equation and must be incorporated into a gene pool with other currently useful factors. For instance, in the model which Duncan used to develop the graphs in Figure 2.13, he included the following nine variables: leaf area, leaf angle, leaf position, reflectivity, transmissivity, the light effect curve relating photosynthesis to leaf illumination, solar intensity, skylight brightness, and the position of the sun.

NONLAMINAR PHOTOSYNTHESIS

Leaf area calculations commonly include one surface of the leaf blades and not leaf sheaths. However, the photosynthetic efficiency of the leaf sheaths and the inflorescence of monocotyledons is considered to range from 50 to 100% that of a blade. The sheath of small grains may contribute from 15 to 40% to grain yield, and the ear (head) contribute from 9% in awnless to 40% in awned varieties (Thorne, 1959). The advantage of awned wheats over awnless types is especially apparent in drouth years but also holds during years favorable for wheat production (Grundbacher, 1963). Awned lines have an even greater advantage for kernel weight and test weight as compared to yield. Corn leaf sheaths contribute about 6 to 10% of the total photosynthate. One of the major differences in yielding ability between normal and brachytic type dwarf corn (short internodes) has been suggested to be the reduced leaf sheath area in the dwarf. Stickler and Pauli (1961) concluded the leaf sheath of grain sorghum did not contribute significantly to yield except where severe defoliation might have occurred.

Graham and Lessman (1966) found that a 3-dwarf parent (one carrying three of the four dwarfing genes) in a sorghum line had a larger leaf area than the taller entries but the 3-dwarf line yielded less grain. Like Quinby (1961), they found leaves per plant, days to bloom, and tillering not to differ, and thus they speculated that perhaps leaf arrangement was involved in yield differences. They suggested that less leaf shading may be the cause for more efficient light utilization by tall plants.

CARBON DIOXIDE SUPPLY

Under field conditions, the direct addition of CO_2 seems impractical. Therefore, the yield improvement related to CO_2 may be expected to come primarily from continuous replenishment of air near the leaf surface (Fig. 2.6). There are potential changes in a canopy that might provide improved air movement of more natural stirring. An uneven crop surface resulting from species of different heights may have merit in enhancing mixing. Radke (1967) obtained this effect when he grew corn as "windbreaks" in a soybean field and boosted bean yields 25%. While the increase was likely due to several factors, the air turbulence and therefore a more rapidly replenished CO_2 supply most likely contributed to the greater soybean yield. Wright and Lemon (1966) provide a very thorough analysis of this point, and they describe in quantitative terms the CO_2 supplied by such turbulence. Numerous comments throughout this discussion have implied that, given the proper environmental conditions of air turbulence and a canopy in which it can be moved rapidly, there is sufficient CO_2 naturally present in the air for crop production purposes.

WATER SUPPLY

Nichiporovich (1960) emphasizes what the practical agriculturist would commonly understand: that is, since leaves transpire water, the growth or amount of leaves must be adjusted to the moisture available for the crop. However, to a large measure, water transpired and evaporated equals radiation received. This holds true especially in humid regions and dense crops (Evans, 1963).

In considering water evaporation and radiant energy utilization by the plant community, it is useful to apply the law of the conservation of energy to the energy balance at the surface of plants and soil. Of the net radiant energy absorbed by plant and soil surfaces through radiation exchange with the sun and sky, part goes toward sensible heat, part to latent heat, and some is fixed by photochemical processes. The sensible heat is composed of three major units. There is sensible heat gained or lost from "storage" in the soil, convected by the air, or exchanged as latent heat. A small but significant portion of the net radiation available during the daytime goes into photosynthetic processes in plants. All of these parts add up to zero in the energy conservation law application. For the simplest situation—a short, uniformly dense surface in the actively growing stage, without border effects and supplied with sufficient soil moisture to avoid restrictions on normal plant functions—this set of values is reasonably well defined.

About 75–85% of the net radiation absorbed during the daytime is used to evaporate water; 5–10% goes into sensible heat storage in the soil; 5–10% goes into sensible heat exchange with the atmosphere by convective processes, and about 5% goes into photosynthesis.

The percentage of transmission of light through a corn leaf canopy is described by a straight line decline; as leaf area index increases, percent of transmission declines. Much of the reflected shortwave radiation from vegetation takes place in the infrared range, and 75% of the shortwave radiation transmitted through the corn crop is in the near-infrared range. Presumably this phenomenon plays an important role in providing energy at the soil surface for sensible heating and evaporation under dense vegetation. From studies in this area, values ranging from 17 to 54% of the net radiation have been suggested as being available under varying crop canopies to evaporate water from the soil.

Under adequate soil moisture supplies, the movement of water from the soil to the root is more rapid than movement in the plant. However, in drier soils there does appear to be an impedance of moisture flow in the soil near plant roots.

There is little indication that wind has much effect on evaporation in a humid climate where there is a dense crop. In humid regions where advected sensible heat plays only a minor role in dense crops, evaporation appears to be largely controlled by radiant energy supply. Where advection is important in drier regions when irrigation is practiced, wind may play a role.

A heavier plant population increased water use by corn (Table 2.6), but narrower rows (a more uniform plant distribution) reduced water use for a given population (Yao and Shaw, 1964). Noting the small differences in evapotranspiration in Table 2.6, it is apparent that there are counterbalances under field conditions which do not cause water use to increase as rapidly as the L increases. For example, earlier shading of the soil surface reduces water loss directly from the soil under irrigated conditions. It may be that without irrigation, the

TABLE 2.6. Water use by corn at two populations of irrigated corn, May 31 to Sept. 20 (Yao and Shaw, 1964)

Row Width	Inches of Evapotranspiration	
	14,000 plants/acre	28,000 plants/acre
21″	13.2	15.0
32″	15.5	16.6
42″	16.1	17.4

rate of evaporation could have been reduced significantly in the wider rows. Nevertheless, the water-conserving aspects would likely provide a yield advantage for the narrow rows. In subhumid and humid regions, then, the problems related to increased stand densities, especially in corn, are related to light reduction (light stress) effects on ear, head, or tiller development and not as much to the availability of water or its rate of use. The midday depression in photosynthesis commonly observed under field conditions also is likely the result of water stress. This photosynthetic depression may be up to 50% when the relative turgidity drops below 85% (see Fig. 2.9).

SUMMARY

A summary of the factors influencing gross photosynthesis of a field crop includes the following observations:

1. A relative insensitivity to daytime temperature changes possibly because leaf temperature is poorly correlated to air temperature at a weather station reading.

2. A strong dependence on the intensity of radiation and its distribution within the canopy. Moss et al. (1961) found a correlation between insolation and CO_2 assimilation by corn of 0.95. Baker and Musgrave (1964) report that under adequate water and CO_2 a corn stand with an L greater than 0.6 never reaches light saturation and the theoretical maximum rate of photosynthesis. Shibles and Weber (1965) report the dry matter increase rate for soybeans to be directly related to the percent of interception.

3. An approximate proportionality to the concentration of CO_2 in the air surrounding the leaves.

4. The angle of leaf display (Duncan, 1967).

5. The relative stomatal and mesophyll resistance to CO_2 diffusion as influenced by water supply and genetic differences and the

boundary resistance of the canopy as related to eddy transfer and the leaf surface (Hesketh, 1963).

DRY MATTER DISTRIBUTION IN THE PLANT

There are many phases in a discussion of dry matter distribution in the plant, some of which can be more appropriately treated in the chapters related to growth regulators, seed development, and carbohydrate nutrition. Here the phasic development of the plant and the end result of dry matter moving into different plant parts will be described. Most of the carbohydrate in the grain of barley and other cereals is formed from CO_2 assimilated after the head or ear emerges. The nonblade portions (especially awns) have been known to contribute 15–40% of the final grain dry weight (Thorne, 1959). Careful measurements of the time at which the leaf becomes a carbohydrate exporter are limited; Webb and Gorham (1964) reported that developing leaves of squash begin exporting photoassimilates when they reach 50% of their maximum dry weight. Figure 2.14 describes the fresh and dry weight of various organs of the pea plant. Root weight appears to hold constant after a maximum is reached, but other organs increase, then decline in dry weight, except for the fruiting body or storage organ which has a continuous increase. A somewhat similar pattern occurs for soybeans (Hanway and Thompson, 1967). In his studies on corn, Hanway (1963) did not observe the apparent transfer of dry matter from the leaf and stem to the ear; that is, these organs maintained a constant weight once they reached the maximum. This would appear to be a difference between production of seed by legumes (peas and soybeans) and cereals.

Human and livestock food consumption patterns have dictated that different organs of different plants will be used for nutritional purposes. It is therefore important to know how the plant distributes its dry matter into various organs. Two useful terms used to describe this partitioning of dry matter by the plant are *biological yield* and *economic yield*. The term "biological yield" was proposed by Nichiporovich (1960) to represent the total dry matter accumulation of a plant system. "Economic yield" and "agricultural yield" have been used to refer to the volume or weight of those plant organs that comprise the product of economic or agricultural value. The proportion of biological yield represented by economic yield has been called the "coefficient of effectiveness," the "migration coefficient," and the "harvest index" or K. All of these terms imply the movement of dry

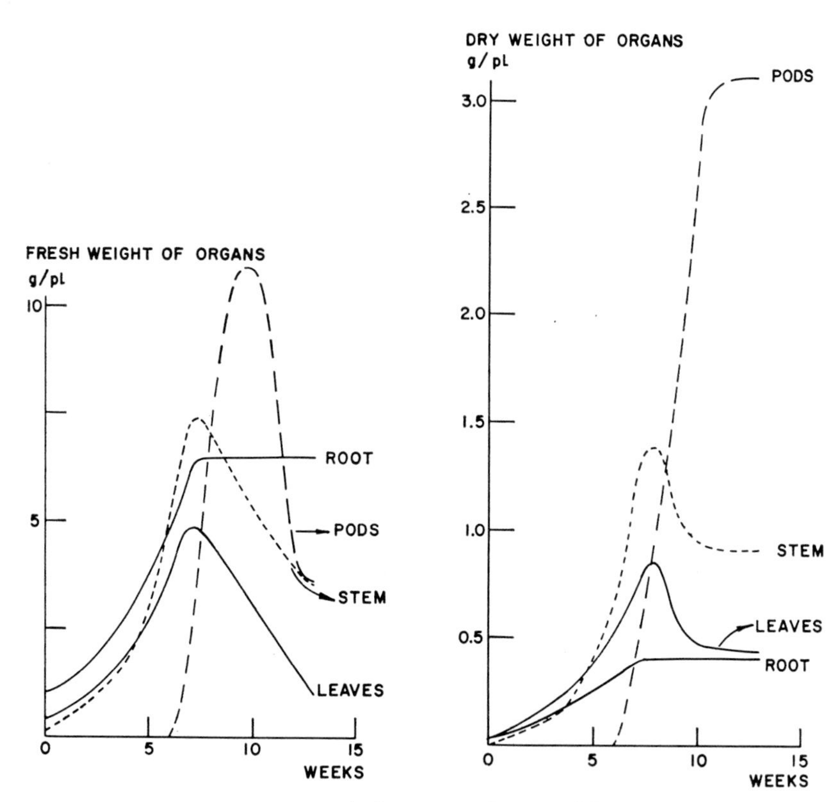

FIG. 2.14 ☙ Fresh and dry weight of the various organs of pea plants plotted against time after emergence (Brouwer, 1962).

matter to a useful organ. In thinking of the dry matter distribution pattern of annuals versus perennials, it is of interest to note that the annual is a plant that places much of its carbohydrate into the seed, an economic yield. In contrast the perennial stores a significant portion of its carbohydrate production in an overwintering organ. When discussing maximum harvestable yield, it is necessary to evaluate the difference between the 1,000–2,000 pounds of carbohydrate stored in the alfalfa root for it to overwinter and the 10–60-pound seeding rate used to establish annuals. Perhaps only where stand establishment is very difficult or expensive can the perennial be justified, assuming that maximum carbohydrate can be well used if available in the harvestable portion.

It is of interest to note that successful breeding of higher yielding varieties has arisen in part from unconscious selection by plant breeders for a higher harvest index. Wallace and Munger (1965) note that five wheat varieties developed in the Netherlands over the period 1902–1955 had a K value increase from 34 to 40. Recent varieties had smaller biologic yields but greater economic yields. The same patterns of change have occurred in wheat, rice, and peas. Semidwarf wheats developed in the United States increased in K values from 32 to 38. Corn silage varieties in the 1920s had K values of 36; in the 1950s they were 44. This factor is another useful example of how the applied biologist cannot hope to look at single factors but must constantly consider several; K by itself is not an adequate measure of a new variety but must be considered along with yield. In the study reported by Wallace and Munger (1965), their highest yielding line was second in K value because the highest K value line produced too little biological yield. Nevertheless, the K value and an understanding of what it means is a useful guide in crop production.

LIFE PERIOD OPTIMAL FOR SEED YIELD

In theory it would be expected that the longer the stage of maximum net assimilation per unit of soil surface lasts, the higher the total dry matter production will be, and the more fruits or other parts the plant will be able to produce. Some data support this view. For example, the best late-maturing varieties of a crop normally outyield the best early-maturing varieties in the temperate zone.

In this respect the tropics might seem to have an advantage over the temperate zone, for in the tropics the duration of the vegetation period is not limited by temperatures, and varieties with very long vegetative periods could be used which would produce high yields. Best (1962) suggests that this is incorrect. He states that many other factors come into play, most of which may probably be considered under the heading of senility of the plant. In the examples in Figure 2.15, seed yield was studied as a function of the vegetation period. In both rice and soybeans, a variety sensitive to photoperiod was used and variations in the duration of the vegetation were obtained with the help of weak supplementary irradiation, maintaining at the same time a similar amount of daylight per 24 hours in all treatments.

The yield of grain in these figures is plotted against the length of the vegetation period. The graphs show that there are certain optimum vegetation periods for seed production which are not correlated with maximum dry matter production. Best cites other informa-

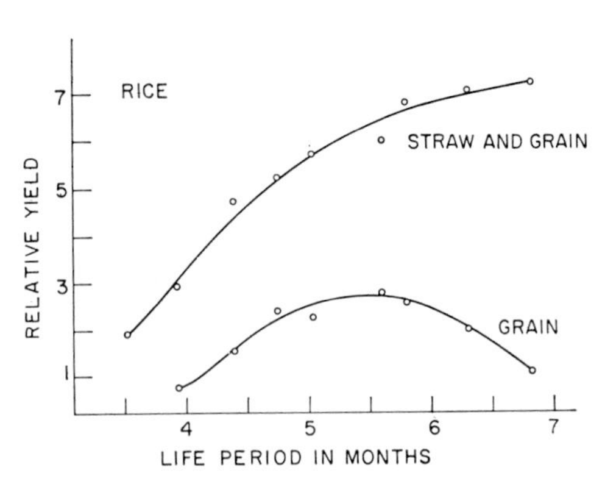

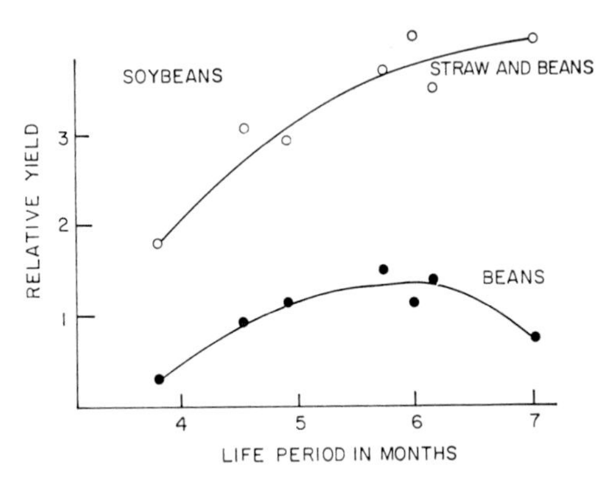

FIG 2.15 ☙ Effect of duration of the vegetative period on yield of soybeans and rice, with and without straw. Duration of the vegetative periods regulated by photoperiodic treatments, the daily amount of photosynthetic light being similar in all treatments (Best, 1962).

tion to suggest that light and duration of particular photoperiods may not favor high rates of photosynthetic productivity in the tropics. From this discussion it may be suggested that grain production with high rates of dry matter production over relatively short periods may best be done in the temperate zone, and forage production which has steadier rates of dry matter production may better be done in the tropics. Best also points out that temperature has a distinct influence on the distribution of dry matter in the crop—cool temperatures favoring carbohydrate transport to the grain. ☙

LITERATURE CITED

Alberda, Th. 1962. Actual and potential production of agricultural crops. *Neth. J. Agr. Sci.* 10:325–33.

Baker, D. N., and R. B. Musgrave. 1964. Photosynthesis under field conditions. *Crop Sci.* 4:127–31.

Best, R. 1962. Production factors in the tropics. *Neth. J. Agr. Sci.* 10:347–53.

Brix, H. 1962. The effect of water stress on the rates of photosynthesis and respiration in tomato and loblolly pine seedlings. *Physiol. Plantarum* 15:10–20.

Brougham, R. W. 1956. Effect of intensity of defoliation on regrowth of pasture. *Australian J. Agr. Res.* 7:377–87.

———. 1960. The relationship between the critical leaf area, total chlorophyll content and maximum growth rate of some pasture and crop plants. *Ann. Botany* 24:463–74.

Brown, R. H., R. B. Cooper, and R. E. Blaser. 1966. Effects of leaf age on efficiency. *Crop Sci.* 6:206–9.

Brouwer, R. 1962. Distribution of dry matter in the plant. *Neth. J. Agr. Sci.* 10:361–76.

Chandraratna, M. F. 1964. *Genetics and Breeding of Rice.* Longmans, Green & Co., Ltd., London.

Crops and Soils. Apr.–May 1963. Leaf area-oriented corn. 15(7):8.

DeWit, C. T. 1959. Potential photosynthesis of crop surfaces. *Neth. J. Agr. Sci.* 7:141–49.

Duncan, W. G. 1965. A model for simulating photosynthesis and other radiation phenomena in plant communities. *Proc. 10th Intern. Grassland Congr.,* pp. 120–25.

Duncan, W. G., R. S. Loomis, W. A. Williams, and R. Hanau. 1967. A model for simulating photosynthesis in plant communities. *Hilgardia* 38:181–205.

Eik, K., and J. J. Hanway. 1965. Some factors affecting development and longevity of leaves of corn. *Agron. J.* 57:7–12.

Evans, L. T. 1963. *Environmental Control of Plant Growth.* Academic Press, New York.

Friend, D. J. C., V. A. Helson, and J. E. Fisher. 1962. Leaf growth in Marquis wheat as regulated by temperature, light intensity and day length. *Can. J. Botany* 40:1299–1311.

Gaastra, P. 1962. Photosynthesis of leaves and field crops. *Neth. J. Agr. Sci.* 10:311–24.

Gardener, C. J. 1966. The physiological basis for yield differences in three high and three low yielding varieties of barley. M.S. thesis, Univ. of Guelph, Ontario, Canada.

Grundbacher, F. J. 1963. The physiological function of the cereal awn. *Botan. Rev.* 29:366–81.

Graham, D., and K. J. Lessman. 1966. Effect of height on yield and yield components of two isogenic lines of *Sorghum vulgare* Pers. *Crop Sci.* 6:372–74.

Hanway, J. J. 1963. Growth stages of corn. *Agron. J.* 55:487–92.

Hanway, J. J., and H. E. Thompson. 1967. How a soybean develops. *Iowa State Univ. Spec. Rept.* 53.

Hesketh, J. D. 1963. Limitations to photosynthesis responsible for differences among species. *Crop Sci.* 3:493–96.

Hesketh, J. D., and D. N. Moss. 1963. Variation in the response of photosynthesis to light. *Crop Sci.* 3:107–10.

Hesketh, J. D., and R. B. Musgrave. 1962. Photosynthesis under field conditions. IV. Light studies with individual corn leaves. *Crop Sci.* 2:311–15.

Hoover, W. H., E. S. Johnston, and F. S. Brackett. 1933. Carbon dioxide assimilation in a higher plant. *Smithsonian Inst. Misc. Collections* 87:1–19.

Hoyt, P., and R. Bradfield. 1962. Effect of varying leaf area by partial defoliation and plant density on dry matter production in corn. *Agron. J.* 54:523–25.

Idso, S. B., D. G. Baker, and D. M. Gates. 1966. The energy environment of plants. *Advan. Agron.* 18:171–218.

International Rice Research Institute. 1965. *The Mineral Nutrition of the Rice Plant.* Johns Hopkins Press, Baltimore.

Kasanaga, K., and M. Monsi. 1954. On the light transmission of leaves and and its meaning for the production of matter in plant communities. *Japan. J. Botany* 14:304–24.

Krizek, D. T., W. J. McIlrath, and B. S. Vergara. 1966. Photoperiodic induction of senescence in xanthium plants. *Science* 151:95–96.

Langer, R. H. M. 1959. Growth and nutrition of timothy *(Phleum pratense).* IV. The effect of nitrogen, phosphorus and potassium supply on growth during the first year. *Ann. Appl. Biol.* 47:211–21.

Laing, D. P. 1966. The water environment of soybeans. Ph.D. thesis, Iowa State Univ. Microfilm No. 66–6988.

Loomis, R. S., and W. A. Williams. 1963. Maximum crop productivity: An estimate. *Crop Sci.* 3:67–72.

Loomis, R. S., W. A. Williams, and W. G. Duncan. 1967. Community architecture and the productivity of terrestrial plant communities. In A. San Pietro, F. A. Greer, and T. J. Army (eds.), *Harvesting the Sun.* Academic Press, New York.

Loomis, R. S., W. A. Williams, W. G. Duncan, A. Dovrat, and A. F. Nunez. 1968. Quantitative descriptions of foliage display and light absorption in field communities of corn plants. *Crop Sci.* 8:352–56.

Milthorpe, F. L. 1956. *The Growth of Leaves.* Butterworth Scientific Publications, London.

Monteith, J. L. 1965. Light distribution and photosynthesis in field crops. *Ann. Botany* 29(113): 17–37.

Moss, D. N., R. B. Musgrave, and E. R. Lemon. 1961. Photosynthesis under field conditions. III. Some effects of light, carbon dioxide, temperature and soil moisture on photosynthesis, respiration and transpiration of corn. *Crop Sci.* 1:83–87.

Moss, D. N., and D. E. Peaslee. 1965. Photosynthesis of maize leaves as affected by age and nutrient status. *Crop Sci.* 5:280–81.

Netherlands Journal of Agricultural Science. 1962. Vol. 10, No. 5, Special Issue. Fundamentals of dry-matter production and distribution.

Nichiporovich, A. A. 1960. Photosynthesis and the theory of obtaining high crop yields. An abstract with commentary by J. N. Black and D. J. Watson. *Field Crop Abstr.* 13:169–75.

Oshima, E. 1962. Studies on photoassimilation of CO_2 by sugar beets with regard to situation of leaves for light receiving (Japanese). Rept. 59. Hokkaido Nat. Agr. Exp. Sta., p. 59. English summary.

Peaslee, D. E., and D. N. Moss. 1968. Stomatal conductivities in K-deficient leaves of maize (*Zea mays* L.). *Crop Sci.* 8:427–30.

Pendleton, J. W., G. E. Smith, S. R. Winter, and T. J. Johnston. 1968. Field investigations of the relationships of leaf angle in corn (*Zea mays* L.) to grain yield and apparent photosynthesis. *Agron. J.* 60:422–24.

Quinby, J. R. 1961. Use of spontaneous mutations in sorghum. *Mutation and Plant Breeding* NAS-NRC 891:183–205.
Radford, P. J. 1967. Growth analysis formulae—Their use and abuse. *Crop Sci.* 7:171–75.
Radke, J. K. 1967. In basic applied research—Scientists study wind—Corn rows as windbreaks. *Agr. Res.* 15:15.
Sakamoto, C. M., and R. H. Shaw. 1967. Apparent photosynthesis in field soybean communities. *Agron. J.* 59:73–75.
San Pietro, A., F. A. Greer, and T. J. Army (eds.). 1967. *Harvesting the Sun.* Academic Press, New York.
Shibles, R. M., and C. R. Weber. 1965. Leaf area, solar radiation interception and dry matter production by soybeans. *Crop Sci.* 5:575–77.
Slayter, R. O. 1967. *Plant-Water Relationships.* Academic Press, New York.
Stalfelt, M. G. 1937. Der Gesaustauch der Moose. *Planta* 27:30–60.
Stickler, F. C., and A. W. Pauli. 1961. Grain producing value of leaf sheaths in grain sorghum. *Agron. J.* 53:352–53.
Tanner, J. W., and N. C. Stoskopf. 1967. The plant resource. *Agr. Inst. Rev.* 22:25–29.
Thorne, G. N. 1959. Photosynthesis of lamina and sheath of barley leaves. *Ann. Botany* 23:365–70.
Thorne, G. N., and D. J. Watson. 1955. The effect on yield and leaf area of wheat of applying nitrogen as a top dressing in April or in sprays at ear emergence. *J. Agr. Sci.* 46:449–56.
Wallace, D. H., and H. M. Munger. 1965. Studies on the physiological basis for yield differences. I. Growth analysis of six dry bean varieties. *Crop Sci.* 5:343–48.
Waggoner, P. E., D. N. Moss, and J. D. Hesketh. 1963. Radiation in the plant environment and photosynthesis. *Agron. J.* 55:36–39.
Watson, D. J. 1952. The physiological basis of variation in yield. *Advan. Agron.* 4:101–45.
Webb, J. A., and P. R. Gorham. 1964. Translocation of photosynthetically assimilated C^{14} in straight-necked squash. *Plant Physiol.* 39:663–72.
Williams, W. A., R. S. Loomis, W. G. Duncan, A. Dovrat, and A. F. Nunez. 1968. Canopy architecture at various population densities and the growth and grain yield of corn. *Crop Sci.* 8:303–8.
Wright, J. L., and E. R. Lemon. 1966. Photosynthesis under field conditions. IX. Vertical distribution of photosynthesis within a corn crop. *Agron. J.* 58:265–68.
Yao, A., and R. H. Shaw. 1964. Effect of plant population and planting pattern of corn on water use and yield. *Agron. J.* 56:147–51.

❧ CHAPTER THREE ❧ MINERAL NUTRITION

❧ IN PRESENT-DAY AGRICULTURE, especially since the work of Liebig, Gilbert and Lawes, and others in the mid-1800s, considerable emphasis has been given to the inorganic nutrition of the plant, in some cases with seeming disregard for the massive role of carbon dioxide and light. Keeping the latter two factors in perspective, it is appropriate now to discuss mineral nutrition. The mineral elements are critical indeed, and this facet of the environment is one readily changed by the agriculturist through soil management and fertilizer application practices.

NUTRIENT REQUIREMENTS

Plants are unique because they are autotrophic. That is, they need no organic molecules to survive, no previously incorporated carbon sources from which to derive energy. Rather they can synthesize the necessary growth constituents from inorganic substances. Plants are ranked in their degree of autotrophism because some can use nitrogen gas (N_2) directly while others require it in an oxidized or reduced form. The bluegreen algae are the most autotrophic, directly fixing N_2 for their use. Many legumes are given an intermediate rank in the autotrophic order because they host the symbiotic bacteria which reduce N_2 to the valence of NH_4^+. Finally, most other green plants can use nitrogen present as NO_3^- or NH_4^+ but not that present as N_2 in the soil air. All other nutrients can be used in the forms found in the soil, although a certain oxidation or reduction state is most favored in absorption, as will be discussed later.

The list of essential plant nutrients has continued to increase over the years. This increased list is due primarily to two factors: (1) improved techniques for the study of mineral nutrition, from soilless water culture to radioactive tracers and very sensitive techniques of paper and gas chromatography; and (2) as Steward (1964) suggests, "There is a real possibility that plants in their native habitat may have managed to encompass all their activities with fewer chemical elements than some plants that are now virtually man-made."

There are basically two criteria by which the essentiality of an element may be established, and both are subject to certain limitations and qualifications.

1. An element is considered essential for a plant if the plant, when grown in a medium devoid of that element, fails to grow and to complete its life cycle, whereas in the presence of a suitable concentration of that element it grows and reproduces normally. Indirect or secondary beneficial effects of an element, such as reversal of the inhibitory effect on some other element, do not qualify an element as essential. It is by this means that every mineral element's essentiality for higher plants has been determined.
2. An element is considered essential for a plant if it is shown to be a constituent of a molecule which is known as an essential metabolite.

In reviewing the progress of knowledge of mineral nutrition, it must first be noted that Aristotle suggested the plant's growth requirements could be met by earth, air, fire, and water. Beginning with the work of van Helmont in the early 1700s, the singular importance of water as a contributor to dry weight was emphasized, almost to the exclusion of soil and air. As measurement techniques became more sophisticated, the soil's direct contribution became apparent. Initially, agriculturists like Jethro Tull conceived the idea that the plant root devoured soil particles directly and that thorough working of the soil enhanced the plant's ability to "feed on the soil." Later, an understanding of chemistry built a foundation for the idea that ions entered the root and were the source of mineral nutrition.

In the early 1800s, the source of carbon was considered to be the humus of the soil, but Liebig is given credit in the mid-1800s for emphasizing that the air and not the soil was the source of the carbon. The other nutrients required in large amounts were also identified during the 1800s. Most nutrients required in small amounts have been known since 1900. The plant's dry weight can now be accounted

TABLE 3.1. Concentrations of nutrient elements in plant material at levels considered adequate (Bonner and Varner, 1965)

Element	Atomic Weight	Concentration in Dry Matter		Relative Number of Atoms with Respect to Molybdenum
		μmole/g	ppm or %	
			ppm	
Molybdenum (Mo)	95.95	0.001	0.1	1
Copper (Cu)	63.54	0.10	6	100
Zinc (Zn)	65.38	0.30	20	300
Manganese (Mn)	54.94	1.0	50	1,000
Iron (Fe)	55.85	2.0	100	2,000
Boron (B)	10.82	2.0	20	2,000
Chlorine (Cl)	35.46	3.0	100	3,000
			%	
Sulfur (S)	32.07	30	0.1	30,000
Phosphorus (P)	30.98	60	0.2	60,000
Magnesium (Mg)	24.32	80	0.2	80,000
Calcium (Ca)	40.08	125	0.5	125,000
Potassium (K)	39.10	250	1.0	250,000
Nitrogen (N)	14.01	1,000	1.5	1,000,000
Oxygen (O)	16.00	30,000	45	30,000,000
Carbon (C)	12.01	35,000	45	35,000,000
Hydrogen (H)	1.01	60,000	6	60,000,000

for very completely. More specialized plants and more exacting analyses, however may demonstrate additional nutrients needed either in minute supply or present but not essential.

Essential elements have come to be divided somewhat arbitrarily into those required by the plant in large or small quantities. Table 3.1 presents a comparison of quantities of the several elements found in plant tissue and ranks their concentration in relation to molybdenum.

The mineral elements represent a vast number of unique functions in the plant and for this reason need to be discussed individually. However, there are some functions which are shared by several elements in a group. The groupings suggested below are not absolute but do provide ways to organize the elements and think about them in categories.

First, nutrient requirements may be grouped in the following quantity requirement sets: carbon, hydrogen, and oxygen are required in tons per acre; nitrogen, sulfur, phosphorus, potassium, calcium, and magnesium are used in tens to hundreds of pounds per acre; and the micronutrients may be required in pounds, ounces, or grams per acre per year (Table 3.2).

TABLE 3.2. Essential nutrients absorbed from soil and representative roles in the plant

Element	Absorption Form	Total Lb/A (elemental form) Present in Soil (general value)*	Available Lb/A (general value)*	PPM of Element in Nutrient Solution (relative amt. needed)†	Representative Role in Plant
Nitrogen (N)	NO_3^- NH_4^+	4,000	1–50	100–200	Amino acid; protein synthesis; nucleic acids
Phosphorus (P)	$H_2PO_4^-$ $HPO_4^{=}$	1,200	0.01–0.10	63	Utilizing energy from food reserves
Sulfur (S)	$SO_4^{=}$	800	1–10	32	Sulfhydryl groups
Potassium (K)	K^+	50,000	5–15	200	Hexokinase
Calcium (Ca)	Ca^{++}	15,000	10–100	120	Calcium pectate
Magnesium (Mg)	Mg^{++}	6,000	5–50	24	Chlorophyll; respiration
Iron (Fe)	Fe^{++}	50,000	Trace	5.6	Cytochromes; ferredoxin
Manganese (Mn)	Mn^{++}	1,600	Trace	0.6	Formation of amino acids
Boron (B)	$BO_3^{=}$	100	Trace	. . .	Possibly in sugar translocation
Copper (Cu)	Cu^{++}	50	Trace	0.02	Nitrate reduction
Zinc (Zn)	Zn^{++}	50	Trace	0.07	Dehydrogenases
Molybdenum (Mo)	$MoO_4^{=}$	Trace	Trace	0.01	Nitrate reductase
Chlorine (Cl)	Cl^-	Trace	Trace	. . .	Photosynthetic phosphorylation

Adapted from: * AG-26 "Agronomy Practice with Understanding." 1965. Cooperative Extension. Iowa State University.
† W. G. Schrenk and J. C. Frazier. Enzymes—like plant nutrients—necessary for life. *Plant Food Rev.* Fall 1964.

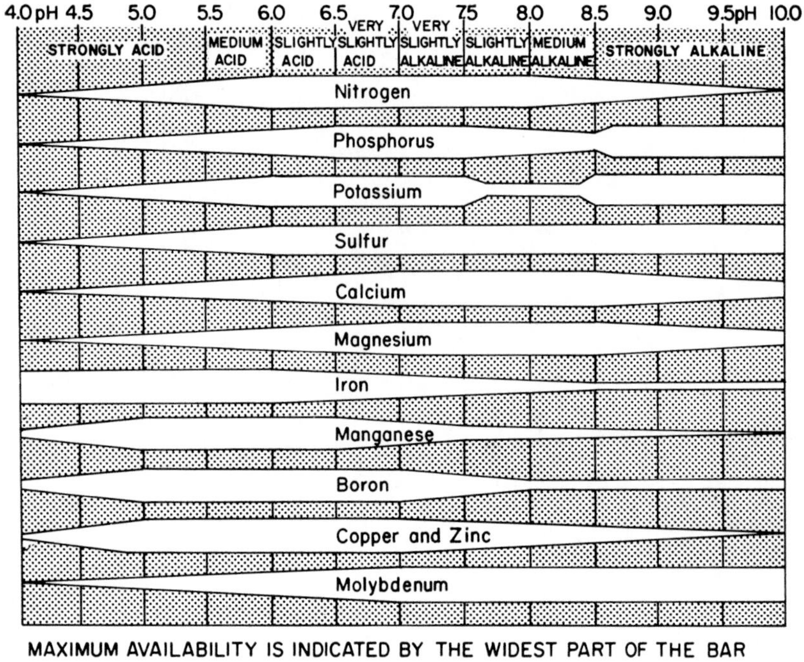

4.0 pH 4.5 5.0 5.5 6.0 6.5 7.0 7.5 8.0 8.5 9.0 9.5 pH 10.0

STRONGLY ACID | MEDIUM ACID | SLIGHTLY ACID | VERY SLIGHTLY ACID | VERY SLIGHTLY ALKALINE | SLIGHTLY ALKALINE | MEDIUM ALKALINE | STRONGLY ALKALINE

Nitrogen

Phosphorus

Potassium

Sulfur

Calcium

Magnesium

Iron

Manganese

Boron

Copper and Zinc

Molybdenum

MAXIMUM AVAILABILITY IS INDICATED BY THE WIDEST PART OF THE BAR

FIG. 3.1 Soil reaction influence on availability of plant nutrients (Truog, 1946).

Secondly, a useful means of grouping these elements is in a use classification to be discussed more fully later. On this basis, the (1) *basic structure* is provided by carbon, hydrogen, and oxygen. (2) *Energy storage, transfer,* and *energy bonding* are provided by nitrogen, sulfur, and phosphorus. (3) *Charge balance* is provided by potassium, calcium, and magnesium; and the micronutrients play a role in (4) *enzyme activation* and *electron transport.*

NUTRIENT AVAILABILITY

The two primary factors affecting the availability of the elements provided by the soil are: (1) the natural supply of the nutrient, which is closely tied to the parent materials of that soil and the vegetation under which it developed; and (2) the soil pH as it affects nutrient availability (Fig. 3.1). Other modifying factors are: (3) the relative activity of microorganisms, which play a very active role in nutrient release (especially nitrogen), and may, as in the case of the mycorrhizae, directly function in nutrient uptake; (4) fertility additions from commercial fertilizer and animal or green manure; (5) soil temperature; (6) soil moisture; and (7) soil aeration.

Even though the essential nutrients are present in the soil, many external and internal factors influence whether or not they will be absorbed. A summary discussion of these factors follows.

EXTERNAL FACTORS AFFECTING ABSORPTION

1. Oxidation-reduction state of the element. Each element (with the exception of N and Mn) is absorbed most effectively at a particular charge level. For many of the elements, the most oxidized state in which they occur naturally is favored for absorption in aerated soils.

2. Concentration of the element. While accumulation is an active process and does not function extensively on a concentration gradient or a diffusion pressure gradient, the relative concentration influences the likelihood that the nutrient will be present at an absorption site. In addition, because movement of the ion to the xylem is passive, it will be affected by the concentration.

3. Water content of the soil. This point interacts heavily with items 4 and 5, for while water must be adequate to keep the elements in solution, the soil must be dry enough so that

4. aeration remains adequate for respiration and energy release, and

5. the temperature is favorable for both root and microbial activity.

6. Effect of pH on root growth. This is most commonly indirect and related to its influence on the availability of beneficial, essential nutrients (Fig. 3.1) as contrasted to elements like aluminum which are toxic in large quantities. The toxic elements are released at lower pH.

INTERNAL FACTORS AFFECTING ABSORPTION

1. Cell wall. The wall is differentially permeable and selectively absorbs particular cations and anions. The cation exchange capacity of the cell wall is discussed in Chapter 9. Cations appear to have some competitive advantage over anions in uptake, according to many reports. Yet three essential nutrients in the plant are absorbed as anions (NO_3^-, $H_2PO_4^-$, and $SO_4^=$), so it is obvious the plant has a general ability to absorb anions in considerable quantities.

2. Respiration. Movement across the plasmalemma and movement across the tonoplast to the vacuole are energy-requiring reactions (Hendricks, 1966). Rapid ion uptake and growth are associated with warm, moist, well-aerated soils, and ion accumulation is linked to aerobic metabolism. Compact soils, for example, can lead to potassium deficiency symptoms even when adequate potassium has been supplied.

3. Type of cell and stage of development. The entry of inorganic materials into the xylem from the soil appears to occur primarily in the root tips. Experiments with isotopes have shown the uptake from the soil to be principally in the apical meristem and the terminal centimeters of root tips. It is generally conceded that absorption occurs through root hair cells and that a particular cell may be effective in absorption for just a few days.

4. Transpiration. This contributes in a minor way to the passive transport of nutrients across the cell membrane, but the dominant factor in nutrient absorption is the active transport made possible by respiratory energy.

Ions of the soil solution seem to have free access to only a part of the root cross section. Entry of solutes into the xylem is metabolically regulated and does not occur freely from the soil solution in large amounts. Once inside the xylem, the solute is relatively unhindered. In addition to the mineral ions, some surprisingly large molecules can enter the xylem. It is useful to note that the xylem solution is considerably more dilute than the phloem sap. The xylem solution is strongly influenced by the transpirational pull. The nutrients do move on a concentration gradient but finally become most concentrated in the areas of greatest meristematic activity. Thus mineral nutrients may be translocated from older to younger tissue if they are in short supply.

Meristematic cells are very active in ion accumulation. With cell expansion there is a dramatic increase in membranes which serve to control the movement of ions. These include not only membranes of the tonoplast and plasmalemma but also the internal cytoplasmic complex and the inner mitochondrial membrane which forms the cristae. Ion accumulation increases with this membrane development, presumably for any one of four reasons: (1) increased cell or vacuole surface, (2) increased mitochondrial activity, (3) increased complex of cytoplasmic membranes, or (4) changing membrane properties. The Ca^{++} ion plays a key role in maintaining and synthesizing these membranes. In addition, high-energy bonds, probably phosphorylated and derived from either respiration or photosynthetic ATP, are utilized in ion uptake.

SUMMARY

The plant cell is bounded by and contains a complex of membranous structures. With cell growth, the amount of membrane and the rate of ion uptake increase in parallel fashion, suggesting a relationship between rate of transport and content of mitochondria and endoplasmic reticulum. The increased salt uptake may also be related to changing membrane properties, which are reflected in higher respiration, phospholipid content, and response to calcium, the element essential to membrane growth and function in ion uptake.

FUNCTION AND USE OF SPECIFIC ESSENTIAL ELEMENTS

In addition to carbon, hydrogen, oxygen, and nitrogen, twelve mineral elements are generally recognized as essential for higher green plants. Most of these are involved in structural features of the plant body as well as being cofactors of numerous enzymatic reactions. A deficiency then may initially cause a lack of one structure or one enzymatic process but rather quickly disrupts a wide array of metabolic processes. The elements will be considered under the four groups mentioned earlier, and then their specific unique involvement will be described. Table 3.2 will serve as a summary of the form in which each element is absorbed, its concentration in soil (a generalized value), how much the plant needs, and some of the uses made of each element.

BASIC STRUCTURE

Carbon. Carbon is the backbone of organic compounds. Its four valence charges serve as points of attachments for other elements. The four charges provide the possibility of asymmetric, mirror-image structures and thereby provide the biological organism a means of selection and sorting of the materials it uses because the structures have unique shapes and configurations. This asymmetry is commonly utilized by a specialized enzyme system in synthesis processes. Carbon may represent approximately 45% of the dry weight of a plant. It is found in carbohydrates (particularly the sugars and starches), the proteins, the lipids, and in the unique chain and ring structures such as carotene and nucleic acids. Carbon is absorbed into the plant through the stomates of the leaf in the form of carbon dioxide.

Hydrogen. Hydrogen, when attached to carbon, produces a mildly reactive hydrocarbon. In addition, the transport of hydrogen and its accompanying electron during photosynthesis represents one of the energy-generating processes in the plant system. Hydrogen is supplied to the plant via water and released for reaction by photolysis. It composes about 6% of the plant's dry weight.

Oxygen. Oxygen, upon insertion between carbon and hydrogen, enhances the reactivity of the compound. The primary source of oxygen is carbon dioxide. Plants contain approximately 43% oxygen.

Thus, these three elements serve as the backbone, regulators, and fillers of the plant's chemical structure.

ENERGY STORAGE AND TRANSFER; ENERGY BONDING

Nitrogen. Nitrogen is quantitatively (next to hydrogen) the second most important root-absorbed essential element. It is present in the air as N_2 gas but must be bonded to hydrogen or oxygen by lightning, bacteria (free-living or symbiotic), or industrial processes (Haber process) before the plant can use it.

Young plants may contain 6% nitrogen on a dry basis; older plants range from 0.5 to 2%. A lack of nitrogen limits cell expansion and, more especially, cell division. The first effect of limited nitrogen is a reduction in the rate of meristematic activity. As a consequence, in the early stages of nitrogen deficiency, carbohydrates may tend to accumulate.

Nitrogen is a key component of protoplasm as a part of the protein molecule and other derivatives. It plays a central role in the *peptide bond* which bonds molecules together at a low energy (Fig. 3.2). This bond makes possible the synthesis of proteins from long chains of amino acids. Nitrogen is also important in ring structures such as adenine (Fig. 3.3) and tryptophan (Fig. 3.4). Adenine in turn represents one of the nitrogen bases in the genetic material of heredity (DNA) and protein synthesis (RNA), as well as being combined with ribose and phosphate groups to form ATP, ADP, and AMP. Nitrogen

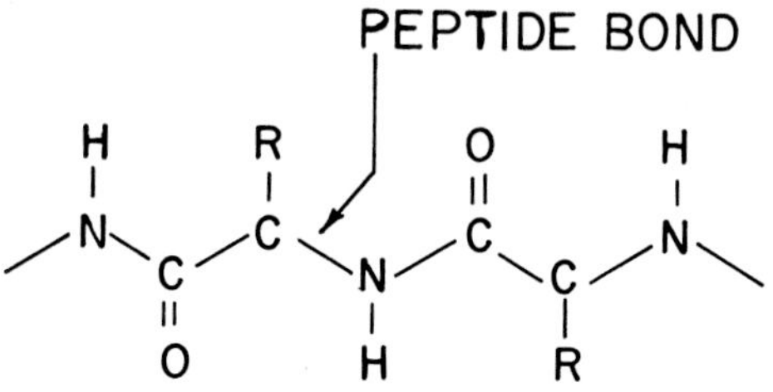

FIG. 3.2 ☙ The peptide bond, a bond between carbon, oxygen and nitrogen which is developed with a low-energy input.

PEPTIDE BOND

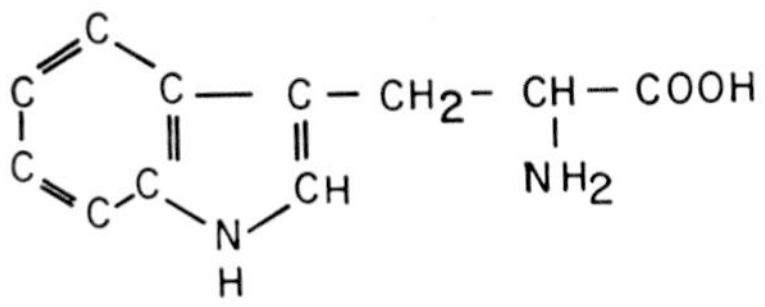

FIG. 3.3 Tryptophan, an amino acid with nitrogen in the ring as well as in the alpha position of the acid.

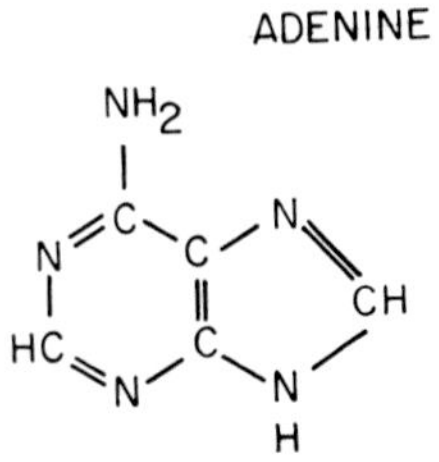

ADENINE

FIG. 3.4 Adenine, a nitrogen-containing ring structure.

thus plays a primary role in many types of low-energy bonding. The proteins in the vegetative cells of plants are largely functional rather than structural in nature. Many of these proteins are enzymes, still others are nucleoproteins, some of which are present in chromosomes. In such combinations proteins serve as catalysts and as directors of metabolism. Functional proteins are not stable entities, for they are continually being broken down and reformed.

In addition to its involvement in many vital plant compounds—ATP, NAD, DNA, RNA, IAA, all of which are important in growth—nitrogen is also integrally involved in differentiation. The following nitrogen compounds, unique to the species in many cases, form under differentiation conditions.

NICOTINE: tobacco leaves
QUININE: Cinchona tree bark
MORPHINE: poppy leaves
CAFFEIN: coffee seeds and tea leaves
COLCHICINE: meadow saffron

While the compounds listed above are referred to as differentiation products and as such have often been considered to be synthesized at the point of storage, research has shown that they are translocatable. For example, nicotine is formed in the roots of tobacco and then transported in the xylem to the leaves. A question commonly asked is, Did these differentiation products have value evolutionarily or are they merely coincidental in their presence? Two general observations might be made relative to the question. First, it is conceivable the differentiation materials cut down on insect attack or pathogen infestation and such plants survived as the most fit members of the species. Secondly, it is of interest to note that man has commonly adapted the differentiation products to his use for medicinal or drug purposes, whether or not they were of value to the plant itself. In this way some species have been perpetuated which might not have otherwise survived.

Absorption of nitrogen by the root occurs both as NO_3^- and NH_4^+. However, the mass of available nitrogen in the soil is converted to NO_3^- by bacterial action; as a result NO_3^- is the most commonly absorbed form. Once in the plant, NO_3^- is rapidly converted to NH_4^+ if there is a normal activity level of nitrate reductase. Water stress or low light conditions may reduce the activity of the nitrate reductase system and result in an accumulation of NO_3^- in the plant.

Sulfur. Sulfur, like nitrogen, serves as a basis of low-energy bonding in protein synthesis, forming thiol bonds analogous energetically to the peptide bonds of nitrogen. Sulfhydryl groups (SH) are thought to be important in the hardening of protoplasm to cold or drouth. Sulfur also functions in energy transfer in a manner similar to phosphorus.

Sulfur is needed for the synthesis of sulfur-containing amino acids—cystine, cysteine, and methionine—and for protein synthesis. This element activates certain proteolytic enzymes and is a constituent of some vitamins, of coenzyme A, and of glutathione. Sulfur is present in the oils of plants in the mustard and onion families. Sulfur fertilization has been shown to increase the oil content of crops such as flax and soybeans. Sulfur-deficient plants may accumulate nitrates as well as amides.

This element enters the plant as $SO_4^{=}$ and is, like nitrogen, quickly reduced, in this case to an SH group.

Phosphorus. Phosphorus is involved in energy storage and release and structurally in the phospholipids. When phosphate groups in plant cells are attached to an acceptor group—phosphorylation—the reactivity of the acceptor is increased. This phosphorylation results in a reduction of the activation energy (see Potassium) and overcomes otherwise unfavorable thermodynamic conditions within the plant system. As a result, the number of reactions in the biological system is increased.

There are three distinct phases of phosphate turnover in the plant. First, inorganic phosphate is absorbed and combined with organic molecules, one of which is ADP. In the second step, these phosphate groups are transferred to other molecules. For example, the first step

PHOSPHATE — RIBOSE — NICOTINAMIDE
|
PHOSPHATE — RIBOSE — ADENINE
|
PHOSPHATE

FIG. 3.5 ☙ Nicotinamide-adenine dinucleotide phosphate, the hydrogen carrier in many oxidation-reduction reactions. Phosphorus serves to make this a high-energy compound.

in the activation of glucose in respiratory breakdown is this energization by acceptance of a phosphate group. Finally, the phosphate group is split from the phosphorylated intermediates. The major source of energy for the incorporation of phosphate into organic combinations is the oxidation-reduction potential made available in oxidative metabolism. Phosphorus is also a key constituent in the electron (and hydrogen) transporting molecules NADP (nicotinamide-adenine dinucleotide phosphate) (Fig. 3.5).

Phosphate is an integral part of the phospholipids, fatty substances believed to play an essential role in the structure of the cell membrane and the structure to which chlorophyll is directly attached by its alcohol side chain phytol in the chloroplast. Nucleic acids (RNA and DNA) contain phosphorus, and while they serve as templates for protein synthesis rather than as energy storage components, their role apparently requires a high energy state. Finally, phosphorus serves to buffer the protoplasm, keeping it in the pH range to which the plant is best adapted.

Absorption of phosphorus occurs primarily as the orthophosphate ion $H_2PO_4^-$, but it may also be absorbed as $HPO_4^=$. These ions are very mobile and translocated freely within the plant. Therefore, phosphorus usually occurs in highest concentration in cells of high metabolic activity. In growing plants, this results in its being highest in the meristems and then later moving into the seed. While its mobility was stressed above, storage may occur for short or long periods of time. One complex storage form, phytic acid, serves as a reservoir for phosphorus that may be reactivated to a mobile form. Phytic acid serves as a source of phosphorus for the germinating seed and the developing seedling.

CHARGE BALANCE

Potassium. Potassium serves a wide range of functions in the plant (Fig. 3.6) but does not enter into structural components. It serves with magnesium and calcium to maintain cellular organization and to

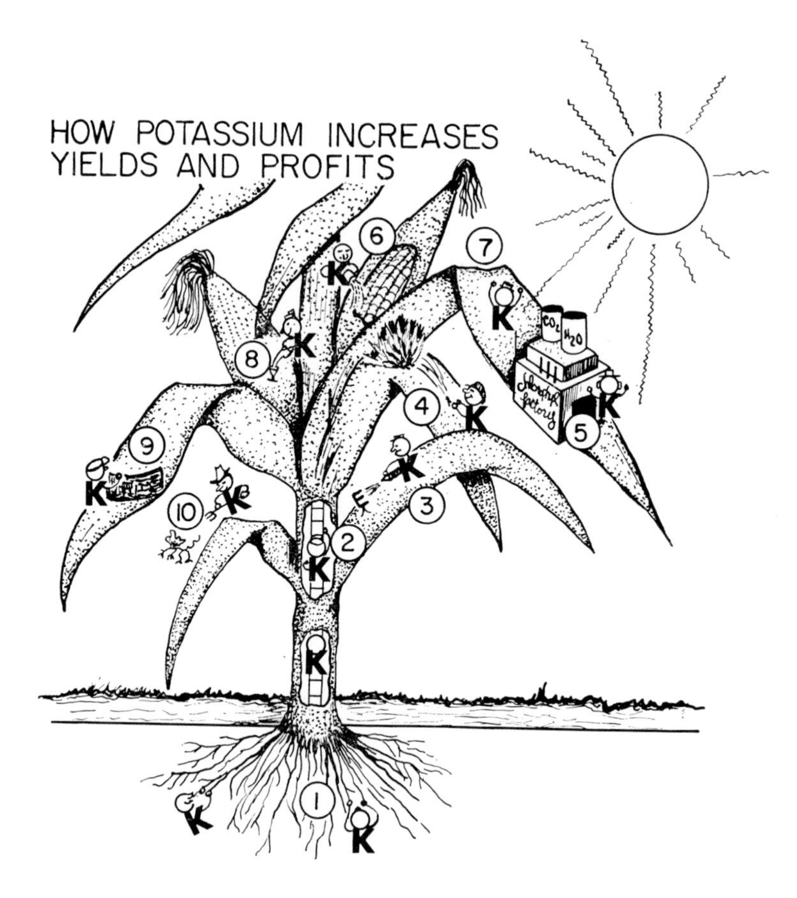

1. INCREASES ROOT GROWTH AND IMPROVES DROUGHT RESISTANCE.
2. BUILDS CELLULOSE AND REDUCES LODGING
3. AIDS MANY ENZYME ACTIONS
4. REDUCES RESPIRATION, PREVENTING ENERGY LOSSES
5. AIDS IN PHOTOSYNTHESIS AND FOOD FORMATION
6. GIVES GRAIN RICH IN STARCH
7. HELPS TRANSLOCATION OF SUGARS AND STARCH
8. INCREASES PROTEIN CONTENT OF PLANTS
9. MAINTAINS TURGOR, REDUCES WATER LOSS AND WILTING
10. HELPS RETARD DISEASES

FIG. 3.6 ☙ The functions of potassium in the plant (courtesy American Potash Institute).

provide an electrical charge balance, hydration, and permeability. For many such reactions, potassium, calcium, and magnesium appear to function as a group and interchangeably.

Potassium, along with the micronutrients, plays a major role as an enzyme activator or a catalytic entity. This activator or cofactor role (Fig. 3.7) is proposed to be one of the most important functions of many essential elements. As a basis for further discussion about enzyme activation, it is useful to review the nature and function of the enzyme itself. Enzymes are proteins that catalyze reactions involving organic substances. As organic catalysts, enzymes increase the rate of reaction of organic substances that commonly require cofactors for activity. Neither the enzyme nor its cofactor is degraded in the catalytic process but may be used over and over. Hence, a little potassium, manganese, or other mineral element "goes a long way" in this function. Of the 13 essential elements absorbed from the soil, apparently only nitrogen and boron do not serve as enzyme activators. The way in which metal ions activate certain proteins is not fully understood. It is known that catalysts lower the *energy of activation* needed for a reaction. All chemical substances require a certain minimum quantity of energy to undergo reaction. If the energy requirement is lowered, the reactions will proceed more readily (see Fig. 3.7).

One suggestion for the metal ion's activity is that it acts as a bridge between the enzyme protein and the substrate. Figure 3.8 is a diagram of such an arrangement. The enzyme A cannot catalyze the reaction of substrate B unless the metal ion M^+ is present.

Potassium is highly mobile and is present in high concentration in meristematic tissue but low in the seed. It serves to maintain a limited amount of iron active in chlorophyll synthesis. In potassium-deficient plants, free amino acids accumulate; where deficiencies are severe, the free acids decline with an increase in the concentration of amides. This results in an accumulation of nonprotein nitrogen. Potassium

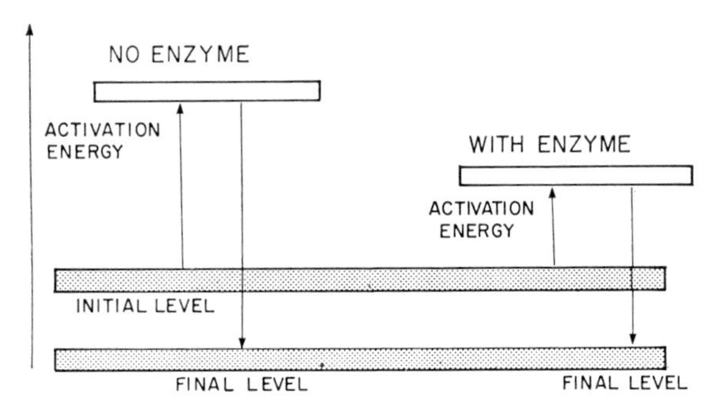

FIG. 3.7 ☙ An illustration of how "activation energy" is reduced by an enzyme, thereby allowing a reaction to occur more rapidly (Schrenk and Frazier, 1964).

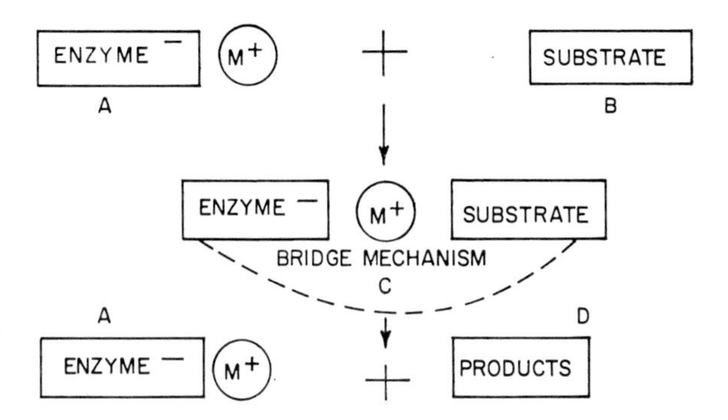

FIG. 3.8 ❧ The enzyme-substrate reaction. The metal ion serves as a bridge to make the linkage between the enzyme and the substrate more effective (Schrenk and Frazier, 1964).

deficiency is associated with a decrease in resistance to certain plant diseases, including powdery mildew and root rot. The winterhardiness of both alfalfa and Bermudagrass is greater with adequate potassium. Potassium is absorbed as K^+.

Calcium. Calcium is essential for maintenance of organization in the protoplasm and has been commonly considered to cement the cell walls together as calcium pectate serving to strengthen the middle lamella of the cell wall. A deficiency of calcium manifests itself in the failure of the terminal buds of plants and the apical tips of roots to develop.

Calcium is related to protein synthesis by its enhancement of the uptake of nitrate nitrogen and is associated with the activity of certain enzyme systems. It is relatively immobile and, unlike phosphorus and potassium, more is present in older than in younger leaves. It is absorbed as the divalent cation Ca^{++}.

Magnesium. Magnesium plays a key role as the center and the only mineral of the chlorophyll molecule. It also acts as a cofactor of several enzymes and serves to activate sulfhydryl groups, particularly those involving phosphorus metabolism. Magnesium frequently plays a role in the formation of an enzyme-substrate complex, for example, as an activator of hexokinase.

Magnesium is a mobile element and is readily translocated from older to younger plant parts in the event of a deficiency. In many species the deficiency results in an interveinal chlorosis of the leaf, in which only the veins remain green. Magnesium is required for the activation of many enzymes concerned with carbohydrate metabolism and is prominent in the Krebs citric acid cycle. Numerous phosphorylation reactions relating to nitrogen metabolism in plants are catalyzed by this element. Magnesium is related also to the synthesis of oil, and with sulfur it brings about significant increases in the oil content of several crops. Absorption occurs as Mg^{++}.

ENZYME ACTIVATION AND ELECTRON TRANSPORT

The remaining elements are most closely associated with electron transfer. These elements appear often to function in a subspecific manner. To explain this observation, Arnon proposed the concept of a "total metal" effect or a "multiple activation" effect. He suggested that in many systems one metal is required, but the other necessary metal might be one of several. It appears that the level of micronutrient requirement is genetically determined. There is a real possibility that plants in their native habitat may have succeeded in encompassing all their activities with fewer chemical elements than present-day plant varieties which have been extensively manipulated by plant breeders.

Before discussing specific activities of each micronutrient, the point should be made of the small range of tolerance allowable for their use under field conditions. Both natural soil conditions and nutrient culture experiments have shown that while small additions may give dramatic growth improvement under deficient conditions, further increments may be very toxic to plant growth. Boron, zinc, molybdenum, manganese, and copper have all been shown to be toxic at relatively low concentrations.

Iron. Iron is present in porphyrin-protein complexes such as cytochrome and catalase. It is at the center of the porphyrin ring in a manner analogous to magnesium in chlorophyll, and it presumably plays a central role by shifting from Fe^{+++} to Fe^{++} in oxidation-reduction reactions.

Iron is present as a constituent of catalase and peroxidase, and these enzymes in turn are hypothesized to degrade any potentially toxic levels of H_2O_2 in cells. The element has been shown to be capable of partly replacing molybdenum as the metal cofactor necessary for the functioning of nitrate reductase. Iron is also necessary for chlorophyll synthesis; a deficiency is not due to its involvement in the final structure but in the synthesis steps. Studies with algae indicate that iron functions in the synthesis of a specific kind of ribonucleic acid which in turn regulates chlorophyll synthesis through a chain of reactions not yet fully understood.

Boron. Boron was one of the first micronutrients identified as necessary for plant growth and yet its specific role is not well understood. There is evidence that it enhances ATP synthesis, plant reproduction, and pollen germination and that it may serve as a complexing agent to enhance the movement of sugars as they are translocated.

Boron is believed to influence cell development by the control it exerts on polysaccharide formation. Another function attributed to this element is its inhibition of the formation of starch by combining with the active site of phosphorylase. In this manner boron may function in a protective way by preventing the excessive polymerization of sugar at the sites of sugar synthesis. Further, it appears that boron may determine whether sugars are decomposed for energy release via the glycolytic pathway or the pentose phosphate shunt, two alternate pathways of sugar decomposition to pyruvic acid. Requirements for boron and calcium often go hand in hand, and this has suggested that boron, like calcium, may be needed in wall formation and in the metabolism of pectic compounds. It is of interest to note that numerous physiological (nonpathogenic) diseases, such as brown heart of turnips, heartrot of sugar beets, and leaf roll of potato, have been traced to boron deficiency. Boron is absorbed as $BO_3^{=}$.

Zinc. If zinc is absent, inorganic phosphate accumulates. Zinc is very likely needed for tryptophan or indoleacetic acid synthesis because less IAA synthesis occurs when zinc is deficient. In addition, many of the enzymes active in hydrogen transfer are activated by zinc, and it is a constituent of the protein which forms the enzyme carbonic anhydrase. This anhydrase catalyzes the reaction of carbonic acid with water and vice versa. Zinc is absorbed as Zn^{++}.

Manganese. Manganese plays a key role in nitrate assimilation, in the further conversion of soluble nitrogen compounds to protein, and in the splitting of peptide bonds. It is the predominant metal involved in general enzymatic decarboxylation and hydrolysis reactions and activates peptidases, such as the leucine aminopeptidase in malt and in spinach and cabbage extracts. Like iron and others in the heavy metal group, manganese functions in the activation of numerous enzymes concerned with carbohydrate metabolism. It functions in certain photochemical processes, including the Hill reaction. Hill reaction activity of isolated chloroplasts was found to be proportional both to the level of manganese supplied in the external nutrient solution and to the manganese content of the isolated chloroplasts.

Sometimes symptoms of iron deficiency may be traceable to an excess of manganese, most likely because manganese interferes with the reduction of the iron from the ferric to the ferrous state. Manganese is relatively immobile and is absorbed as Mn^{++}.

Molybdenum. Molybdenum is part of the flavoprotein molecule nitrate reductase which is one of the enzymes involved in nitrate reduction. Lacking molybdenum, growth is limited by the inability to use nitrate; thus nitrate may accumulate in the plant. If furnished with nitrogen in the form of ammonium salts or as organic nitrogen compounds, molybdenum-deficient plants may grow quite satisfactorily. Molybdenum is absorbed as $MoO_4^{=}$.

Copper. Copper-containing proteins are the enzymes which bring about certain oxidation reactions, like those which oxidize ascorbic acid and many phenolic compounds (polyphenol oxidases). Copper has been shown to facilitate the movement of calcium in barley and wheat. Phosphate accentuates a copper deficiency. Copper is absorbed as Cu^{++}.

Chlorine. Little is known of the role of chlorine in plant nutrition but it is required for the growth and development of several plants including barley, corn, cotton, sugar beets, tobacco, and tomatoes. Bromine at somewhat higher concentrations than chlorine may substitute for it, in part at least, in a way similar to that in which sodium substitutes for potassium. Chlorine is absorbed as Cl^{-}.

OTHER ELEMENTS PRESENT IN PLANTS

Cobalt plays a definite role in bacterial nutrition. It is essential in the formation of vitamin B_{12}, which in turn is essential to the formation of leghemoglobin needed for nitrogen fixation. The role of cobalt in higher plants is in question at present. Vanadium has not been universally accepted as essential for the growth of higher plants, but it may replace molybdenum to a certain extent in the nutrition of *Azotobacter*. Both sodium and silicon have recently been considered to be of possible importance in plant nutrition. Sodium has been suggested by workers in England to be essential on its own and not just as a substitute for potassium. In earlier years the primary emphasis was on sodium's ability to substitute for up to 60% of the potassium needs of root crops. Silicon, according to work in Japan, has been suggested to be essential for plant growth, but this view is not widely accepted and silicon's proposed action in plant metabolic and physiologic processes has not been determined.

FIELD PRACTICE

Mineral nutrition under field practice is the subject of many monographs and texts. This is an area regularly modified by the agriculturist and about which encyclopedic information has been generated. Throughout this text reference will be made to the impact of mineral nutrition on plant growth. For example, in Chapter 2 the influence of nitrogen, phosphorus, and potassium on leaf growth and photosynthesis was discussed. In Chapter 8 the effect of nitrogen on tillering is described, and the primary influence of nitrogen and phosphorus on top growth and their secondary effect on root growth is supported in Chapter 9. Further, in Chapter 11, the effect of fertilization on maintaining seeding mixtures is considered, and in Chapter 12 the increased winterhardiness of perennials fertilized with potassium is discussed as it relates to carbohydrate and protein synthesis. In addition to these examples, a few field problems related to mineral nutrition are described below.

SPECIES DIFFERENCES

Species have distinctly different relative requirements for the mineral nutrients. The grass family is characterized by a heavy demand for nitrogen in some fixed form, and it is common also to observe complementary benefits from potassium additions under intensive management. This potassium effect may be specifically tied to the activation effect on phosphoenolpyruvate carboxylase and protein synthesis. In a grass-legume mixture, the lack of potassium favors the grass, pointing up the ability of this family to compete for limited supplies—an ability ascribed by some researchers to be due to the lower cation exchange capacity of grass as compared to legume roots. Legumes have a heavy demand for phosphorus and potassium. With nitrogen fixation functional, legumes can obtain adequate nitrogen supplies symbiotically. This fixation process and the heavy nitrogen uptake appear to cause unique demands for the micronutrients molybdenum and cobalt. Further, boron deficiencies in legumes commonly result in internode shortening.

SPLIT APPLICATIONS

Forage crop fertilization must take into account the objective of supplying a relatively constant supply of production throughout the season where the management system calls for harvest by grazing. A grass-legume mixture will provide a rather uniform production of dry matter over the season. However, if the legume content of the forage mixture is low or lacking, grass may be fertilized with high rates of nitrogen to seek the same high yields as obtained with the grass-legume mixture. To obtain a similar uniformity of production over the season, the nitrogen should be applied in several applications. That is, equal applications might be made in April, June, and August. If high rates of potassium are to be applied, they also should be made in split applications. This split method is recommended as a means of minimizing the crop's tendency to absorb more potassium than necessary for maximum yield (that is, the tendency of the crop to take up certain elements as luxury consumption).

NITROGEN IMMOBILIZATION

The perennial grasses have a tendency to become nitrogen deficient, a condition the layman may refer to as "sod-bound." The difference between increasing nitrogen rates the first year and in later years is described in Figure 3.9. No single factor to explain this observation has been established but it might be explained by the large mass of nitrogen in the dead roots which is mineralized slowly because of the high ratio of energy to nitrogen in roots and a deficiency of mineral nitrogen in the soil for decomposition. The observation may also be related to the lack of cultivation and the large mass of living roots under grass.

POTASSIUM-NITROGEN INTERACTIONS

Numerous examples of nutrient interactions and the need for elemental balance in plant nutrition may be cited. For example, increased nitrogen fertility may cause a higher incidence of stalk rot in corn. The problem lies partly in an imbalance between nitrogen and potassium (Liebhardt and Murdock, 1965). Liebhardt et al. (1968) further propose that stalk and root parenchymas senesce prematurely

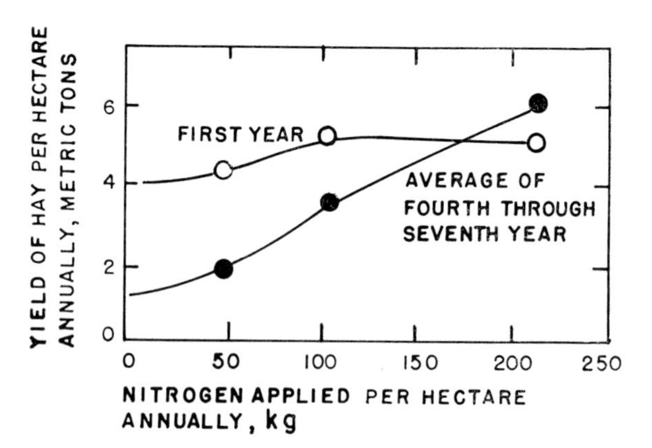

FIG. 3.9 ❧ Yield of perennial grasses (*Chloris guyana* and *Paspalum dilatatum*) with different quantities of ammonium sulfate (expressed as N) during a seven-year experiment in South Africa (Haylett and Theron, 1955).

when potassium is deficient, probably as a result of abnormal metabolism in the plant. There is lower carbohydrate production, the parenchyma disintegrates, and a weakened stalk results. It has been suggested (Younts and Musgrave, 1958) that the reduction of stalk rot was related to the chloride in KCl additions and that K_2SO_4 or KPO_3 did not give the same benefit. To date these observations cannot be fully explained, but the nitrogen-potassium balance does appear important in minimizing this disease.

ZINC-PHOSPHATE INTERACTIONS

Still another representative interaction of nutrients is described by zinc and phosphate additions to corn. Soils containing a high available supply of phosphorus produced less grain with phosphorus applied in a band along the row at planting time than without phosphorus additions. The yield-depressing effect of phosphorus was accentuated by additions of nitrogen fertilizer. The conclusion is drawn from experiments like the one in Table 3.3 that an antagonism between zinc and phosphate occurs at the root surface and results in a zinc-phosphate precipitate. Therefore, if phosphorus is added, an increased supply of zinc must also be added to overcome the partial tieup of zinc occurring in the zinc-phosphate precipitate. High levels of available soil phosphorus induced zinc deficiencies in fiber flax (Burleson and Page, 1967). The addition of phosphorus followed by an addition of zinc increased the total phosphorus in the upper and lower roots and decreased total phosphorus in the tops. When zinc was added first, phosphorus increased the total zinc in the lower roots

TABLE 3.3. Influence of nitrogen, phosphorus, and zinc on yields of corn and grain sorghum (Olson et al., 1965)

Treatment*	Yield, Bu/A When Soil P Test† Was	
	10 ppm or less (8 experiments)	More than 10 ppm (18 experiments)
Check	59	84
N only	77	95
N + P	81	93
N + P + Zn	81	89

* N sidedressed when plants were 6–18 inches in height at rates of 60–120 pounds N/A depending on the moisture supply; P and Zn applied at planting as starter, usually at the rate of 10 pounds of P and 5 or 10 pounds Zn/A.

† Bray and Kurtz No. 1 Test Solution.

and decreased total zinc in the tops. These workers conclude that phosphorus and zinc react together within the root in a manner that reduces either their mobility or solubility. ✻

LITERATURE CITED

Barber, S. A., and R. P. Humbert. 1963. Advances in knowledge of potassium relationships in the soil and plant. In M. H. McVicker, G. L. Bridger, and L. B. Nelson (eds.), *Fertilizer Technology and Usage.* Soil Sci. Soc. Am., pp. 231–68.

Black, C. A. 1968. *Soil-Plant Relationships,* 2nd ed. John Wiley and Sons, Inc., New York (esp. Chaps. 7–9).

Bonner, J., and J. E. Varner. 1965. *Plant Biochemistry.* Academic Press, New York.

Brouwer, R. 1962. Nutritive influences on the distribution of dry matter in the plant. *Neth. J. Agr. Sci.* 10:399–408.

Burleson, C. A., and N. R. Page. 1967. Phosphorus and zinc interactions in flax. *Soil Sci. Soc. Am. Proc.* 31:510–13.

Clark, R. B. 1966. Effect of metal cations on aldolase from leaves of *Zea mays* L. seedling. *Crop Sci.* 6:593–96.

Crops and Soils. March 1965. Corn lodging: Does K affect it? 17:7–9.

———. March 1965. Don't overlook the importance of micronutrients. 17:15–16.

DeWit, C. T., P. G. Tow, and G. C. Ennik. 1966. Competition between legumes and grasses. *Verslag. Landbouwk. Onderzoek.* No. 687.

Hageman, R. H. 1964. Corn's enzyme sets pace for yield. *Crops and Soils* 16(8):13–14.

Haghiri, F. 1966. Influence of macronutrient elements on the amino acid composition of soybean plants. *Agron. J.* 58:609–12.

Haylett, D. G., and J. J. Theron. 1955. Studies on the fertilisation of a grass ley. *Union S. Africa Dept. Agr. Bull.* 351.

Hendricks, S. B. 1966. Salt entry into plants. *Soil Sci. Soc. Am. Proc.* 30:1–7.
Hewitt, E. J. 1963. The essential nutrient elements: Requirements and interactions in plants. In F. C. Steward (ed.), *Plant Physiology: A Treatise.* Vol. III. *Inorganic Nutrition of Plants.* Academic Press, New York.
Jones, J. B., Jr. 1965. Molybdenum content of corn plants exhibiting varying degrees of potassium deficiency. *Science* 148:94.
Lawton, K., and R. Cook. 1954. Potassium in plant nutrition. *Advan. Agron.* 6:253–303.
Leonce, F. S., and M. H. Miller. 1966. A physiological effect of nitrogen on phosphorus absorption by corn. *Agron. J.* 58:245–49.
Liebhardt, W. C., and T. J. Murdock. 1965. Effect of potassium on morphology and lodging of corn. *Agron. J.* 57:325–28.
Liebhardt, W. C., P. J. Stangel, and T. J. Murdock. 1968. A mechanism for premature parenchyma breakdown in corn (*Zea mays* L.). *Agron. J.* 60:496–99.
Miller, R. J., J. T. Pesek, J. J. Hanway, and L. C. Dumenil. 1964. Soybean yields and plant composition as affected by P and K fertilizers. *Iowa Agr. Exp. Sta., Res. Bull.* 524.
Moore, A. W. 1966. Non-symbiotic nitrogen fixation in soil and soil-plant systems. *Soils and Fertilizers* 29:113–28.
Nelson, L. B. 1965. Advances in fertilizers. *Advan. Agron.* 17:1–84.
Nutman, P. S. 1965. Symbiotic nitrogen fixation. *Agronomy* 10:360–83.
Olson, R. A., D. D. Stukenholtz, and C. A. Hooker. 1965. Phosphorus-zinc relations in corn and sorghum production. *Better Crops with Plant Food* 49–50:19–24.
Ozburn, J. L., R. J. Volk, and W. A. Jackson. 1965. Effect of potassium deficiency on photosynthesis, respiration and utilization of photosynthetic reductant by immature bean leaves. *Crop Sci.* 5:69–75.
Schrenk, W. G., and J. C. Frazier. 1964. Enzymes—like plant nutrients—necessary for life. *Plant Food Rev.* Fall.
Sprague, H. B., ed. 1964. *Hunger Signs in Crops,* 3rd ed. David McKay Co., New York.
Steward, F. C. 1964. *Plants at Work.* Addison-Wesley Publishing Co., Inc., Reading, Mass.
Truog, E. 1946. Soil reaction influence on availability of plant nutrients. *Soil Sci. Soc. Am. Proc.* 11:305–8.
Vose, P. B. 1963. Varietal differences in plant nutrition. *Herbage Abstr.* 33:(1)1–13.
Wallace, T. 1961. *The Diagnosis of Mineral Deficiencies in Plants by Visual Symptoms: A Colour Atlas and Guide,* 2nd ed. Chemical Publ. Co., New York.
Watson, P. J. 1956. The physiological basis of the effect of potassium on crop yield. *Potassium Symposium 1956,* pp. 109–19.
Younts, S. E., and R. B. Musgrave. 1958. Chemical composition, nutrient absorption and stalk rot incidence of corn as affected by chloride in K fertilizer. *Agron. J.* 50:426–29.
Zieserl, J. F., W. L. Rivenbark, and R. H. Hageman. 1962. Nitrate reductase activity, protein content and yield of four maize hybrids at varying plant populations. *Crop Sci.* 3:27–32.

☙ CHAPTER FOUR ☙ CARBOHYDRATE, PROTEIN, AND LIPID NUTRITION

☙ THE PREVIOUS CHAPTERS have developed the ideas related to light utilization and photosynthesis as well as a consideration of the mineral nutrients necessary to drive reactions and build more complex substances. The dry matter thus formed is initially present as carbohydrate and then interconverted to a range of plant constituents which will conserve the energy for later use. These interconversions from carbon skeletons as first formed in photosynthesis to the structures of storage and structural carbohydrates, proteins, and lipid materials will be the topic of this chapter.

The term "carbohydrate resources" refers to all available carbohydrate present at a given time in the plant. Whether this carbohydrate was derived from accumulated reserves or from current photosynthesis is not readily distinguishable. Very early in the study of plants the botanist Sachs demonstrated that starch was formed in the chloroplast, and he considered starch to be the first product of photosynthesis. Simpler carbohydrates have now been identified using radioactive tracers, and in many plants 3-phosphoglyceric acid is considered to be the first formed structure from which all others are built.

CARBOHYDRATE INTERCONVERSIONS

It is convenient to regard photosynthesis as coming to an end with the formation of a hexose sugar (Fig. 4.1), although a decision about the point or structure with which to stop is relatively arbitrary. The hexose may immediately undergo further changes and interconvert from the 6-carbon glucose to the 6-carbon fructose. These two sugars

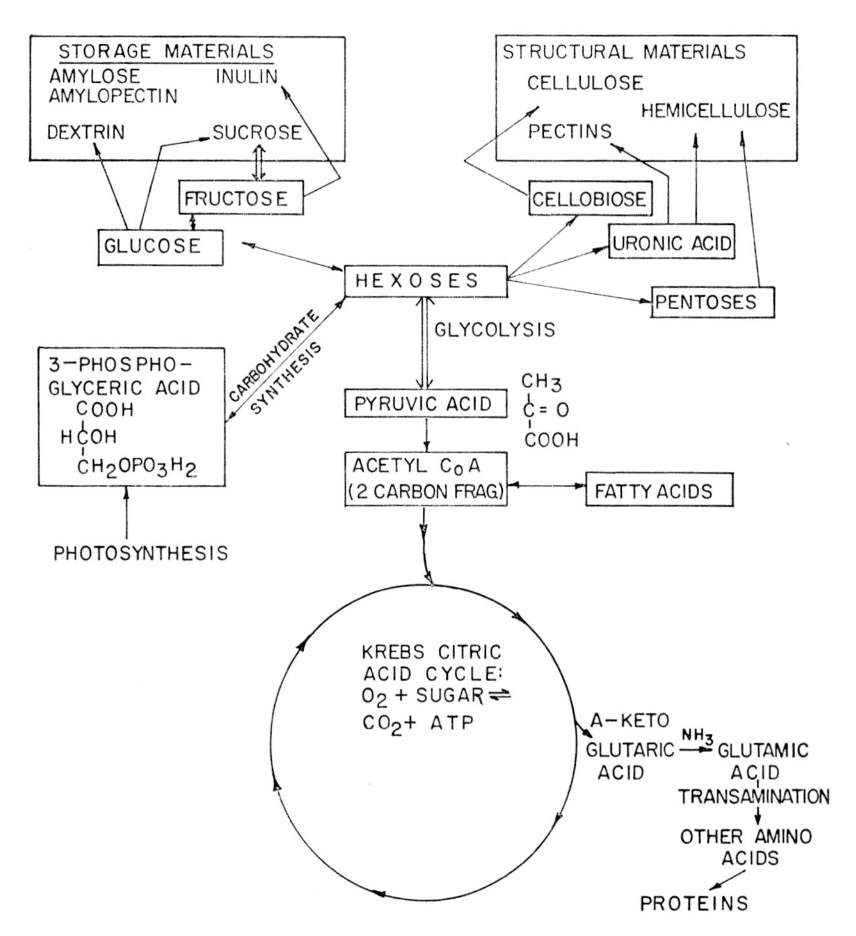

FIG. 4.1 ☙ Carbohydrate interconversions from 3-phosphoglyceric acid to storage and structural components or to fatty acids and proteins (adapted from Harlan, 1956).

may combine to form sucrose or the glucose may form the starch which Sachs first observed. When a fresh leaf is examined all of these carbohydrates are found, and if any one of such carbohydrates is fed to a starved leaf in a radioactive form, rapid interconversion occurs and the radioactive tracer is found quickly in all three. The equilibrium between starch ⟵⟶ glucose ⟵⟶ sucrose appears to give the plant a means of temporarily holding photosynthate (as starch) until this carbohydrate can move into the translocation stream as sucrose. Whether the photosynthate is retained in the leaf for further growth or is translocated depends on leaf age. During early leaf expansion, translocation is all import, shown in Figure 4.2 as radioactivity translocated from older leaves. When the leaf has reached 50% of its final size it is largely self-sufficient, and then the outward transport begins at a high rate. The result of these activities is to form a labile pool of carbohydrates in plant tissues from which carbon units may be drawn for respiration or synthesis of storage or structural components (Fig. 4.1).

Other plant organs may serve as important temporary storage sites for sucrose, as described by Duncan et al. (1965) for the stalk of corn. While photosynthate production was distinctly diurnal, directly correlated with solar radiation, and temperature independent, kernel growth was steady and uniform and appeared to be distinctly temperature dependent, as was indicated also by Vernon and Aronoff (1952) from soybean studies.

The several carbohydrate constituents have been emphasized as being in ready equilibrium. After the plant's respiratory needs are met, accumulation products are formed that later may be resynthesized into new growth. Plant tissues vary widely in the efficiency of carbohydrate use. A dark-grown test of this efficiency shows that plants may lose 20–67% of their dry weight, and only 9–16% of that dry weight can be accounted for in new growth.

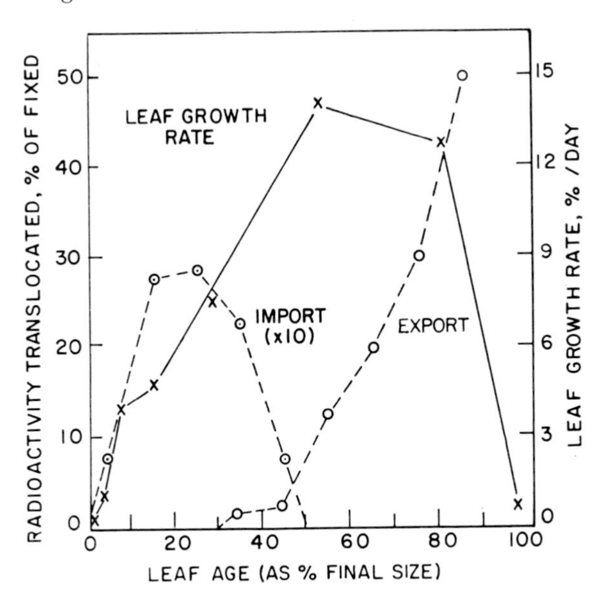

FIG. 4.2 ☙ Translocation into and out of a soybean leaf as a function of stage of development (Thrower, 1962).

As long as the production of short-chain sugars is occurring, the reaction is driven toward accumulation products. If these early products of photosynthesis are in short supply, the accumulation products break down.

The accumulation products (sometimes called "food reserves") and the site of accumulation are quite different for annual and perennial species. In annuals, seed formation often becomes the site of primary storage to nourish the new generation. The annual is less likely to store significant reserves in vegetative parts for extended periods although the *Sorghum* species and so-called "high-sugar" corn are notable exceptions. Perennials may store carbohydrates in the seed, but the perennial existence of the plant is completely dependent upon accumulation in the rhizomes, crowns, leaf sheaths, roots, and stems. Much of the management of perennial forage species is tied to a knowledge of these constituents, as will be discussed later.

Considerable discussion has occurred as to the relative importance of the accumulation products and leaf area when a forage crop recovers from defoliation. In 1927 Graber summarized his work at the Wisconsin Agricultural Experiment Station and stated that organic reserves are essential for normal top growth and root development (Smith, 1962). Carlson (1966) observed that the rate of leaflet unfolding was more rapid for defoliated than for nondefoliated plants and suggested growth regulators were also a factor in regrowth. He observed that root carbohydrate disappeared from shaded and defoliated plants at the same rate. This suggests that stored carbohydrate may be mobilized whenever the top is not forming more carbohydrate.

Work by Ward and Blaser (1961) on orchardgrass suggests the rate of regrowth is dependent on both carbohydrate reserves in the stubble and the leaf area remaining after cutting. For a crop like alfalfa, dependence is apparently more on root reserves. Recent studies on birdsfoot trefoil suggest that if it is allowed to grow for an extended period and then fully defoliated, severe stand losses may result. This suggests that regrowth is heavily dependent on foliage remaining after cutting, because birdsfoot trefoil appears to store low amounts of carbohydrates in the roots in the summer months (Fig. 4.3). As a result, this species is much better suited to partial defoliation and use as pasture, in contrast to alfalfa's usefulness as hay.

STORAGE AND STRUCTURAL CARBOHYDRATES

To be classified as a storage carbohydrate, the constituent commonly has: (1) a primary period of accumulation; (2) a period during

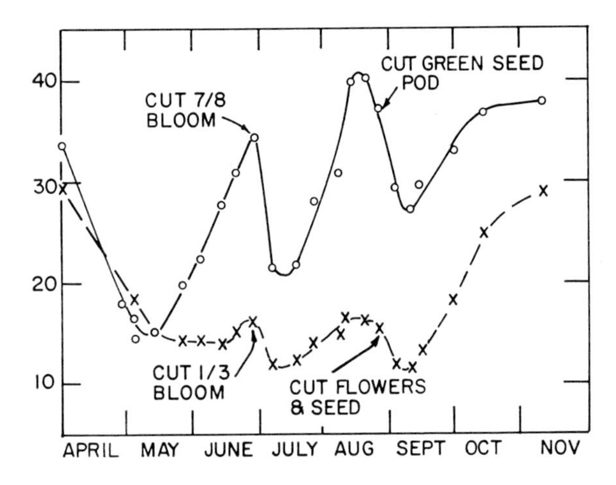

FIG. 4.3 ☙ Percent of total available carbohydrates in the roots of alfalfa (0-0) and birdsfoot trefoil (x-x) with two cuttings (Smith, 1962).

which the substance is maintained at a relatively high concentration; and (3) later on, in association with physiological processes, a decline in that constituent's concentration.

The following carbohydrates are commonly classified as storage materials (Fig. 4.1). Sucrose ($C_{12}H_{22}O_{11}$) and its components glucose and fructose (C_6 sugars) are all important. Dextrin is a storage carbohydrate of intermediate chain length. Starch is probably the most abundant organic compound on the earth and is known to be of two types: amylose and amylopectin. Amylose is a straight chain of 100–300 glucose molecules. Amylopectin has a branch off the main chain at approximately every twentieth glucose molecule. The main chain of glucose molecules is in 1-4 linkages with 1-6 cross links in amylopectin. Inulin is the only commonly known fructosan and usually occurs in shorter carbon chains than does starch.

The structural carbohydrates are considered to be cellulose, hemicellulose, and the pentosans. They are classified as structural because their concentration in the plant is subject to less variation and their rate of decomposition is slower when a plant is grown in the dark. Table 4.1 gives an example of this.

TABLE 4.1. Composition of alfalfa roots before and after growth in the dark (Smith, 1962)

Components	Before	After
Dry matter	34.2 g	26.4 g
Starch	10.8%	0%
Dextrins and soluble sugars	3.3%	1.8%
Total sugars	7.9%	1.4%
Hemicelluloses	10.1%	16.5%
Total nitrogen	2.6%	2.3%

The most important structural carbohydrate, cellulose, is a water-insoluble polysaccharide which upon degradation yields cellobiose. Cellulose is similar to amylose except that the glucose molecules are linked by beta linkages. As the plant ages, cellulose becomes impregnated with lignin, a complex nitrogen-containing molecule, and suberin, thus reducing the digestibility of the cellulose. The term hemicellulose is applied to certain polysaccharides in the cell walls composed of 5-carbon sugars (often xylose or arabinose) plus uronic acid. Hemicelluloses stored in seeds serve as a source of food for the young seedling, but it becomes available at a relatively slow rate.

Pentosans are long chains of 5-carbon sugars found primarily in the cell walls. They surpass the hexosans in their capacity to imbibe water, and the water-holding capacity of drouth-enduring plants such as the cacti is traceable to the presence of pentosans in mucilagenous materials. Pentosans may be drawn on as reserve food after the more readily utilizable starch and dextrins have been consumed. Pectic acid is the fundamental unit of the pectic compounds and represents a polymer of galacturonic acid. There are roughly 100 galacturonic acid residues in a chain and they are soluble in water. Calcium and magnesium salts of pectin form a key component of the middle lamella in the plant cell wall.

The accumulation products in temperate and tropical grass species appear to be distinctly different from each other. DeGugnac (Chap. 12 in Troughton, 1957) classified these plants as *"les sacchariferes"* and *"les levuliferes." Les sacchariferes* (warm-season plants) never form fructosan; they contain sucrose and the reducing sugars glucose and fructose. *Les sacchariferes* may be further subdivided according to the presence or absence of starch. *Les levuliferes* do contain fructosan and are commonly native to cool and temperate regions. They can be further subdivided on the basis of whether or not the fructosan is hydrolyzed by sucrase. Okajima and Smith (1964) note that fructosan was the highest in stem bases of orchardgrass, reed canarygrass, Kentucky bluegrass, and timothy. Tall fescue had about equal supplies of sucrose and fructosan, and bromegrass deviated from the pattern by having predominantly sucrose in its stem base (indicative that the classification scheme is not absolute but is a useful guide). The type of storage carbohydrate may vary depending on the organ assayed. For example, fructosan is usually the predominant reserve carbohydrate in orchardgrass stubble while sucrose is present in significant amounts in the roots.

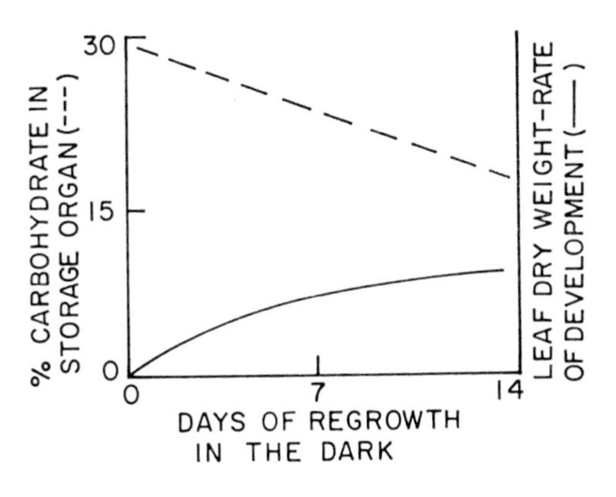

FIG. 4.4 Carbohydrate content of the storage organ as related to leaf regrowth in the dark.

FACTORS AFFECTING CARBOHYDRATE ACCUMULATION

Carbohydrates are of prime importance because they are the most readily available source of energy for growth and other metabolic processes. Because of the importance of storage carbohydrates in many management practices, the patterns of reserve accumulation and use have been studied with reference to a wide variety of modifying factors.

INTERNAL CONTROL BY THE PLANT

Carbohydrate reserves are assumed to be used for the growth of new leaves after defoliation, as described in Figure 4.4 when regrowth occurred in the dark. In the light this same decline occurs, but after a certain quantity of leaf growth is redeveloped, the reserve carbohydrates are restored (as represented by fructosan in Fig. 4.5), most

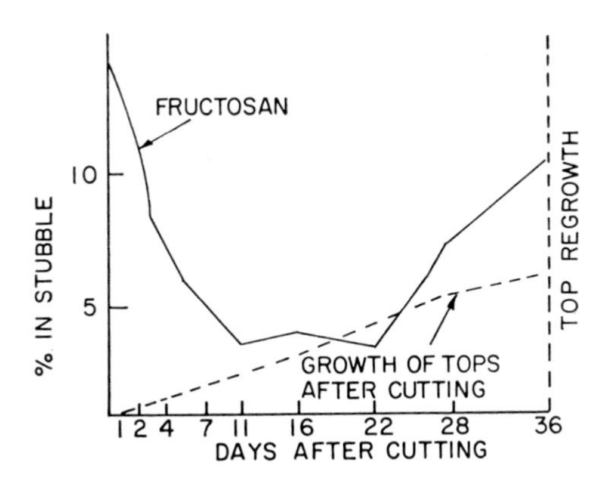

FIG. 4.5 Percent of fructosan in ryegrass stubble at different intervals following cutting and top growth of the same intervals (Hughes et al., 1962).

often to a value near the original level. The carbohydrate reserves are translocated to the aerial parts and used in the production of new photosynthetic area. When sufficient new leaf tissue is formed, the carbohydrate is replenished in the reserve organ. On a smaller scale, roots follow this same pattern even though they may not be the major storage organ. Nitrogen and the minerals show much narrower ranges of fluctuation following defoliation. Their pattern of decline is more analogous to the mild fluctuations of the reducing sugars.

The following priorities often occur for the use of carbohydrate storage materials after the plant has been partially or completely defoliated.

1. New leaf growth
2. Restoration of carbohydrate reserve
3. Root growth
4. Tillering

In this hierarchy of demand, the meristems of the leaf appear capable of controlling any new photosynthate to its own use. If the idea of the organ nearest the limiting factor being favored under stress is applied (the stress of defoliation in this case), it seems logical that the leaf would be favored over the root. Ranking the restoration of stored carbohydrates over new root growth is more difficult to rationalize. More and more evidence would suggest that the storage organ, if it is a modified stem, is not a strong "sink" and would be expected to be low in its competitive ability. The root, in contrast, has active meristems and should compete more vigorously for carbohydrate. Nevertheless, the storage organ does appear to recover first. Tillering usually occurs after the main shoot needs are met or when the main shoot loses its apical dominance.

The position of flowering in the above ranking depends upon the growth stage of the plant and the various environmental stresses to which the plant may be subjected. Defoliation usually favors vegetative growth at the expense of flowering, but defoliation by disease or insect attack may throw the plant into a survival response of reproductive physiology and flowering takes precedence over many other growth processes. When considering a plant under optimum growth conditions, however, the flowering response comes after all the other needs of the plant are met.

ENVIRONMENTAL EFFECTS

Seasonal Variation. In the Netherlands and Great Britain seasonal variation of stored reserves has been studied intensively for perennial

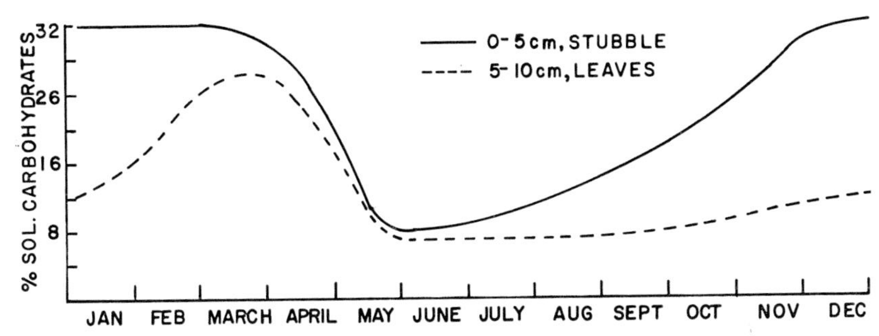

FIG. 4.6. ☙ Schematic curves of the percentage of soluble carbohydrates over the whole year in the dry matter of stubble and leaves of a grass sward (Sonneveld, 1962).

ryegrass, while in the United States alfalfa has been studied most frequently. The behavior of the reserves in autumn, winter, and spring is of first importance in the temperate climates. Figure 4.6 represents the seasonal variation in the stubble and leaves, schematically showing that ryegrass accumulates carbohydrates in the fall, uses very little over winter, but has a rapid use in spring as new growth begins.

For alfalfa the reserve pattern was first described by Graber (1927) at Wisconsin. The fall buildup, decline over winter, sharp decline until a minimum regrowth is established in spring, and then a regeneration of root storage is one of the classic curves in crop management (Fig. 4.7).

Grandfield (1943) showed that as the content of the carbohydrate declined in the root of two legumes, alfalfa and red clover (and it began to do this in autumn), the crown content rose in autumn; then the crown content declined in March. Red clover appears to follow a similar pattern, but it has a more rapid drop in carbohydrate content over winter, suggestive of a relationship to its lesser winterhardiness and lesser disease resistance. Very commonly the peak of carbohydrate storage occurs just before anthesis, as with alfalfa, but this is not true for all plants. With Canada thistle the reserve carbohydrates are lowest when the bud is expanding.

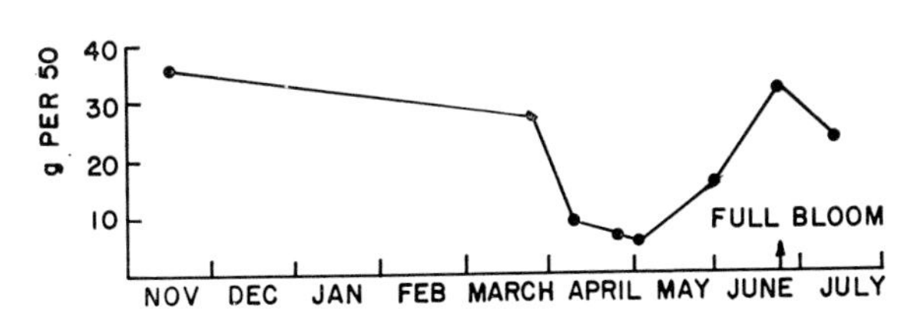

FIG. 4.7 ☙ Total amount of available carbohydrates in the roots of alfalfa during winter and spring (Sonneveld, 1962).

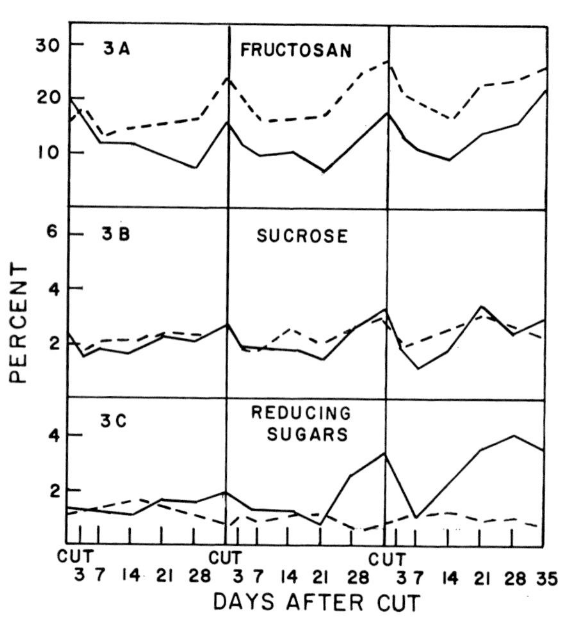

FIG. 4.8 ❦ The percent of fructosan, sucrose, and reducing sugars following cutting. Fructosan content is less under high nitrogen (—) and greater under low nitrogen (- - -) (Hughes et al., 1962).

Nitrogen Fertilizer. The addition of nitrogen fertilizer favors top growth and, combined with defoliation, places a heavier demand on the carbohydrate reserves than defoliation alone. Thus nitrogen can temporarily put a perennial plant at a disadvantage by being relatively more exhausting on its reserves, as shown in Figure 4.8. This effect can be offset by balanced fertility, and, as Figure 4.9 describes, one of the chief reasons that potassium fertilization enhances the winterhardiness of certain species is that increasing the potassium supply increases the carbohydrate content in both alfalfa and orchardgrass.

Clipping Frequency. The number of cuts per season directly affects the reserve supply. An example for orchardgrass by Klapp (1937) is cited in Table 4.2. Recent studies have been directed toward evaluating how much the more photosynthetically efficient young leaves can compensate for total leaf mass in producing yield. There is quite obviously a breaking point at which too frequent defoliation results in insufficient photosynthetic area to provide a good yield. Conversely, a dense leaf canopy left for a season may give a high level of reserves, as described in Table 4.2, but will not provide much harvestable yield above ground.

Light Intensity Following Defoliation. As shown in Figure 4.10, for perennial ryegrass an increase in light intensity simultaneously in-

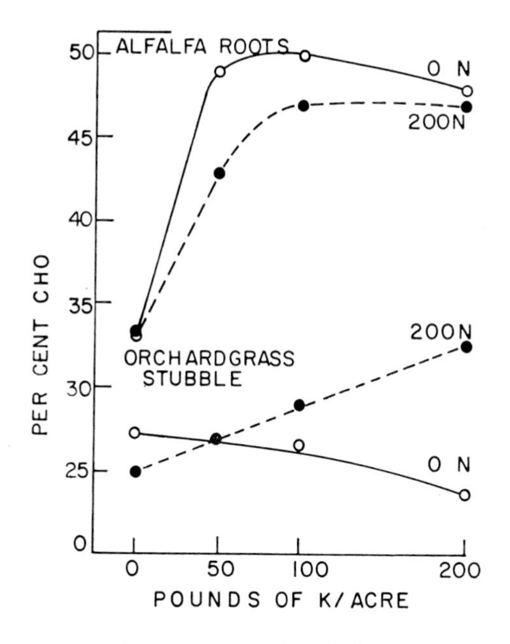

FIG. 4.9 ☙ Effect of potassium at 0- and 200-pound rates of nitrogen on the total available carbohydrate content of alfalfa roots and orchardgrass stubble (MacLeod, 1965).

creases the quantity of carbohydrate accumulated in the storage organ. A period of cloudy weather or shading by a competitor that recovered more rapidly following defoliation could have detrimental effects on the rate and quantity of storage carbohydrates.

Temperature. For a given species, the temperature plays a significant role in the rate of carbohydrate accumulation after defoliation. Higher temperatures favor both more rapid leaf growth and an increased respiration rate in the stubble and root, thus reducing the level of storage. Sullivan and Sprague (1949) placed perennial ryegrass under four temperature regimes, as shown in Figure 4.11. New top growth was most rapid under the 70–60° regime. Fructosan, the major reserve carbohydrate, was more strongly depleted under the higher temperatures. Sullivan and Sprague concluded that the roots were more

TABLE 4.2. Effect of clipping frequency on weight of leaf stubble reserves (Klapp, 1937)

Number of Cuts	Weight of Reserves in Leaf Stubble (g/pot)	Relative Percent
4	450	100
3	620	138
2	1203	267
1	1579	351

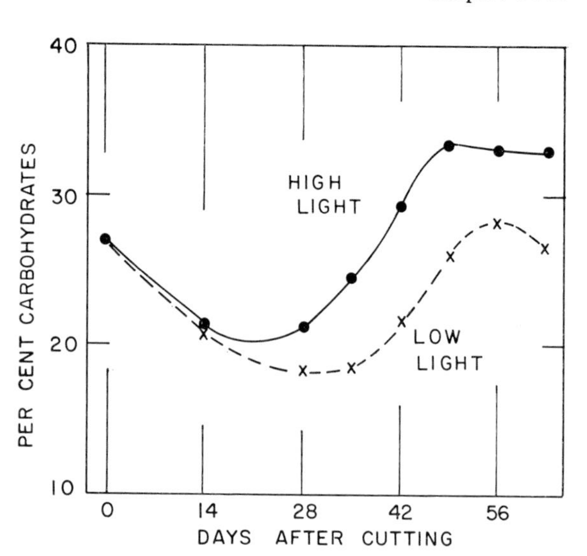

FIG. 4.10 ❧ Changes in percentage of total available carbohydrates in the roots of alfalfa following cutting when grown in the greenhouse under high light intensity and low light intensity. The carbohydrates did not reach as high a level of storage at maturity under low light as under high light. (Smith, 1962).

seriously injured by heat than the stubble and that root death occurred in part as the result of carbohydrate exhaustion. For perennial ryegrass, a cool-season grass, a relatively cool temperature is necessary to favor accumulation. This temperature range will change from species to species, but the generalization of relatively cooler temperatures favoring storage over growth will hold.

Irrigation. Water applied as irrigation improves the conditions for growth under the conditions where irrigation is practiced. This addition stimulates vegetative growth of the plant, improves the rapidity of leaf tissue development, and, as a result, depletes the carbohydrate reserves in the storage tissue.

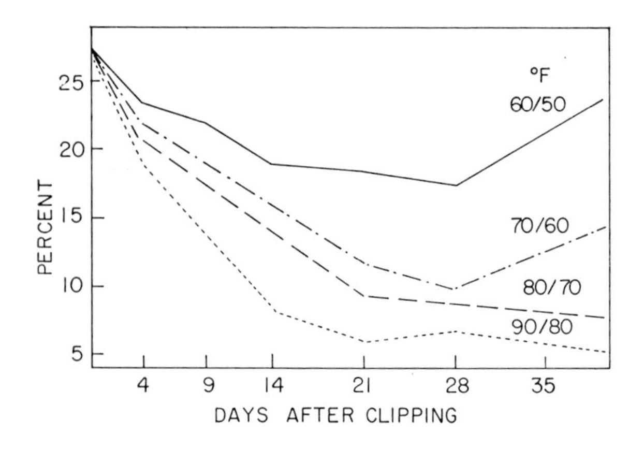

FIG. 4.11 ❧ Changes in the fructosan content of the stubble of perennial ryegrass at four temperature conditions (Laude, 1964, after Sullivan and Sprague, 1949).

As a generalization, any factor favoring rapid top growth will in turn reduce the supply of stored carbohydrate, especially in the first growth period following defoliation.

In summary, a consideration of carbohydrate nutrition in the plant must evaluate those aspects of (1) rapid turnover and use for respiratory energy, (2) accumulation as stored material (Fig. 4.1), (3) whether this stored material may (as in the case of starch in alfalfa roots or may not (as in the case of corn seed) be remobilized and used by the parent plant. The remainder of this discussion has centered around the factors affecting the rapidity of use and regeneration in the storage organs of perennials.

NITROGEN METABOLISM

The nitrogen metabolism of plants results in the synthesis of soluble nitrogen compounds, proteins, alkaloids, chlorophylls, and other complex substances. The soluble nitrogen constituents and protein synthesis will be discussed here. As with carbon compounds, the nitrogen constituents exist in two forms: (a) large, polymeric molecules and (b) monomers or smaller building units. Thus the chemistry of nitrogen compounds is concerned first with the intermediate forms in reduction to the amino level and the formation of amino acids and second with the distribution between the long-chain polymers (proteins) and the smaller molecules (amino acids and their derivatives).

Research culminating in the early 1900s indicated that a dozen amino acids were free in the plant and available for protein synthesis. With research techniques then available, the commonly supported pathway of nitrogen metabolism was:

$$NO_3 \longrightarrow NH_4 + \text{organic acid} \longrightarrow \text{amino acid} \longrightarrow \text{protein.}$$

Amino acids or other soluble forms of nitrogen were given little consideration. However, with the advent of paper chromatography in the late 1940s, many additional soluble, free nitrogen compounds were described. Some of these were amino acids not known to occur in proteins. They were found to be stored in the vacuole as β-alanine, γ-amino butyric acid, and many other γ-glutamyl compounds. Therefore, in addition to nitrogen's role in low energy bonding and its involvement in key ring structures, it was found to be widely distributed in the plant in both insoluble protein forms and soluble nitrogen forms like glutamine, arginine, and asparagine. These latter forms are especially common in succulent tissues.

The plant proteins as constructed from amino acids are composed of the following elements:

C: 50–55%; H: 7%; O: 20–25%; N: 16–18%; S: 2%; P: trace

The smallest known protein has a molecular weight of 17,600 in comparison to a glucose molecule with a weight of 180. The largest proteins have molecular weights in the millions.

NITROGEN FIXATION

Nitrogen gas may be fixed in the light by blue-green algae and photosynthetic bacteria, using the sunlight energy for the reducing power. *Azotobacter, Clostridium* (both free-living bacteria), and *Rhizobium* (nodule bacteria) can fix N_2 in the dark by using energy from organic materials. Whether ammonium or hydroxylamine (see Fig. 4.12) is the first major fixation product is not agreed upon.

NITROGEN ABSORPTION AND CONVERSION

Nitrate is the most readily absorbed form of nitrogen in higher plants, although certain microorganisms cannot utilize NO_3^- and must be fed ammonium. The reduction of NO_3^- is energized by reduced nicotinamide dinucleotide ($NADH_2$) generated by light. With $NADH_2$ and energy stored in carbohydrates, the plant can reduce NO_3^- in the light or in the dark. The enzyme involved, nitrate reductase, is a light-adaptive enzyme and is favored by adequate light. Nitrate reductase is a flavoprotein containing molybdenum and catalyzes the conversion of NO_3^- to NO_2^-. The remaining steps have not been fully elucidated, but considerable evidence supports the reduction pathway shown in Figure 4.12.

Ammonium is converted readily into organic combination while nitrate must first be reduced. Nitrate often accumulates in consider-

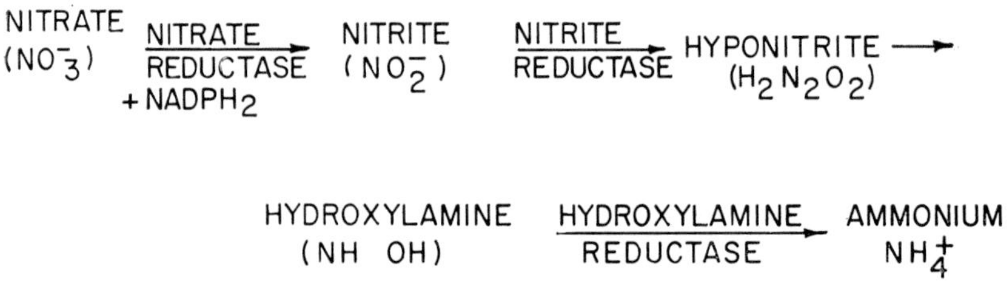

NITRATE (NO_3^-) —NITRATE REDUCTASE + $NADPH_2$→ NITRITE (NO_2^-) —NITRITE REDUCTASE→ HYPONITRITE ($H_2N_2O_2$) →

HYDROXYLAMINE (NH OH) —HYDROXYLAMINE REDUCTASE→ AMMONIUM NH_4^+

FIG. 4.12 ❧ Stepwise scheme for nitrate reduction.

able quantities in plant tissues without apparent deleterious effects on plant growth. (Tissues with extremely high levels of nitrate may be toxic to animals. However, nitrate poisoning is very difficult to reproduce experimentally, and it may be that many animal deaths are blamed on nitrate poisoning but are produced by something else.) In contrast, the presence of ammonium ions is frequently associated with toxic effects on plant growth and development. If ammonium supply is continuous, reactions which incorporate NH_4^+ into organic combinations may occur at the expense of other vital growth processes. Nitrate, in contrast, does not impose a continuous stress on the carbon reserve for synthesis of organic nitrogenous compounds. Leaves from plants grown with ammonium nutrition have a greater rate of oxygen consumption than leaves from plants receiving nitrate nutrition. This agrees with the hypothesis (Barker et al., 1965) that nitrate oxidizes part of the respiratory reductant in nitrate-grown plants, whereas molecular oxygen serves as the primary oxidant in ammonium-grown leaves.

AMINO ACID AND SOLUBLE NITROGEN SYNTHESIS

The ammonia derived from the above series unites with a carbon-structured acid to form an α-amino acid. This means the amino acid has an NH_2 group in the α position, adjacent to the carboxyl. The amino acids in higher plants have the same spatial configuration about the α-carbon as L-serine and thus belong to the L-series. The most important first step is the reductive amination of α-ketoglutaric acid to form glutamic acid. This reaction is catalyzed by the enzyme glutamic dehydrogenase. While other pathways have been noted, this one appears to be the most significant in producing amino acids. Reductive amination of α-ketoglutaric acid to form glutamic acid is well established. Such amination is probable in the use of pyruvate to form alanine and may occur with oxaloacetate to form aspartate. It appears that most other amino acids are formed from their α-keto analogues by transaminations directly or indirectly from glutamic acid.

Glumatic acid may form a peptide bond with free ammonia at its distal carboxyl group to form glutamine (Fig. 4.13). This reaction apparently plays an important role in preventing the accumulation of free ammonia which could be toxic to the plant. In earlier years most plant researchers implied that all nitrogen was incorporated into protein forms. When the storage of nitrogen takes place in organs of low

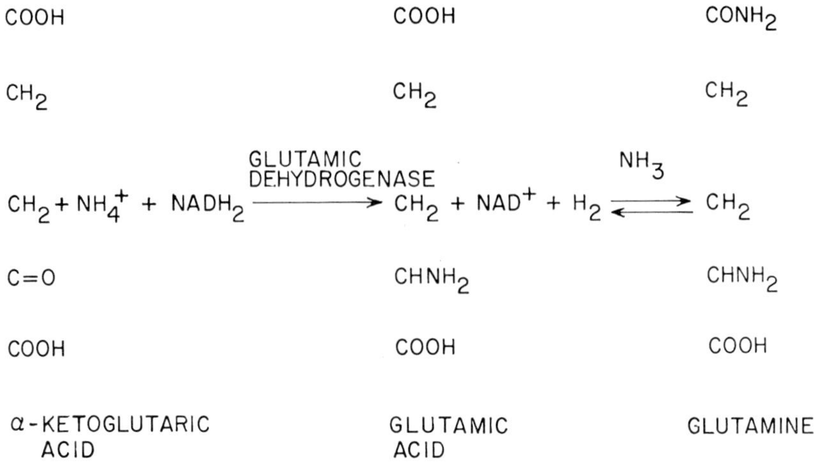

$$
\begin{array}{llllll}
COOH & & & COOH & & CONH_2 \\
| & & & | & & | \\
CH_2 & & & CH_2 & & CH_2 \\
| & & \text{GLUTAMIC DEHYDROGENASE} & | & NH_3 & | \\
CH_2 & + NH_4^+ + NADH_2 & \longrightarrow & CH_2 & + NAD^+ + H_2 \rightleftarrows & CH_2 \\
| & & & | & & | \\
C{=}O & & & CHNH_2 & & CHNH_2 \\
| & & & | & & | \\
COOH & & & COOH & & COOH \\
\alpha\text{-KETOGLUTARIC ACID} & & & \text{GLUTAMIC ACID} & & \text{GLUTAMINE}
\end{array}
$$

FIG. 4.13 ☘ Formation of glutamic acid and glutamine.

water content, more of it may be in the form of protein (e.g. in the cotyledons or endosperm of seeds). However, it is now agreed that succulent or fleshy tissues contain much of what may be called nitrogen-rich storage products. Glutamine, asparagine, and arginine are the major representatives in this group, but Steward (1964) outlines an extensive group of amino acids and other soluble materials studied chromatographically and shown to be widespread in plants. These are β- and γ-amino compounds in contrast to the α-amino acids of proteins.

Under conditions of protein breakdown, amino groups may be liberated. These may be evolved as NH_3 or formed into amides. When seeds germinate normally, protein is broken down in the seed, translocated to the growing point, and then reformed as protein. If kept in the dark, amides will form if carbohydrate is supplied and if respiration is active to supply ATP. Otherwise, free ammonia is liberated and the death of the plant tissue follows.

PEPTIDES AND PROTEIN SYNTHESIS

As in the case of amide synthesis, the formation of protein is dependent on a supply of ATP and thus tends to run parallel to the rate of respiration. The fundamental reaction is a peptide linkage, enzymatically completed after amino acids have been placed in standard order on the ribosome "template" by RNA activity. When two amino acids condense between the carboxyl group of one and the amino group of the other, a peptide link is formed. When this process is repeated many times a protein is formed. The constituent amino acid residues form side chains that determine the properties of the protein. Since a protein contains at least one free carboxyl group which can ionize with a negative charge and one amino group which can ionize positively, proteins can exist with a net negative, a positive, or a neutral charge; i.e., they are amphoteric.

Protein synthesis occurs most vigorously in young healthy tissues and is abundant in the several meristematic regions of the plant. Another prime organ of protein synthesis is the green leaf in the light, since energy in the forms of ATP and reduced pyridine nucleotides ($NADH_2$) is formed there and because the green leaf has available the carbon compounds with which inorganic nitrogen combines. By contrast, green leaves in the dark are organs in which protein breaks down with the release of soluble nitrogen compounds. Thus in the growing cell there is a continuous cycle of protein synthesis and degradation. In other organs protein synthesis occurs wherever protein is stored. Such storage proteins are globulins (potato tuber) or gliadins and glutelins (cereal grain endosperm).

In summary, the following characteristics of plant proteins are noted. Those of a reserve nature are primarily the seed proteins—the gliadins and the glutelins. Functional proteins may exhibit rapid turnover or they may be of an enzymatic nature. Leaf proteins fit in this category. Of the total weight of the grana in the chloroplast, 33–50% is protein not soluble in water. The cytoplasmic proteins are mostly water soluble. Arginine is the most abundant cytoplasmic protein and composes 14–15% of the total. The nuclear proteins are represented by deoxyribonucleic acid (DNA) and ribonucleic acid (RNA).

LIPID METABOLISM

It is generally accepted that carbohydrates constitute the starting materials for the synthesis of lipids in plants. This synthesis likely occurs in three steps for the fats: (1) synthesis of fatty acids, (2) formation of glycerol, and (3) combination of fatty acids with glycerol to form a triglyceride.

Fatty acid synthesis occurs in the cell in the mitochondria and is closely allied to the endoplasmic reticulum. One scheme to describe this synthesis is outlined in Figure 4.14.

a. ACETYL CoA $\xrightarrow[+\ CO_2]{+ATP+BIOTIN}$ MALONYL CoA (ENERGY-RICH INTERMEDIATE)

b. MALONYL CoA + ACETYL CoA $\xrightarrow[+\ BIOTIN\ +\ CO_2]{+NADP+Mn^{++}\ \ +ATP}$ FATTY ACID

FIG. 4.14 Fatty acid synthesis via malonyl CoA.

LIPID CLASSIFICATION

The lipids which occur naturally can be separated into three groups.

1. True fats which result from the esterification of fatty acids with glycerol.
2. Waxes formed by esterification of fatty acids with long-chain alcohols.
3. Phospholipids, the derivatives of fat-containing phosphate groups.

True fats may be found in physiologically active cells of the leaf and stem, and they also accumulate as reserve materials in seeds and rhizomes. The fatty acids are most commonly 12 to 18 carbons in length. Leaf fats appear to contain primarily the C_{18} unsaturated group, in which linoleic and linolenic acids frequently predominate. The seed fats contain a wide range of specific acids but oleic acid is the most widespread of all natural fatty acids, in some cases making up 50% of the total fatty acid composition (Fig. 4.15).

The fatty acids occur in nature in varying degrees of saturation. One with no double bonds (that is, all the available carbon valences are filled by hydrogen) is saturated. Then, as described in Figure 4.15, a varying number of double bonds may be present; the number and placement of these bonds give the fatty acid its uniqueness. Linoleic has 18 carbons and two double bonds at the 9th and 12th carbon, counting from the COOH end group. Since these double bonds of fatty acids are relatively unstable, they can be easily hydrogenated or

a. 18:1(9).OLEIC: $CH_3(CH_2)_7\ CH{=}CH(CH_2)_7\ COOH$

b. 18:2(9,12).LINOLEIC: $CH_3(CH_2)_4\ CH{=}CHCH_2\ CH{=}CH(CH_2)_7\ COOH$

c. 18:3(9,12,15).LINOLENIC: $CH_3CH_2\ CH{=}CHCH_2\ CH{=}CHCH_2\ CH{=}CH(CH_2)_7\ COOH$

FIG. 4.15 Three widespread fatty acids.

hydrolyzed. A different fatty acid may be attached to each of the three hydroxyl groups of glycerol. These two characteristics—(a) different degrees of saturation in the several acids and (b) different acids attached to glycerol—serve as the basis for differences between the plant fats.

Nearly all plant fats are liquids at ordinary temperatures and are thus referred to as oils. Climate greatly affects the percentage of oil in the plant, especially that laid down in the seed. High temperatures may decrease the iodine number (high iodine number represents a high degree of unsaturation) as much as 20 to 40 points on a 200-point scale. Little oil synthesis occurs the first days of seed development, then there is a mid-period of rapid development and finally a tapering off. The iodine number continues to rise after the oil reaches a maximum percentage, so unsaturation is a separate process from fat formation. In effect, early oil formation is a reduction process from carbohydrates, while the continued increased unsaturation involves oxidation (Eyre, 1931). There is also an effect of environment on the specific fatty acids. In soybeans Collins and Sedgwick (1959) reported that 18 varieties from 43 locations ranged from 5 to 11% linolenic, 43–56% linoleic, 15–33% oleic, and 11–26% saturated acids. Much of this variation was related to temperature and other effects.

The enzyme lipase catalyzes the hydrolysis of fats. Fats have a higher heat of combustion than the polysaccharides, and consequently when they are oxidized, they yield a greater amount of energy per unit weight of material consumed. Fatty acid decomposition also occurs in the mitochondria via the process known as β-oxidation. This means the fats are broken down two carbons at a time, the reverse of their buildup (Fig. 4.16).

Waxes are found as structural materials in the cell wall and as a

a. $\text{FAT} \xrightarrow{\text{LIPASE}} \text{FATTY ACID} + \text{GLYCEROL}$

b. $\text{GLYCEROL} + \text{ATP} \longrightarrow \text{TRIOSE PHOSPHATE}$

c. $\text{FATTY ACID: } (CH_2)_n\ \text{COOH} + \text{COENZYME A} \xrightarrow[\text{NAD}]{Mg^{++}} (CH_2)_n\ \text{CO-S-CoA}$

d. $(CH_2)_n\ \text{CO-S-CoA} \xrightarrow{-2H} (CH_2)_n^{-2H}\ \text{Co-S-CoA} \xrightarrow[+\text{CoA}]{+H_2O-2H}$

e. $\longrightarrow (CH_2)_{n-2}\ \text{CO-S-CoA} + CH_3\ \text{CO-S-Co-A} \longrightarrow$ KREBS CITRIC ACID CYCLE

FIG. 4.16 ❧ Fatty acid catabolism.

$$
\begin{array}{l}
CH_2 - O - OC - \text{FATTY ACID} \\
| \qquad\qquad\qquad\qquad\qquad\qquad (\text{HYDROPHOBIC}) \\
CH - O - OC - \text{FATTY ACID} \\
| \qquad\qquad O \\
| \qquad\qquad | \\
CH_2 - O - P - CH_2 - CH_2 - N^{+}(CH_3)_3 \quad (\text{HYDROPHILIC}) \\
\qquad\qquad\quad | \\
\qquad\qquad OH \qquad \text{CHOLINE}
\end{array}
$$

FIG. 4.17 ❧ Phospholipid (example–lecithin).

protective coating of cuticle overlying the epidermis.

Phospholipids are found in the cytoplasm, the chloroplast, and in plant membranes. The simple phospholipids are similar to fat and have one of the three fatty acids replaced by phosphoric acid. In more complex derivatives, the third fatty acid is replaced by an organic nitrogenous phosphate derivative like choline to form lecithin (Fig. 4.17). Thus phospholipids have a hydrophobic group in the fatty acid and a hydrophilic group in the phosphate radicle. Phospholipids are therefore well suited to occupy an interface separating aqueous and other phases as in the chlorophyll structure. ❧

LITERATURE CITED

Auda, H., R. E. Blaser, and R. H. Brown. 1966. Tillering and carbohydrate contents of orchardgrass as influenced by environmental factors. *Crop Sci.* 6:139–43.

Barker, A. V., R. J. Volk, and W. A. Jackson. 1965. Effects of ammonium and nitrate nutrition on dark respiration of excised bean leaves. *Crop Sci.* 5:439–44.

Bryant, H. T., and R. E. Blaser. 1964. Yield and persistency of an alfalfa-orchardgrass mixture as affected by cutting treatment. *Va. Agr. Exp. Sta. Bull.* 555.

Carlson, G. E. 1966. Growth of clover leaves after complete or partial leaf removal. *Crop Sci.* 6:419–22.

Colby, W. G., M. Drake, D. L. Field, and G. Kreowski. 1965. Seasonal pattern of fructosan in orchardgrass stubble as influenced by nitrogen and harvest management. *Agron. J.* 57:169–73.

Collins, F. I., and V. F. Sedgwick. 1959. Fatty acid composition of several varieties of soybeans. *J. Am. Oil Chemists' Soc.* 36:641–44.

DeMan, T. J., and J. G. Detteus. 1949. Carbohydrates in grass. I. The soluble carbohydrate. *Rec. Trav. Chim.* 68:43–50.

Dobrenz, A. K., and M. A. Massengale. 1966. Changes in carbohydrate in alfalfa (*Medicago sativa* L.) roots during the period of floral initiation and seed development. *Crop Sci.* 6:604–7.

Duncan, W. G., A. L. Hatfield, and J. L. Ragland. 1965. The growth and yield of corn. II. Daily growth of the corn kernels. *Agron. J.* 57:221–23.
Eyre, J. V. 1931. Notes on oil development in the seed of a growing plant. *Biochem. J.* 25:1902–8.
Fergason, V. L., and M. S. Zuber. 1962. Influence of environment on amylose content of maize endosperm. *Crop Sci.* 2:209–11.
Fuwa, H. 1957. Formation of starch in young maize kernels. *Nature* 179:159–60.
Graber, L. F. 1927. Organic food reserves in relation to the growth of alfalfa and other perennial herbaceous plants. *Univ. of Wis. Res. Bull.* 80.
Grandfield, C. O. 1943. Food reserves and their translocation to the crown buds as related to cold and drought resistance in alfalfa. *J. Agr. Res.* 67:33–47.
Harlan, J. R. 1956. *Theory and Dynamics of Grassland Agriculture.* D. Van Nostrand Co. Inc., Princeton, N.J.
Hughes, H. D., M. E. Heath, and D. S. Metcalfe. 1962. *Forages.* Iowa State University Press. Ames.
Hojjati, S. M., R. A. McCreery, and W. E. Adams. 1968. Effects of nitrogen and potassium fertilization, irrigation and clipping interval on chemical composition of coastal Bermudagrass. *Agron. J.* 60:617–18.
Homer, M. G., and S. N. Fertig. 1960. Relationships between control of quackgrass *(Agropyron repens)* and carbohydrate content of rhizomes. *Proc. Northeast. Weed Control Conf.* 14:357–62.
Klapp, E. 1937. Über einige Wachstumregeln mehrjähriger Pflanzen unter der Nachwirkung verschiedener Nutzungsweise. *Pflanzenbau* 14:209–24.
Laude, H. M. 1964. Plant response to high temperature. In *Forage Plant Physiology and Soil-Range Relationships.* Am. Soc. Agron. Spec. Publ. No. 5, pp. 15–31.
MacLeod, L. B. 1965. Effect of nitrogen and potassium fertilization on the yield, regrowth, and carbohydrate content of the storage organs of alfalfa and grasses. *Agron. J.* 57:345–50.
Miller, E. V. 1957. *The Chemistry of Plants.* Reinhold Publishing Corp., New York.
Moran, C. H., V. G. Sprague, and J. T. Sullivan. 1953. Changes in the carbohydrate reserves of Ladino white clover following defoliation. *Plant Physiol.* 28:467–74.
Nelson, C. D., and P. R. Gorham. 1957. Translocation of radioactive sugars in the stems of soybean seedlings. *Can. J. Botany* 35:703–13.
Okajima, H., and D. Smith. 1964. Available carbohydrate fractions in the stem bases and seed of timothy, smooth bromegrass and several other northern grasses. *Crop Sci.* 4:317–20.
Priestly, C. A. 1962. Carbohydrate resources within the perennial plant (Their utilization and conservation). *Commonwealth Bur. Hort. Plantation Crops, Tech. Commun.* 27.
Smith, D. 1962. Carbohydrate root reserves in alfalfa, red clover, and birdsfoot trefoil under several management schedules. *Crop Sci.* 2:75–78.
———. 1962. *Forage Management in the North.* Wm. C. Brown Publishing Co., Dubuque, Iowa.
Smith, D., and R. D. Grotelueschen. 1966. Carbohydrates in grasses. I. Sugar

and fructosan composition of the stem bases of several northern adapted grasses at seed maturity. *Crop Sci.* 6:263–66.

Sonneveld, A. 1962. Distribution and redistribution of dry matter in perennial fodder crops. *Neth. J. Agr. Sci.* 10:427–44.

Sprague, V. G., and J. T. Sullivan. 1950. Reserve carbohydrates in orchardgrass clipped periodically. *Plant Physiol.* 25:92–102.

Steward, F. C. 1964. *Plants at Work.* Addison-Wesley Publ. Co., Inc., Reading, Mass.

Sullivan, J. T., and V. G. Sprague. 1949. The effect of temperature on the growth and composition of the stubble and roots of perennial ryegrass. *Plant Physiol.* 24:706–19.

Thrower, S. L. 1962. Translocation of labelled assimilates in the soybean. II. The pattern of translocation in intact and defoliated plants. *Australian J. Biol. Sci.* 15:629–49.

Troughton, A. 1957. The underground organs of herbage grasses. *Commonwealth Bur. Pastures and Field Crops, Bull.* 44.

Vernon, L. P., and S. Aronoff. 1952. Metabolism of soybean leaves. IV. Translocation from soybean leaves. *Arch. Biochem. Biophys.* 36:383–98.

Ward, C. Y., and R. E. Blaser. 1961. Carbohydrate food reserves and leaf area in regrowth of orchardgrass. *Crop Sci.* 1:366–70.

Whittingham, C. P. 1964. The chemistry of plant processes. Chap. 7. *Nitrogen Metabolism.* Metheun & Co., Ltd., London.

Younis, M. A., A. W. Pauli, H. L. Mitchell, and F. C. Stickler. 1965. Temperature and its interaction with light and moisture in nitrogen metabolism of corn (*Zea mays* L.) seedlings. *Crop Sci.* 5:321–26.

❧ CHAPTER FIVE ❧ LIMITING FACTORS AND CONCEPTS RELATED TO THEIR INFLUENCE ON CROP GROWTH

❧ THE DISCUSSION of plant growth in the preceding chapters on light utilization and mineral and carbohydrate nutrition has considered many plant growth requirements individually. This single-factor study is helpful but must finally be placed into a system of interactions if it is to be useful agriculturally. Agronomists over the years have considered the interaction of many factors as they sought to increase yield per acre and to identify why yields were not reaching an expected maximum.

Plant growth factors are a heavily interrelated and complex group, and growth analysis curves are helpful in understanding the general pattern of plant development. The plant is the product of its genetic composition and its environment. The genetic pattern is relatively fixed for a given plant, but the environment influences the development and interacts with the genetic composition. In the end the result is very definitely a product of the two and not just one decelerated or accelerated by the other. The plant breeder tends to emphasize that the genetic potential sets the upper limit beyond which no environmental conditions can take the plant. The climatologist suggests that the environment sets the limit of production. But the phenomenon is one of interaction to set an upper limit.

Many factors are recognized as limiting to crop growth (Fig. 5.1). Some factors (weeds, insects, diseases) directly detract from the yield potential; that is, they reduce the yield that might have been achieved with the particular crop under that environment. A second group of factors (nutrients, stand densities, leaf orientation of varieties, row width) can add to the yield potential and thus affect yield ability. Most

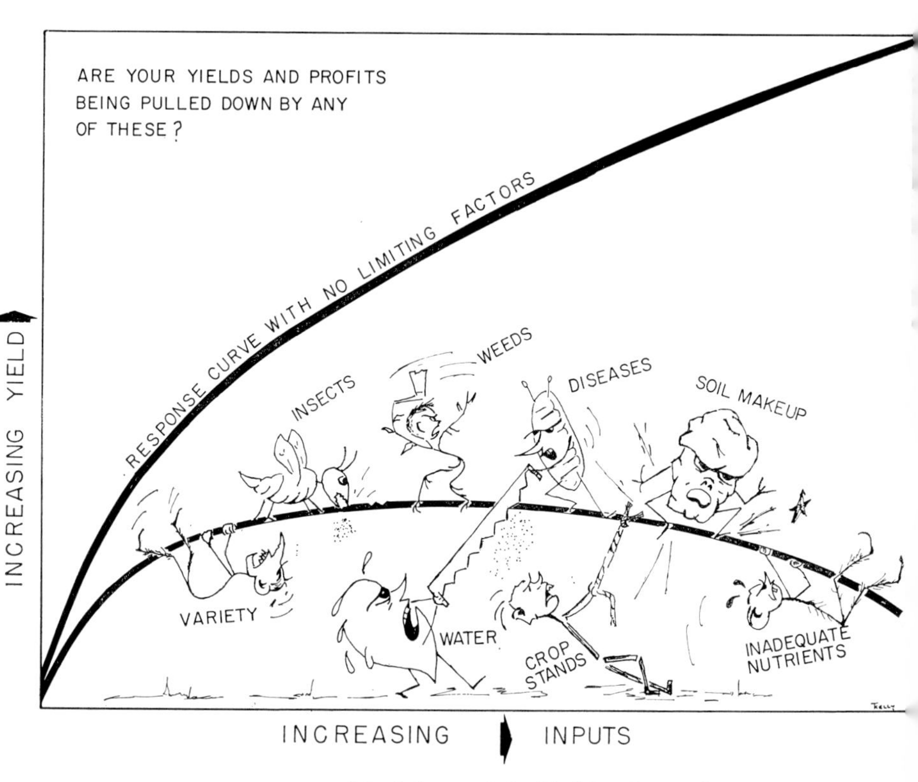

FIG. 5.1 Some of the limiting factors in crop production (American Potash Institute, 1960).

crop management practices are directed toward balancing these factors in order to obtain a range of yields up to the maximum. Sometimes students and other observers misunderstand what each subject matter has as its primary responsibility. The agronomist (like the animal scientist and the agricultural engineer) provides information on how to develop steps to maximum production—how to obtain the response curve with no limiting factors. Then the farm management specialist seeks to explain the basis for deciding where to stop on a particular response curve—how to utilize scarce resources of time, money, and management skill to maximize income and other goals. But the farm management specialist can make useful recommendations only when he has sound parameters with which to work. In developing these parameters, the agronomist must not be stifled initially by the notion that a new system is "uneconomical"; it may be so only in certain contexts or until it is perfected.

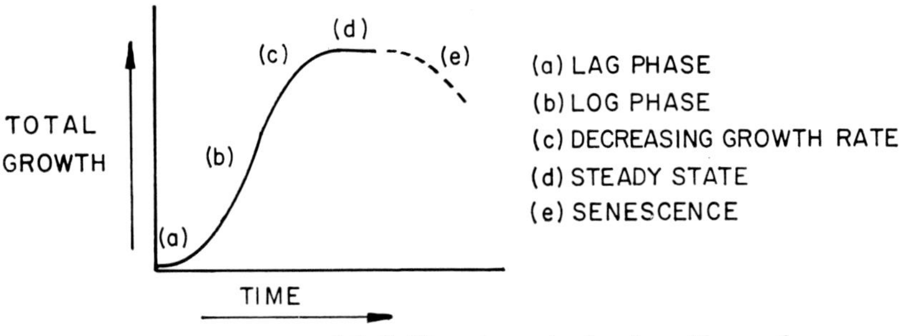

FIG. 5.2 ☙ Five phases in the sigmoid growth curve.

SIGMOID GROWTH CURVE

Several useful concepts have been developed which generalize about plant response to a variety of growth-promoting factors. The S-shaped, or sigmoid, curve (Fig. 5.2) is typical of the growth pattern of individual organs, of a whole plant, and of populations of plants. It can be shown to consist of at least five distinct phases: (a) an initial lag period during which internal changes occur that are preparatory to growth; (b) a phase of ever-increasing rate of growth (Since the logarithm of growth rate, when plotted against time, gives a straight line during this period, this phase is frequently referred to as the log period of growth or "the grand period of growth."); (c) a phase in which growth rate gradually diminishes; (d) a point at which the organism reaches maturity and growth ceases. If the curve is prolonged further a time will arrive when (e) senescence and death of the organism set in, giving rise to another component of the growth curve.

One can think of the log phase (b) of plant growth as an example of the operation of compound interest (Blackman, 1919). The growth increment in any period adds to the total capital available for subsequent growth. This is not completely accurate because the "capital invested" in root may not contribute the same as in leaves, but it is a useful idea with which to begin.

LIEBIG—LAW OF THE MINIMUM

Very frequently in studies of plant growth and response to varied inputs, section b through d of this curve are analyzed in detail. Three men—Liebig, Blackman, and Mitscherlich—have their names attached to interpretations of this response curve, and their ideas provide a useful consideration of several key points. Justus von Liebig stated his *Law of the Minimum* in 1840 as follows: "A deficiency or absence of one necessary constituent, all others being present, renders the soil barren for crops for which that nutrient is needed." This has sometimes been restated as the "barrel concept." The barrel is suggested to have staves of different lengths—the barrel cannot contain anything above the height of the "shortest stave" (the most deficient or limiting factor). In like manner, growth can be no greater than that allowed by the factor lowest in availability.

BLACKMAN—OPTIMA AND LIMITING FACTORS

"Optima and Limiting Factors" were discussed by F. F. Blackman in 1905. Blackman suggested that "when a process is conditioned as to its rapidity by a number of separate factors, the rate of the process is limited by the pace of the slowest factor." His ideas have been interpreted frequently as described in Figure 5.3. This figure suggests that as the two factors, light and carbon dioxide, interact there is an abrupt cutoff when one becomes limiting. Such a distinct break is referred to as a "Blackman" response. When trying to confirm this concept the expected straight lines and angles are rarely found. For example, the values are curvilinear and do not break sharply in most response curves, as shown in Figure 5.4. As a result, Blackman's and Liebig's ideas have been seriously criticized. However, the Liebig and Blackman concepts are still useful as a guide, even though they are not absolute boundary conditions.

MITSCHERLICH—LAW OF DIMINISHING RETURNS

In 1909 a German soil scientist, E. A. Mitscherlich, developed an equation which related growth to the supply of plant nutrients. He

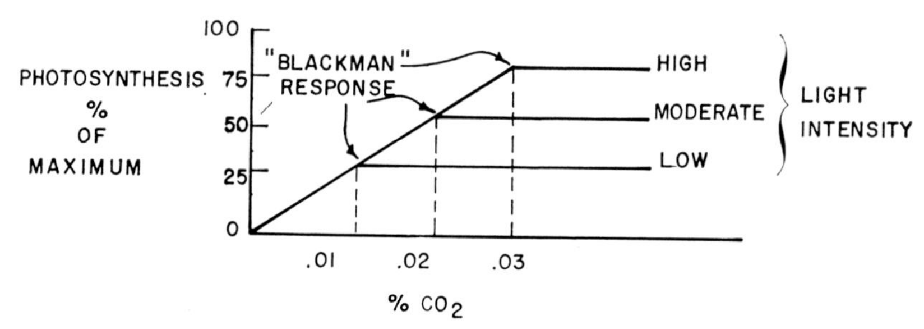

FIG. 5.3 ❧ Assimilation of CO_2 and its interaction with light intensity (Blackman, 1905).

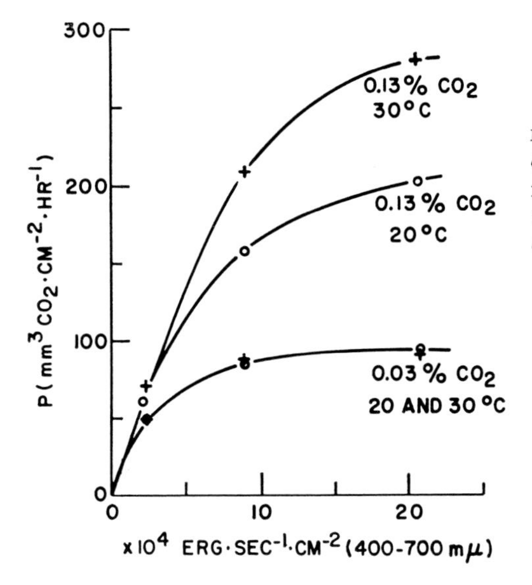

FIG. 5.4 ⁂ Photosynthesis (P) of a cucumber leaf in relation to light intensity, temperature, and CO_2 concentration (Gaastra, 1963).

observed that when plants were supplied with adequate amounts of all but one limiting element, their growth was proportional to the amount of this one limiting element as it was supplied to the soil. Plant growth increased as more of this element was added, but not in direct proportion to the amount of the growth factor added (Fig. 5.5). Mitscherlich thus proposed the concept of the "law of diminishing returns" in which he stated: "The increase in any crop produced by a unit increment of a deficient factor is proportional to the decrement of that factor from the maximum." The effect of an added growth factor was taken to depend upon the difference between the amount available and the amount necessary to obtain maximum yield. This condition led to an exponential function instead of Liebig's straight line.

Mitscherlich developed an equation as follows:

$$dy/dx = C(A - y)$$

where dy is the increase in yield resulting from an increment of the growth factor dx; dx is an increment of the growth factor x; A is the maximum possible yield obtained by supplying all growth factors in optimum amounts; y is the yield obtained after any given quantity of the factor x has been applied; and C is a proportionality constant which depends on the nature of the growth factor.

The rate of increase in y is greatest at the lowest level of x. The rate of increase in yield with respect to the amount of x added is proportional to the decrement from maximum yield. This means that

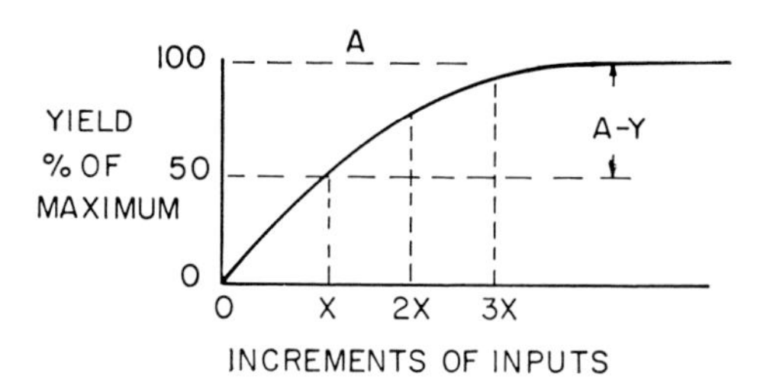

FIG. 5.5 ☙ Response curve as described by the Mitscherlich equation.

factors limiting yield at the maximum also have an effect at all other levels.

The value 0.301 replaces the constant C when yields are expressed on a relative basis of A = 100. If the maximum yield A is 100 percent, then the equation may be written and solved by taking antilogs.

If $x = 0$ $\quad A = 100$

$$\begin{aligned} \log (100 - y) &= \log 100 - 0.301x \\ \log (100 - y) &= 2 - 0 \\ 100 - y &= 100 \\ y &= 0 \end{aligned}$$

If $x = 1$ $\quad A = 100$

$$\begin{aligned} \log (100 - y) &= \log 100 - .301x \\ &= 2 - .301 \\ &= 1.699 \\ 100 - y &= 50 \\ y &= 50\% \text{ increase} \end{aligned}$$

If $x = 2$ $\quad A = 100$

$$\begin{aligned} \log (100 - y) &= \log 100 - .301(2) \\ &= 1.398 \\ 100 - y &= 75 \\ y &= 25\% \text{ increase} \end{aligned}$$

If $x = 3$, $\quad y = 12.5\%$ increase
If $x = 4$, $\quad y = 6.25\%$ increase

The amount of a nutrient or other factor required to give one-half yield is called a *Baule* unit. Where one unit is present, the yield is 0.5 A. Therefore:

$$\begin{aligned} 1 \text{ unit} &= 50\% \ A \\ 2 \text{ units} &= 75\% \ A \\ 3 \text{ units} &= 87.5\% \ A \\ \text{many units} &= 100\% \ A \end{aligned}$$

Willcox (1937) considered it possible to determine a C value that was constant for all crops, but this idea is not generally accepted. Nevertheless, the overall concept of this curvilinear and diminishing response curve is very useful.

MACY–POVERTY ADJUSTMENT; CRITICAL PERCENTAGE

Macy (1936) adds an additional clarification that is very helpful in placing the work of Liebig, Blackman, and Mitscherlich in perspective. He suggests a relationship between the sufficiency of a nutrient and its percentage content in the plant. The central concept assumes that there is a critical percentage of each nutrient in each kind of plant (Fig. 5.6B). Above that point there is luxury consumption and below that point there is poverty adjustment. This poverty adjustment is proportional to the deficiency until a minimum percentage is reached.

He suggests that the Liebig law holds at the minimum percentage because there is not enough of that growth factor to allow normal plant growth. Liebig's law holds again in luxury consumption because, although there is a large quantity of this element, some other element has now become limiting. Mitscherlich's law of diminishing returns holds during the exponential or poverty adjustment stage. According to Macy, Liebig's is the basic law and Mitscherlich's holds during the transitional stage (Fig. 5.6).

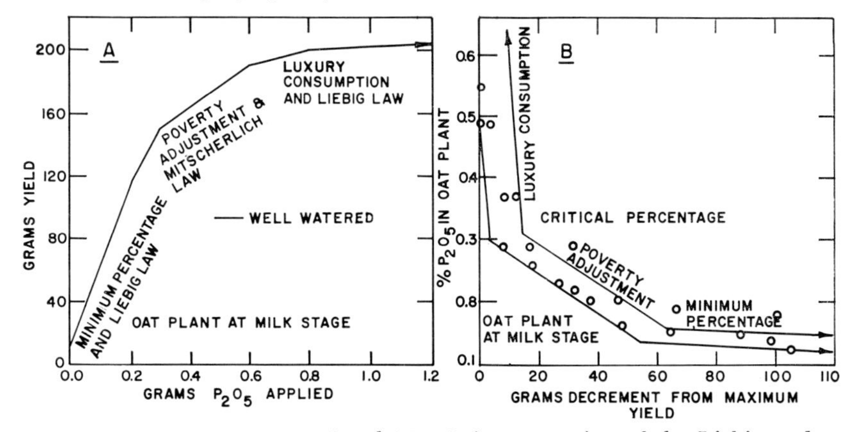

FIG. 5.6 Macy's interpretation of the Liebig and Mitscherlich concepts. Macy coordinated these with the minimum percentage, poverty adjustment, and luxury consumption. He separated poverty adjustment and luxury consumption at the critical percentage (Macy, 1936).

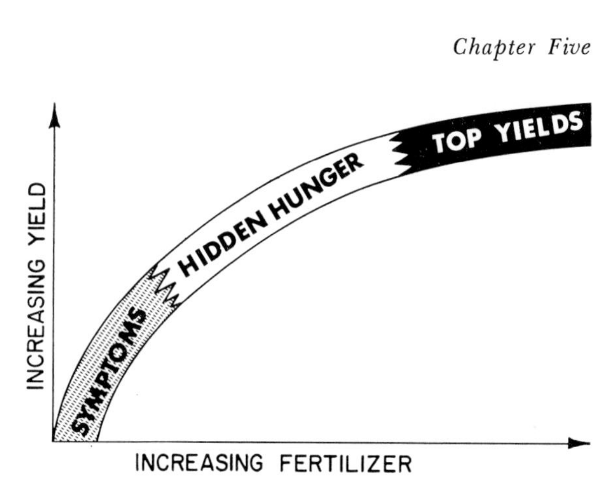

FIG. 5.7 The zone of poverty adjustment may not have visible deficiency symptoms but may represent "hidden hunger" (American Potash Institute, 1964).

One commercial fertilizer organization speaks of the upper portion of this poverty adjustment area as "Hidden Hunger," a time when the crop is not yielding its maximum but does not yet exhibit marked deficiency symptoms (Fig. 5.7).

FACTORS MODIFYING THE LIMITING FACTORS CONCEPT

The ideas presented above have often been referred to as laws but might better be described as concepts. There are several reasons why they do not hold as laws, some of which may be described as follows.

1. Biological reactions are complex. One may have compound A going to compound B and B to C, or it is possible that A can go directly to C. In other words, there are several different pathways within the plant that will allow the same end product. Therefore, slowing one pathway or reducing the concentration of one element may not reduce the plant's rate of growth.
2. Substitution of factors. Sodium can substitute for potassium up to 60% in cotton and up to 90% in sugar beets and mangolds. The use of a different element allows normal growth even though a primary element is limiting.
3. Addition of one factor makes another more efficient. When nitrogen is added to corn, more dry matter is produced per pound of water transpired. This may be due to either more roots or more leaf area. In this same regard, addition of nitrogen gives greater root growth and benefits phosphorus uptake.

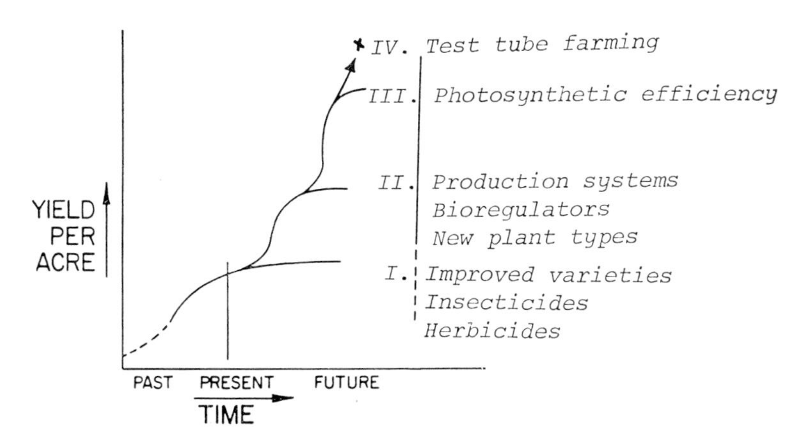

FIG. 5.8 ☙ Yied plateaus in crop production. Agricultural research is now taking us into Phase II where yields can be doublted by using new cropping systems. Phase III can be reached only after we have been able to alter the basic metabolic processes of photosynthesis. Phase IV may occur prior to, or simultaneously with, the attainment of Phase III (Army and Greer, 1967).

4. Addition of one factor has a direct effect on another factor. Increased light causes increased temperature and this in turn causes less efficient use of water.

5. Plants affect factors as well as factors affecting plants. An application of nitrogen causes a chain reaction. More nitrogen gives greater plant growth, greater shade, a lowered soil temperature, and a reduced soil moisture loss.

6. More than one factor may be limiting.

SUMMARY

Crop production yield responses seem to progress with relatively sharp increases for a span of time and then to level out. Several of the characteristics which have contributed to yield advances to date and some of the possibilities for breakthroughs in the future are described in Figure 5.8. Each major advance may require several pieces of new knowledge (plant design, fertility rates, hormone balance) and yet builds upon all the practices used in the past.

In summary, the concept of limiting factors is useful as a means to think broadly about plant growth and yield potential. The modifiers of this concept are many, but the underlying expectation of a curvilinear response continues to be basic and predictable in a large number of cases. ☙

LITERATURE CITED

American Potash Institute. 1960. Know your limiting factors in crop production. Reprint A–1–60.

———. 1964. Fighting hidden hunger with chemistry. Midwest Potash Newsletter.

Army, T. J., and F. A. Greer. 1967. Photosynthesis and crop production systems. In A. San Pietro, F. A. Greer, and T. J. Army (eds.). *Harvesting the Sun.* Academic Press, Inc., New York.

Bertrand, D. 1962. Logical mathematical expression of the law of the optimum nutrient concentration: Its application to agriculture (French). *C. R. Acad. Agr. Fr.* 48(9):422–28.

Blackman, F. F. 1905. Optima and limiting factors. *Ann. Botany* 19:281–95.

Blackman, V. H. 1919. The compound interest law and plant growth. *Ann. Botany* 33:353–60.

Gaastra, P. 1963. Control of photosynthesis and respiration. In L. T. Evans (ed.). *Environmental Control of Plant Growth.* Academic Press, Inc., New York.

Hanway, J. J. 1960. Growth and nutrient uptake by corn. Iowa State Univ., Ames, Pam. 277.

Justesen, S. H., and P. M. L. Tammes. 1962. Random variation and the interpretation of biological response curves. *Neth. J. Agr. Sci.* 10:23–26.

Liebig, J. von. 1862. *Die Chemie in ihre Anwendung auf Agrikultur und Physiologie.* 7. Aufl. Braunschweig.

Loomis, R. S., and W. A. Williams. 1963. Maximum crop productivity: An estimate. *Crop Sci.* 3:67–72.

Macy, P. 1936. The quantitative mineral nutrient requirements of plants. *Plant Physiol.* 11:749–64.

Mitscherlich, E. A. 1909. Das Gesetz des Minimums und das besetz des abnehmenden Bodenertrages. *Landwirtsch Jahrb. Schweiz* 38:537–52.

———. 1922. *Das Wirkungsgesetz der Wachstumsfaktoren Zeitscher.* F. Pflanzenernahrung und Dingung, pp. 49–95.

Sayre, J. D. 1948. Mineral accumulation in corn. *Plant Physiol.* 23:267–81.

Steenbjerg, F., and S. T. Jakobsen. 1963. Plant nutrition and yield curves. *Soil Sci.* 95:69–88.

Willcox, O. W. 1937. *ABC of Agrobiology.* W. W. Norton & Company, Inc., New York.

Van der Honert, T. H. 1928. Koolzuurassimilateae en beperkende factoren. Thesis. Utrecht (cited in Justesen and Tammes, 1962).

Voss, R., and J. Pesek. 1962. Generalization of yield equations in two variables. III. Application of yield data from 30 initial fertility levels. *Agron. J.* 54:267.

❧ CHAPTER SIX ❧ STAND DENSITY, PLANT DISTRIBUTION, AND CROP YIELD

❧ SOLAR RADIATION is the factor governing the ultimate yield of a plant community, as has been discussed in Chapter 2. If water and nutrients are available in adequate supply so that competition for these factors ceases, then light becomes the sole limiting factor to production. Irrigation and fertilizers are making this situation more and more common in world agriculture. Therefore, usefully intercepting all the light by greater plant density seems to be a logical goal. A prominent Corn Belt production specialist has commented that every generation of agronomists gets excited about the idea that if you increase the plants per acre you get higher yields of corn. His statement implies that increased plant population does not automatically insure grain yield increases. Yet a higher plant population or a shift in plant distribution is frequently the way to improve light interception. The plant's response to these practices will be the subject of this chapter.

GENERALIZED POPULATION-YIELD INTERACTIONS

Holliday (1960a, b) summarized a large volume of literature which emphasizes the two different population-yield interactions that occur when crop density is increased. These differences depend on whether the yield is a product of the plant's growth in the reproductive phase or a product of growth in the vegetative phase. That is, is the economic yield equal to the seed yield (a product of reproduction) or to the entire above ground dry matter (biologic yield)?

For the case of yield from the reproductive stage, Holliday cites work with wheat at Cambridge, England. These data describe a para-

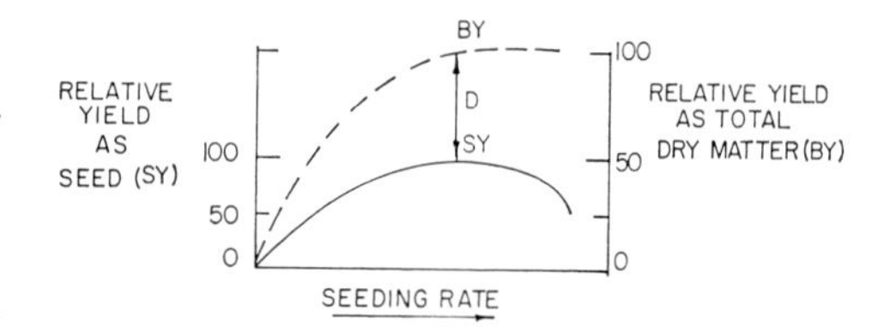

FIG. 6.1. ☙ The effect of increased seeding rate on seed yield (SY) and dry matter or biological yield (BY) of a crop. The BY curve levels off where the SY curve reaches a peak or plateau (D).

bolic response curve (Fig. 6.1—SY) which is typically a flat-topped one with decreases in yield on both sides of the optimum. Therefore, when seed yield is the desired product there is an optimum population and the population can become too high. Holliday suggests the curve can be fitted to the quadratic equation:

$$Y = a + bx - cx^2$$

where:

Y = yield per unit area
x = plant population
a, b, and c are regression constants.

In situations where yield is the product of growth of vegetative material, the yield response to increasing plant density is asymptotic (Fig. 6.1—BY). In this case, it is important to get a stand dense enough to obtain maximum yields; but if the stand is too dense, the only loss is from greater seeding expense. This curve can be defined by the expression:

$$Y = Ax \frac{1}{1 + Abx} \text{ (a rectangular hyperbola),}$$

where:

Y = yield of dry matter per unit area
A = apparent maximum yield per plant
x = number of plants per unit area
b = linear regression coefficient.

In this expression the term $\frac{1}{1 + Abx}$ reoresents the manner in which the maximum plant yield (A) is reduced by the increasing competition resulting from greater plant density. It may be termed the "competition" factor.

While density and dry matter or density and grain yield have been determined in many studies, the three parameters of density, yield of dry matter, and yield of grain have not often been measured together. However, six such studies are cited by Donald (1963). From these data Donald suggests that in each instance the peak of the grain curve occurs approximately at the density at which the yield of dry matter (the biologic yield) is leveling off (Fig. 6.1—D). At this density any gain in total yield per acre due to the addition of extra plants is offset by a loss due to the decrease in the weight per plant. No doubt these relationships represent conditions of either limiting light or limiting nutrients and would not hold, for example, under conditions in which water supply becomes exhausted before grain is formed.

EFFECT OF GREATER CROP DENSITY ON PLANT GROWTH AND REPRODUCTION

In the chapter on shoot and bud development, the yield components of several grain crops are discussed. Corn's limited ability to vary the ears per plant in response to different environmental conditions is emphasized. For this reason agronomists have encouraged farmers to increase corn plant population as a means of increasing yield. Early efforts to increase yields by increased plant population were not always successful, because plants selected to grow at populations of 12,000–16,000 plants per acre would exhibit barrenness at densities of 20,000–24,000 or more plants per acre. Five possible causes may be described for barrenness: poor silk development, low nitrate reductase activity, absence or sterility of the tassel, date of planting, and plant distribution. These causes will serve to characterize the type of problems that arise when plant density is increased for many crops.

Several workers (e.g. Moss and Stinson, 1961) suggest that barrenness is a direct result of shading causing poor corn silk development. For a hybrid selected at a low population, this would mean the shading was more intense as the population of plants increased; as a result the silk could be either stopped entirely or slowed in development enough that it was poorly pollinated. The small or missing ear not only reduces the potential grain yield but may slow the dry matter production of the entire plant. There is evidence that the size of the "sink" or the availability of a storage site affects the production and movement of carbohydrates (Moss, 1962). Early Nebraska work demonstrated a 27% decrease in fodder yield when the ear was removed at pollination. Moss cites other work to suggest that removal of the storage organ causes a carbohydrate accumulation in the leaves and a suppression of photosynthesis. In his own work Moss observed that dry matter accumulation from flowering to harvest in barren plants was only 50% of that for plants with normal ears.

A second possible explanation of lower yields with increased shading relates to nitrate reductase activity. Zieserl et al. (1963) suggest that while reduction of light does reduce carbohydrate synthesis, this is not the first limiting factor. They observe, as Moss did, that shading has a more marked effect on grain yield during the reproductive phase than at other phases of plant development. Short periods of shading drastically reduce the level of nitrate reductase in corn plants. These workers argue that hybrids possessing a high level of nitrate reductase would be better adapted to the maintenance of protein synthesis during critical stages of development and under conditions of environmental stress. This observation provides a possible basis for selection of lines with a greater tolerance to shading.

A third factor associated with silk and ear development at increased plant densities relates to the development of the tassel. It is regularly observed that male-sterile plants are less likely to be barren,

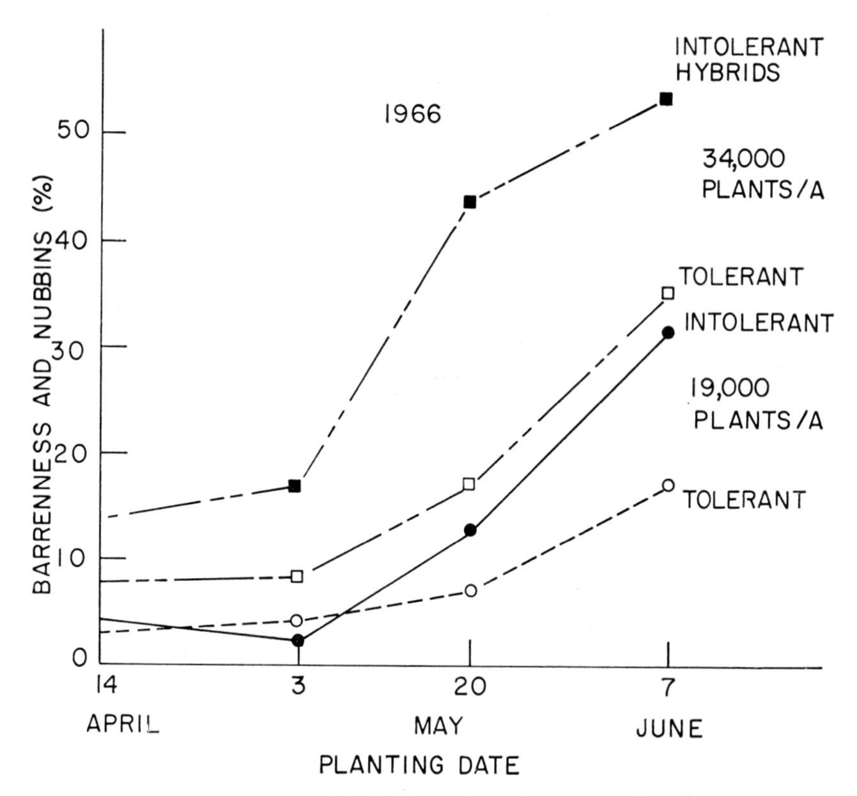

FIG. 6.2 ☙ The effect of planting date on percent of barren stalks or nubbin ears for hybrids tolerant or intolerant to high stand densities (Cardwell, 1967).

as there is a lower auxin activity in the tassel of male-sterile plants. One suggested result of this is that the ear competes more successfully for the carbohydrate supply in the plant when it has less competition from the tassel, a condition which exists when the tassel is low in auxin production or is removed entirely. Based on these experiences, commercial seed corn producers provide the farmer with a genetic (see Chapter 10) or mechanical blend of male-fertile and male-sterile plants to improve tolerance to population and moisture stress. The fertile plants pollinate all the plants. The male-sterile plants are more population tolerant, and the combination yields more in properly selected lines than an all male-fertile field. If very high populations (e.g. 90,000 plants per acre) are desired, it may be feasible to plant 5% of the plants as pollinators and the remaining 95% as male-sterile ear-bearing parents.

A fourth observation is that corn plants are less barren when planted relatively early. By planting in mid-to-late April corn was lower in its percentage barrenness than when planted in mid-May (Fig. 6.2). It may be that daylength favors the ear over the tassel in such cases, but whatever the reason the final effect of reduced barrenness is a very useful one to the farmer.

The ratio of ears per one hundred mature plants was the major contributor to increased yields which resulted from drilled as compared to hill-dropped or checked corn under irrigation (Fig. 6.3). Thus a more uniform distribution of plants over a unit area of ground is a fifth possible way to reduce barrenness and improve yields. Drilling corn resulted in 16 bushels per acre higher yields than checking corn at their respective optimum populations. Yield superiority of 5–6% in another study for corn grown in 20-inch versus 40-inch rows was associated mainly with more two-eared and fewer barren plants.

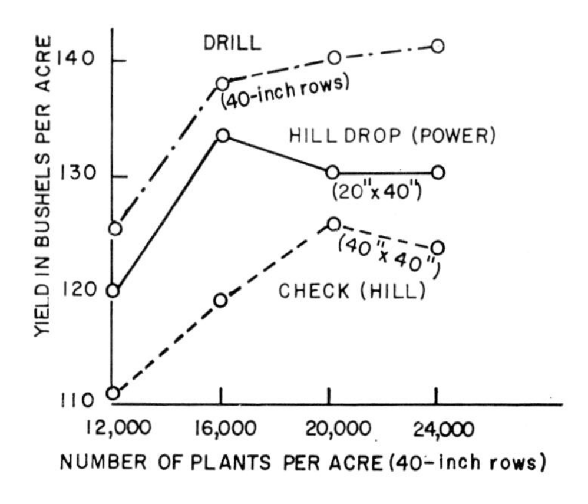

FIG. 6.3 The effect of seed distribution and population on corn yields (irrigated experiments, 1958–60) (Colville, 1964).

Corn breeders have used numerous selection procedures to improve the plant's ability to bear at least one ear under stress conditions and more than one ear when conditions are optimal. Lines are available which produce two ears, but as yet no striking increases in yield have resulted from using such material (Russell and Hallauer, 1968). The primary benefit sought is a broader range of tolerance to population. This tolerance would allow the manager to plant an intermediate population and expect the plants to adjust. If the season were favorable, each plant might set two ears and respond with high yields; if the season were less favorable, the plant's ability to bear ears would ensure at least one ear on each plant. At present, corn varieties have a tendency to bear no ears under a stress as described above; the proposed multiple-eared types should minimize this.

POPULATION RESPONSES IN DIFFERENT ENVIRONMENTS

VARIETY-POPULATION INTERACTIONS

A variety-population interaction is obtained regularly if the range of genetic material used is great enough. Figure 6.4 describes

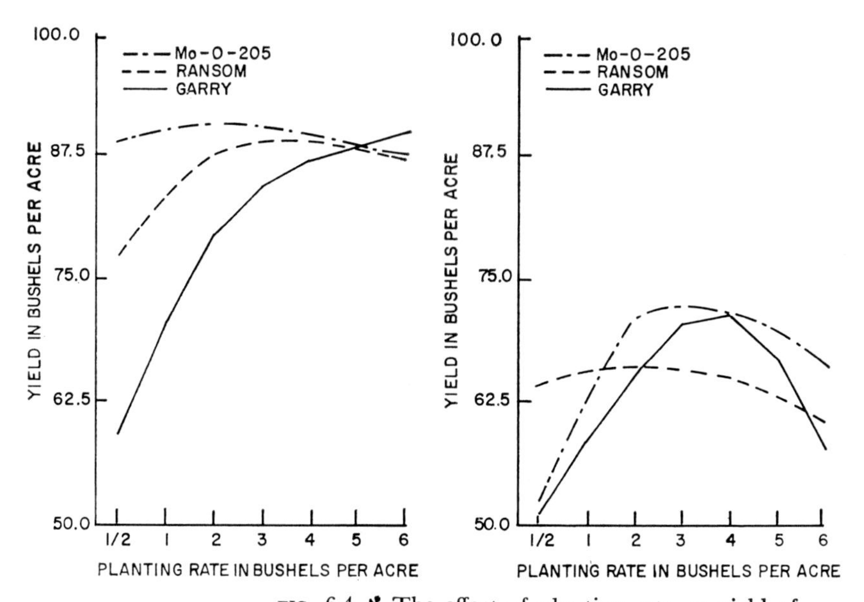

FIG. 6.4 ☙ The effect of planting rate on yield of three oat varieties at two locations: (left) east central South Dakota; (right) south central South Dakota (Shubeck et al., 1967).

the yield responses of three oat varieties to increasing population at two localities in South Dakota. Mo-0-205 and Ransom each exhibit stability of yield over a range of seeding rates in one of the location-by-year combinations. All the varieties vary widely as location and seeding rate are changed. From this information it is possible to select a most desirable seeding rate for a variety and to recommend the range of seeding rates that are acceptable for a particular variety at a location. The key point to recognize here is that varieties do respond differently to differing environments.

SOYBEAN RESPONSE TO PLANT DENSITY

Work by Shibles has provided an outstanding explanation of soybean responses to shifting plant density which may be used to generalize about many species. He describes how a change in plants per unit area changes the time necessary to achieve interception of all the available sunlight, the amount of leaf area necessary to accomplish this interception, and the quantity of the total dry matter that is translocated to the beans under these conditions.

As shown in Table 6.1, the more uniform distributions of corn (5 x 12 or 10 x 6) required fewer days to accomplish 95% light interception (L_{95}), had the largest leaf area index to intercept 95% of the sunlight, and gave the largest yield increases (126 and 132%) over the 40-inch row.

As the population is increased (row width being held constant), the leaf area to intercept 95% of the light (L_{95}) is increased, but days to L_{95} are decreased (Table 6.2). However, the low plant density allows excessive branching in the soybean and makes harvest by machine difficult. From Table 6.2 it is observed that the 50,000 population gives the greatest yield increase over 100,000 plants in 40-inch rows. However, for standard combine harvest, it would be difficult to

TABLE 6.1. Influence of row width and plant spacing on amount of leaf area and time required for 95% light interception in soybeans (R. M. Shibles, Iowa State University)

Row Width (inches)	Plant Spacing (inches)	Plants/Foot of Row	L_{95}	Days to L_{95}	Yield, % of 40″ Rows
5	12	1	3.6	54	126
10	6	2	3.5	55	132
20	3	4	3.4	55	115
40	1½	8	3.2	66	100

TABLE 6.2. Influence of soybean plant population on amount of leaf area and time required to reach 95% light interception (R. M. Shibles, Iowa State University)

Population (plants/acre)	Plant Spacing in Inches (20″ rows)	Plants/ Foot of Row	L_{95}	Days to L_{95}	Branches/ Plant	Yield, % of 40″ Rows*	% of Bean Weight to Total
25,000	12	1	2.5	62	9.0	108	30.7
50,000	6	2	2.9	60	4.0	122	27.3
100,000	3	4	3.4	55	0.5	115	27.2
200,000	1½	8	4.0	52	0.1	100	21.0

* Percent of 100,000 plants/acre in 40″ rows (20.8 bu).

actually get this yield due to excessive branch breaking and branch lodging. Further, it is of interest to note that the harvest index (right column—% of bean weight to total) is highest for the 25,000 population. As stated in the section discussing the harvest index (Chapter 2), this is not necessarily the best management level, because there was a less than optimum biological yield. The harvest index of the intermediate populations provides the maximum economic yield.

The recommended density of seeding of 3–4 beans per square foot provides a desirable balance of several factors: (1) the emerging seedlings help each other break through the soil; (2) stem branching is reduced to a level that allows nearly maximum dry matter yield and provides maximum harvestability; (3) the pods form higher off the ground; and (4) the weeds in the row are shaded. Since a 5–20% loss of this initial stand is expected from rotary hoeing, insect and disease loss, and other environmental causes, final stands will be 2–3 plants per square foot, from drilled to wide-row plantings. Farmers commonly make the mistake of seeding soybeans too heavily, thus getting over-vegetative viny plants and lowered seed yields (Table 6.2—200,000 plants).

Shibles and Weber (1966) obtained higher yields with drilled soybeans (with good weed control) than with row-planted soybeans. Comparisons of drilled, 20-inch, and 40-inch row beans gave an average increase of 16% for 20-inch rows and 30% for drilled beans when compared to 40-inch rows. They concluded the advantages of the drilled beans were due to (1) an earlier ground cover, and (2) a higher percentage of dry matter conversion to beans (25% of dry matter as beans in drill; 21% in 40-inch rows). Decreasing row width caused an increase in plant height and had a greater effect on plant height than did seeding rate.

SEED COMPOSITION

In addition, seed composition changes occur in response to plant density as reported by Williams (1962). He observed a highly significant regression of oil content on plant spacing for safflower. Average oil content of the seed from each flower position decreased significantly as plant spacing increased.

PLANT HEIGHT

Row width had a greater effect on sorghum plant height than did seeding rate. This response at first appears contradictory to some of the earlier observations. However, with sorghum, spreading the plants out more resulted in more tillering and thus more mutual shading. The tillering effect was more distinct than the seeding rates used, and as a result plant spacing had a greater effect on height.

THEORY OF PLANT RESPONSE TO STAND DENSITY CHANGES

Donald (1963) presents an inciteful explanation of plant response to stand density changes. He suggests that the greater seed weight and number of seeds per inflorescence at intermediate densities are due to the timing of interplant and intraplant competition. At the widest spacing competition is absent during the early stages of growth. Flower primordia are laid down by each plant in large numbers. As growth proceeds, interplant competition becomes progressively operative until, when flowering and seed setting occur, the load of inflorescences is so great as to lead to competition among inflorescences on an intraplant basis and thereby reduce the efficiency of seed production in the individual inflorescence. This loss of efficiency at the widest spacing is evident in fewer seeds per inflorescence and reduced seed size compared with somewhat more dense stands. Thus intraplant competition may be intense at low densities. It seems that in moderately dense stands, interplant competition becomes operative at the time of flower initiation or formation. The number of floral primordia laid down by each plant is considerably reduced, and this reduced load lies more closely within the capacity of the plant as interplant competition intensifies. Seeds per inflorescence and seeds per unit area achieve maximum values. In extremely dense stands the competition at the time of laying down of primordia is presumably already intense; plants in such dense communities suffer both interplant and intraplant competition. For example, the weight of any corn plant in the row tends to be inversely related to the weight of its immediate neighbors and directly related to the weight of the "second neighbor" in each direction. There is thus a trend toward alternation of heavier and lighter plants along the row. This observation suggests that in denser corn populations plant size will be more variable and uniformity of both the plants and their inflorescences will be reduced.

ROW WIDTH-PLANT SPACING EFFECTS ON SOIL TEMPERATURE

Narrow rows and more evenly distributed plants in the row both give cooler soil temperatures, as shown in Figure 6.5 (Shubeck et al.,

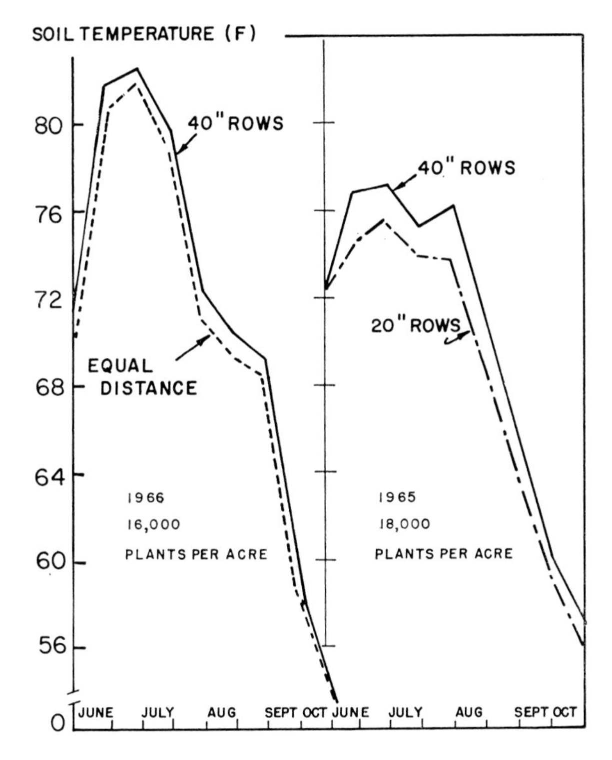

FIG. 6.5 ✤ The effect of space arrangements of corn plants on soil temperature. Cooler soil temperatures result from closer spacings (Shubeck et al., 1967).

1967). This effect is a result of earlier uniform shading of the soil as suggested in the days to 95% light interception (Table 6.1) with soybeans. The uniform shading results in a cooler soil temperature and reduces the rate of evapotranspiration from the soil. Finally, the soil water is conserved for transpiration. While the total quantity of water lost in evapotranspiration from a unit area of ground changes little (10–15%) under a standard supply of radiation, the common observation is made that the more of the water used for transpiration, the better the yield.

DATE OF PLANTING-POPULATION INTERACTIONS

As the question of variety response to population has been considered more intensively, numerous confounding factors have appeared. As shown by Figure 6.6, early planted corn gave increased yields over a span of years. In studying plant response to population and date of planting, Cardwell (1967) observed that these early dates

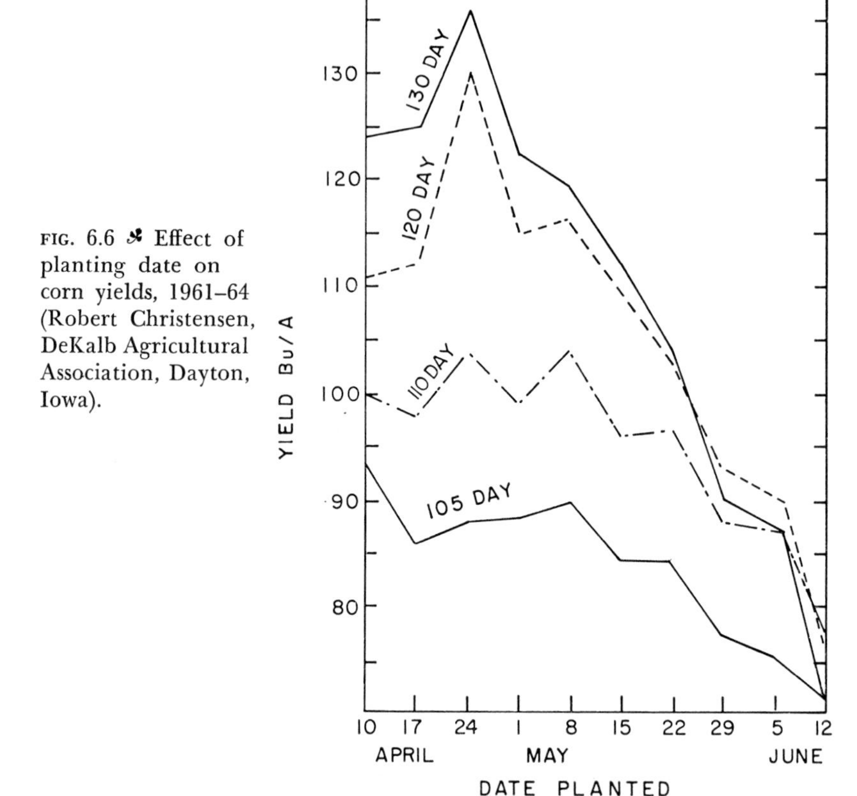

FIG. 6.6 ❧ Effect of planting date on corn yields, 1961–64 (Robert Christensen, DeKalb Agricultural Association, Dayton, Iowa).

of planting shifted the previous expectations of variety response and that varieties considered intolerant of dense population when planted in May became population-tolerant with April plantings (Fig. 6.7). Thus a change in one cultural practice gives an interaction resulting in a change of another factor; that is, planting earlier reduced the plant's sensitivity to high population.

FERTILITY-POPULATION INTERACTIONS

There is sufficient evidence to derive the general principle that as the fertility status is improved, the density required to give maximum yield, especially for annual crops, will increase. Such a population-nitrogen interaction for corn is described in Figure 6.8. Conversely, as density is increased the response to an added nutrient will continue

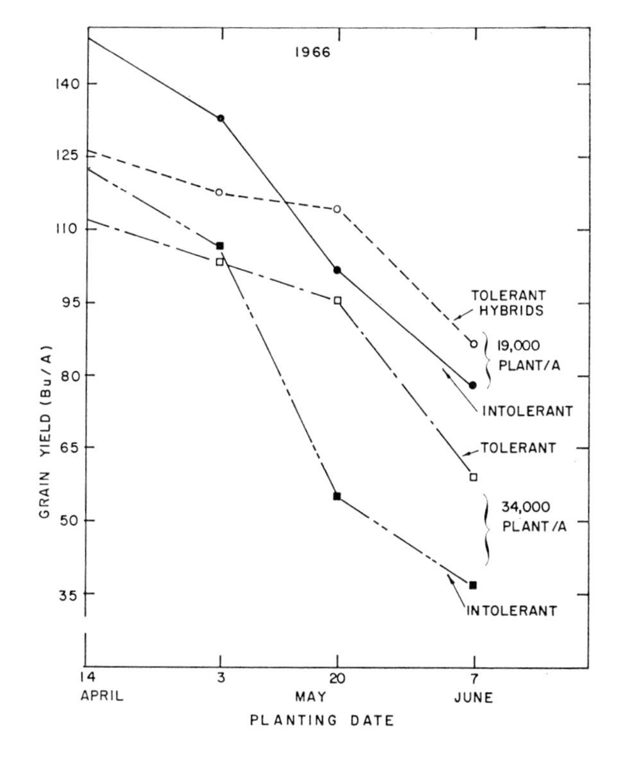

FIG. 6.7 Varieties considered as "tolerant" and "intolerant" to high population when planted in mid-May shift positions; the "intolerant" hybrids give the higher yields when planted in April or early May (Cardwell, 1967).

to a higher level of application. These observations point up the numerous interactions occurring with changes in crop culture. They suggest that a package of practices (i.e. date of planting, population level, fertility rate, and variety selection) may all need to be adopted simultaneously for maximum benefit.

MOISTURE-POPULATION INTERACTIONS

The optimum density of any annual crop will be less in a dry environment than a wet one, again illustrating the point that the more favorable the environment for any reason whatever, the higher will be the optimum population. As the annual precipitation rates increase in Kansas from west to east, the seeding rate recommended for sorghum is also increased. Likewise, wheat grown over a very wide range of rainfall and fertility conditions is sown at rates as low as 40 pounds per acre at its arid limits and as high as 200 pounds in high-precipitation areas.

Karper (1929), studying grain sorghum and its relationships to

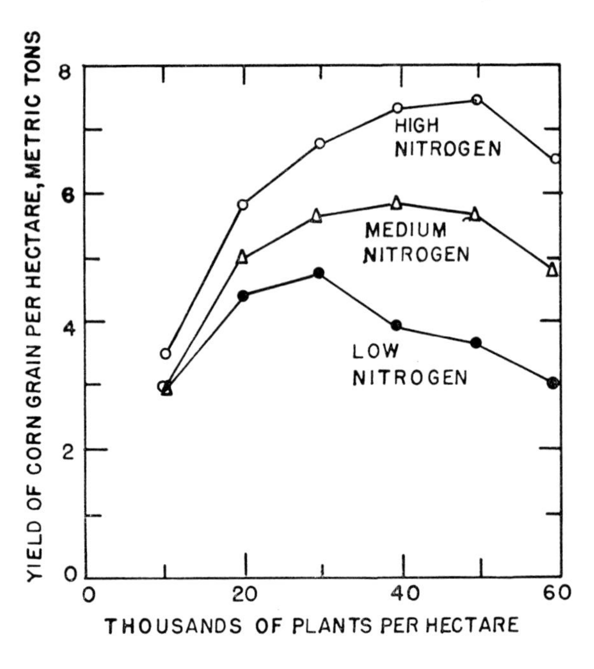

FIG. 6.8 ❧ The yield of corn at different stand densities and different levels of nitrogen (Lang et al., 1956).

rainfall and optimum density, emphasized the difference in rainfall-density relationships between genotypes. The optimum density for kafir was ten times greater in wet seasons compared to dry, but milo gave maximum yield at wide spacings in either wet or dry seasons. These results were directly associated with the degree to which each of the varieties tillered. Other field experiments and long-term weather data indicate that in the southern Great Plains, sorghum plant populations of about 18,000 plants per acre in 40-inch rows would result in the best grain yields in most seasons. Forage yields were higher with reduced row widths and increased seeding rates, but these conditions did not improve grain yields. Havelka (1965) found that regular narrowing of rows was not significant in increasing grain yield of sorghum. He did, however, get yield increases of 400 to 700 pounds of grain per acre by seeding in paired rows 12 inches apart, with 40 inches between the pairs, as compared to single rows uniformly spaced. In his studies forage production was not affected by these methods of planting. Corn plant production reached a maximum at 12,000–16,000 plants per acre in eastern South Dakota. Grain production reached a maximum at the same level and fell off drastically if populations exceeded this optimum (as suggested in Fig. 6.1).

SUMMARY OF INTERACTIONS

Milthorpe (1961) states the general principle that the greater the amount of leaf growth made before plants come into contact with each other, the more extensive will be the root system and the less likely is the plant to suffer drouth. Root systems of corn under various spacings were studied by Haynes and Sayre (1956). If widely spaced, a more or less circular distribution of roots occurred. When the plants were more crowded, lateral penetration out in the row greatly increased. Yao and Shaw (1964b) report that with narrower row spacings, corn roots do not develop as deeply, perhaps because the surface layer remains moist longer.

Average ear weight per stalk has been used to evaluate how well the factors of stand, fertility, and moisture have been balanced for corn. This assumes a minimum quantity of barren stalks. At stands that give the optimum yields, the average ear weight is about 0.5 pounds at 15.5% moisture. Table 6.3 describes a potential range of ear weights, suggested reasons for that weight, and the estimated yield loss when the ear weight differs from the suggested 0.5 pounds. This table has been a useful guide at plant populations used to date, but a wider range of populations and plant types may shift the emphasis to looking at several factors and not just ear weight.

Stands of corn may need to be adjusted in response to predicted moisture conditions. Since forecasting a given season's precipitation is not yet feasible, only that water already stored in the soil can be measured as a basis for planning under nonirrigated conditions. The effective rooting zone of corn may be considered to be five feet in depth. Due to different texture, soils range in water-holding capacity from 0.5 to 3.0 inches of water per foot. One set of population adjustments is suggested in Table 6.4 for Iowa conditions.

TABLE 6.3. Ear weight and yield potential for corn

Ear Weight (lb/stalk @ 15.5% H_2O)	Interpretation	Yield Loss (bu/A) Due to Nonoptimum Stand
.60–.75	stand much too low	12–40
.52–.60	stand too low	0–12
.48–.52	stand balanced to fertilizer and H_2O	0
.40–.48	stand high or fertilizer and H_2O limiting	0–6
below .40	stand too high or fertilizer and H_2O severely limiting	6–20

TABLE 6.4. Suggested stand levels at varying stored water levels

Water Stored in Upper Five Feet	Final Stand (plants/acre)
(inches)	*(thousands)*
less than 5	12–14
5–8	14–16
more than 8	16 and up

ROW WIDTH AS A MEANS OF CHANGING PLANT DISTRIBUTION

Throughout this discussion the desirability of uniform plant population has been mentioned repeatedly. In practical terms, the two ways to achieve uniform distribution are to drop the seeds at equidistant spacing within the row or to narrow the rows. Thus narrowing row width is discussed extensively as a practical means of more uniform plant distribution. Some agronomists suggest rhomboid or hexagonal plant spacing should give optimum light interception and yield. A square placement is a close practical approximation, although planting equipment for corn can very nearly accomplish the hexagonal spacing.

Row width studies on soybeans from several states in the north central United States show a yield increase for narrower rows (Fig. 6.9). Very likely the narrowest row widths yielded less because of poorer

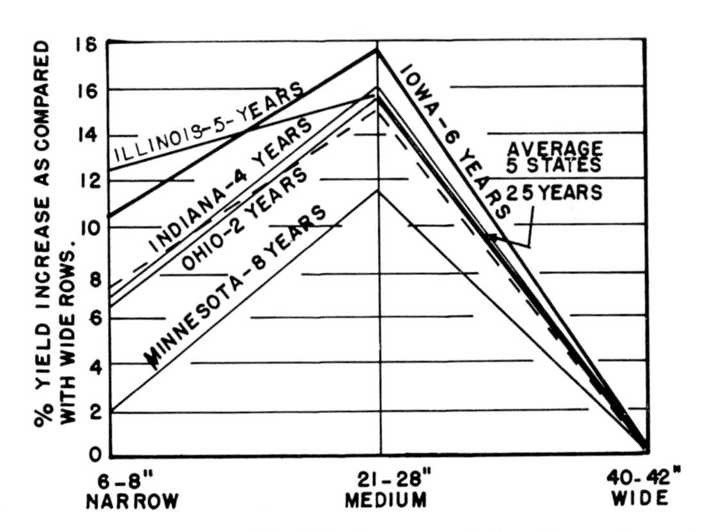

FIG. 6.9 ✤ Various row widths affect soybean yields about the same in five states (after Weber, 1962).

weed control. However, certain genotypes may not produce higher yields under dense plantings since they were selected under wider row widths; more recently, plant breeders are shifting their selection procedures to a range of plant spacings.

Under favorable environmental conditions, then, narrowing the rows of most crop plants will increase the yield. Peanuts gave an 11% yield increase for two years of a Virginia study. When a third dry year was included, the increase averaged only 4–5%. The importance of row spacing diminishes from northerly to southerly latitudes for soybeans. Row width has a greater effect of increasing soybean plant height than does the seeding rate. Eilrich et al. (1964) reported that narrow-rowed sorghum was superior to 40-inch rows in forage yields per acre. Small grains have been planted in a range of drill widths over the years. There has recently been renewed interest in planting wheat in narrower drill widths. Other workers report studies of broadcast planting in cotton instead of row culture. These studies are commonly done in close conjunction with agricultural engineers, as were the peanut studies, because new machinery for planting, weed control, and harvesting is often needed when row width is changed. Narrow-rowed corn (30-inch rows) gives 5–6% yield increase over 40-inch row spacings (Shibles et al., 1966).

For years agronomists and agricultural engineers have known that 40-inch rows were not the most desirable for corn, but 40-inch machinery was available and at intermediate management levels, the advantages for narrower rows were modest. It is emphasized again and again that before the farmer goes to the expense of changing row widths, he must be sure he is getting all the yield increase possible from adapted varieties, fertilization, weed control, insect control, timely cultural practices, *uniform plant distribution in the row* (Table 6.5), and *optimum population.* This last point is underlined by comparing acre yields at three row spacings (Table 6.6) with acre yields at four population levels (Table 6.7). These results were obtained with varieties considered to be population tolerant and under moisture conditions ranging from dry to optimum. Therefore, while population can provide high yields under optimal conditions, narrower row width and population properly balanced can give an even better yield.

NARROW ROW SUMMARY

In summary, narrow rows appear to be one of the final steps in a series leading toward maximum crop yield per acre, and narrow rows

TABLE 6.5. Effect of plant distribution in 30-inch rows on corn yield (Shibles et al., 1966)

Plant Spacings	Hybrid A	Hybrid B	Hybrid C	Average
1 every 8.7″	195	205	209	203
2 every 17.4″	181	192	204	192
3 every 26.1″	185	187	202	191
4 every 34.8″	169	179	190	180

precede the ultimate step of distributing the seed in an equidistant spacing. Shifting to (1) even distribution of single plants in the row, (2) appropriate higher populations, and (3) different plant structural types will be desirable before making the final shift to narrow rows for corn, but narrow rows may well be an early step in crop production improvement in crops like soybeans and sorghum. In Ontario studies, where corn is at the edge of its ecological optimum, 64 hybrids averaged an 18% increase in yield by narrowing the rows. Other areas show less consistent varietal differences to row spacing; it is suggested that population tolerance is the primary factor to consider.

The advantages and disadvantages of narrowing corn row width under Corn Belt conditions will serve to highlight the questions surrounding row width changes. Narrow rows are suggested to give: a higher yield of grain or silage; improved moisture stability by less evaporation, improved infiltration, and earlier ground cover; a better canopy for weed control, less bird damage, and greater tolerance to an early frost.

But there are obvious potential disadvantages: the choice of hybrid and population is not clear, more lodging may result, it is expensive to shift to narrow row equipment, there is a higher investment in seed and fertilizer, and it is absolutely necessary but more difficult to get early season weed control. If the farmer does change to narrow rows the following combination of practices is suggested: early planting, an early hybrid, a responsive hybrid, the right population, early chemical weed control, and proper harvesting equipment.

TABLE 6.6. Yields of corn planted in three-row spacings at three Iowa locations (average of all populations in Table 6.7) (Shibles et al., 1966)

Row Width	Ames (dry)	Castana (moderate)	Columbus Jct. (optimum)
(inches)	*(bushels)*	*(bushels)*	*(bushels)*
20	131.7	159.4	179.0
30	125.7	158.5	161.6
40	133.9	146.4	154.6

TABLE 6.7. Yields of corn at four populations, average of all row spacings (Shibles et al., 1966)

Population (plants/acre)	Ames (dry)	Columbus Jct. (optimum)
	(acres)	*(acres)*
15,500	137.1	144.2
19,500	136.2	164.4
23,500	122.6	175.0
27,500	125.7	176.6

FORAGES

In perennial forage crops the economic yield is the biologic yield, hence the yield response to increasing plant population is asymptotic. After stand establishment (the year of seeding) virtually complete light interception is attained early in the growing season because of tillering and branching. Very frequently the stand count will be reduced by half the first winter and then will stabilize for several years. Cowett and Sprague (1963) verify this observation for alfalfa grown in New Jersey where a wide range in plant density did not give distinctly different yields because tillering offset plant numbers. However, yield can be affected if stands are very low and conditions do not favor extensive tillering, as shown in Table 6.8.

When forages are grown for seed production, they are put in rows to facilitate weed control and increase seed set (see Chapter 10).

SUMMARY

It is reasonable to conclude that economic crops fall into the two yield-plant population relationships described at the beginning of this chapter. In most cases the equation $Y = Ax \frac{1}{1 + Abx}$ provides a suitable description of the asymptotic relationship between dry matter

TABLE 6.8. Average hay yields for Ranger alfalfa at four populations harvested 1961–62, Brookings, S. Dak. (Rumbaugh, 1963)

Spacing (inches)	Plants/Acre	Tons/Acre
42	3,556	.70
21	14,224	1.94
10.5	56,895	2.78
5.25	227,579	5.68

yield and plant population when the yield is a product of vegetative growth. For grain yields, $Y = a + bx - cx^2$ often describes the parabolic nature of yield response to plant population. Weed competition, diseases, and insects can give rise to exceptions to these generalizations. Thus the production manager must always keep these relationships in mind as he plans to shift population level or density. He must determine what genotypes (varieties) will respond favorably to the changes proposed.

The plant breeder has not been able to study his germ plasm under a wide range of conditions but has frequently used only the most common field practices of row width and plant spacing for his selection pressures. This has been due to limited time or funds. Hence, while his varieties are some of the best possible for current conditions, they may not be suited to new practices. The production manager thus must plan thoroughly as he changes production practices such as different plant density or distribution. ☙

LITERATURE CITED

Bond, J. J., T. J. Army, and O. R. Lehman. 1964. Row spacing, plant populations, and moisture supply as factors in dryland grain sorghum production. *Agron. J.* 56:3–6.

Cardwell, V. B. 1967. Physiological and morphological responses of corn genotypes to planting date and plant population. Abstr. No. 3789. Ph.D. thesis, Iowa State Univ.

Carmer, S. G., and J. A. Jackobs. 1965. An exponential model for predicting optimum plant density and maximum corn yield. *Agron. J.* 57:241–44.

Colville, W. L. 1964. Fish and horse . . . or modern corn methods. *Better Crops with Plant Food.* 48(1): 4–9.

Colville, W. L., and D. P. McGill. 1962. Effect of rate and method of planting on several plant characters and yield of irrigated corn. *Agron. J.* 54:235–38.

Cowett, E. R., and M. A. Sprague. 1963. Effect of stand density and light intensity on the microenvironment and stem production of alfalfa. *Agron. J.* 55:432–34.

DeWit, C. T. 1960. On competition. *Verslag. Landbouwk. Onderzoek.* 66:8.

Donald, C. M. 1963. Competition among crop and pasture plants. *Advan. Agron.* 15:1–118.

Duke, G. B. 1964. Effects of close-row spacings on peanut yield and on production equipment requirements. *USDA Prod. Res. Rept.* 77.

Dungan, G. H., A. L. Lang, and J. W. Pendleton. 1958. Corn plant population in relation to soil productivity. *Advan. Agron.* 10:436–71.

Eilrich, G. L., R. C. Long, F. C. Stickler, and A. W. Pauli. 1964. Stage of maturity, plant population and row width as factors affecting yield and chemical composition of Atlas forage sorghum. *Kans. Agr. Exp. Sta., Tech. Bull.* 138.

Goodall, D. W. 1960. Quantitative effects of intraspecific competition: An experiment with mangolds. *Bull. Res. Council Israel,* Sect. D 8:181–94.
Harpstead, D. D., and V. A. Dirks. 1966. How many bushels of seed? *S. Dakota Farm and Home Res.*
Havelka, U. D. 1965. Comparison of single row and double row methods of planting grain sorghum in central Texas. *Texas Agr. Exp. Sta., Prog. Rept.* 2349.
Haynes, J. L., and J. D. Sayre. 1956. Response of corn to within-row competition. *Agron. J.* 48:362–64.
Hinson, K., and W. D. Hanson. 1962. Competition studies in soybeans. *Crop Sci.* 2:117–23.
Hoff, D. J., and H. J. Mederski. 1960. Effect of equidistant corn plant spacing on yield. *Agron. J.* 48:295–97.
Holliday, R. 1960a. Plant population and crop yield: Part I (Review Article). *Field Crop Abstr.* 13:159–67.
———. 1960b. Plant population and crop yield: Part II (Review Article). *Field Crop Abstr.* 13:247–54.
———. 1963. The effect of row width on the yield of cereals (Review Article). *Field Crop Abstr.* 16:71–81.
Hozumi, K., H. Koyama, and T. Kira. 1955. Intraspecific competition among higher plants. IV. A preliminary account on the interaction between adjacent individuals. *J. Inst. Polytech. Osaka City Univ. Ser. D.* 6:121–30.
Karper, R. E. 1929. The contrast in response of kafir and milo to variations in spacing. *Agron. J.* 21:344–54.
Lang, A. L., J. W. Pendleton, and G. H. Dungan. 1956. Influence of population and nitrogen levels on yield and protein and oil contents of nine corn hybrids. *Agron. J.* 48:284–89.
Leffel, R. C., and G. W. Barber, Jr. 1961. Row widths and seeding rates in soybeans. *Md. Agr. Exp. Sta., Res. Bull.* 470.
Lehman, W. F., and J. W. Lambert. 1950. Effects of spacing of soybean plants between and within rows on yield and its components. *Agron. J.* 52:84–86.
Milthorpe, F. L. 1961. The nature and analysis of competition between plants of different species. *Symp. Soc. Exp. Biol.* 15:330–55.
Moss, D. N. 1962. Photosynthesis and barrenness. *Crop. Sci.* 2:366–67.
Moss, D. N., and H. T. Stinson, Jr. 1961. Differential response of corn hybrids to shade. *Crop Sci.* 1:416–18.
Pesek, J., E. A. Heady, J. P. Doll, and R. P. Nicholson. 1959. Production surfaces and economic optima for corn yields with respect to stand and nitrogen level. *Iowa Agr. Exp. Sta., Bull.* 472.
Prine, G. M., and V. N. Schroder. 1964. Above-soil environment limits yields of semiprolific corn as plant population increases. *Crop Sci.* 4:361–62.
Rumbaugh, M. D. 1963. Effect of population density on some components of yield of alfalfa. *Crop Sci.* 3:423–24.
Russell, W. A., and A. R. Hallauer. 1968. Multiple-eared corn may give farmers more stable yields. *Iowa Farm Science* 23(3): 279–83.
Shibles, R. M., W. G. Lovely, and H. E. Thompson. 1966. For corn and soybeans—narrow rows. *Iowa Farm Science* 20:3–6.
Shibles, R. M., and C. R. Weber. 1966. Interception of solar radiation and

dry matter production by various soybean planting patterns. *Crop Sci.* 6:55–59.

Shubeck, F. E., B. E. Lawrensen, and L. A. Nelson. 1967. Corn plant spacing and populations. *S. Dakota Farm and Home Res.* 17(1):4–8.

Stickler, F. C. 1964. Grain sorghum yields from 20- and 40-inch rows at various stand densities in Kansas. *Kans. Agr. Exp. Sta., Bull.* 474.

———. 1964. Row width and plant population studies with corn. *Agron. J.* 56:438–41.

Tanner, J. 1965. Capture that sunshine. *The Furrow.* John Deere, May-June, p. 203.

Termunde, D. E., D. B. Shank, and V. A. Dirks. 1963. Effects of population levels on yield and maturity of maize hybrids grown on the northern Great Plains. *Agron. J.* 55:551–55.

Warren, J. A. 1963. Use of empirical equations to describe the effect of plant density on the yield of corn and the application of such equations to variety evaluations. *Crop Sci.* 3:197–201.

Weber, C. R. 1962. A quick guide for higher soybean yields. Iowa State Univ., Ames, Pam. 290.

Weber, C. R., R. M. Shibles, and D. E. Byth. 1966. Effect of plant population and row spacing on soybean development and production. *Agron. J.* 58:99–102.

Wiggans, R. G. 1939. The influence of space and arrangement on the production of soybean plants. *Agron. J.* 31:314–21.

Williams, J. H. 1962. Influence of plant spacing and flower position on oil content of safflower, *Carthamus tectorus. Crop Sci.* 2:475–77.

Williams, W. A., R. S. Loomis, and C. R. Lepley. 1965. Vegetative growth of corn as affected by population density. I. Productivity in relation to interception of solar radiation. *Crop Sci.* 5:211–15.

———. 1965. Vegetative growth of corn as affected by population density. II. Components of growth, net assimilation rate and leaf area index. *Crop Sci.* 5:215–19.

Woolley, D. G., N. P. Baracco, and W. A. Russell. 1962. Performance of four corn inbreds in single-cross hybrids as influenced by plant density and spacing patterns. *Crop Sci.* 2:441–44.

Yao, A., and R. H. Shaw. 1964a. Effect of plant population and planting pattern of corn on the distribution of net radiation. *Agron. J.* 56:165–69.

———. 1964b. Effect of plant population and planting pattern of corn on water use and yield. *Agron. J.* 56:147–51.

Zieserl, J. F., W. L. Rivenbark, and R. H. Hageman. 1963. Nitrate reductase activity, protein content and yield of four maize hybrids at varying plant populations. *Crop Sci.* 3:27–32.

CHAPTER SEVEN GROWTH REGULATORS

THE GROSS NEEDS of plants for water, minerals, light, and organic plant food have been evident to agriculturists for centuries. More recently, within a time span of less than 100 years and intensively the past 30 years, plant growth regulation by minute quantities of chemicals (naturally present or applied) has been recognized. The activity of a single cell may be studied to provide information about the biochemical reactions in the plant. However, to make use of plant growth as it is known in agricultural production systems, it becomes important that a large mass of cells which form the plant body are coordinated and organized in their total action. Two types of mechanisms have been observed to bring about coordination in the plant: (1) chemical messenger systems which direct cells to carry out various functions and (2) systems of physical forces. A great deal of information has been accumulated on the chemical control systems, and four groups of growth regulators will be discussed: auxins, gibberellins, cytokinins, and growth inhibitors. Recent information suggests a separation may be made in the inhibitor group for abscisic acid and for ethylene, but they will be discussed here under the general heading of inhibitors. Physical forces are less well understood.

As a result of these chemical controls the plant has localized regions of growth, photosynthetic activities, translocation systems, and sites of accumulation or "sinks" in reproductive structures. The controls are generally needed only in minute quantities and in most instances are produced by the plant itself.

Since 1920 work on growth hormones or regulators has been very active. Agriculturally this work has resulted in exciting developments in chemical weed control and is beginning to provide new possibilities in crop growth control. These developments may well have an impact on agriculture as great as the mechanical harvester or the inorganic chemical fertilizer.

Early work was concerned primarily with internal growth regulators. These were considered to be hormones—substances produced in any one part of an organism and transferred to another part where they influenced a specific physiologic function. Thus they function as chemical messengers. One of the earliest observations of this type of reaction was recorded by the French horticulturist du Monceau, who concluded in 1758 that the formation of roots on the trunk of a girdled tree was caused by descending sap. Sachs, in the late 1800s, inferred that differentiation by roots and flowers was due to minute amounts of chemicals moving up and down through the plant.

The auxin group includes two types of materials: (1) the growth hormones, which are endogenous natural plant constituents and regulate cell enlargement in the manner of indoleacetic acid, and (2) the natural or synthetic materials that are not normal components of the growth regulating system and may be classified as exogenous. The gibberellins also regulate growth, but their action is distinctive and especially related to elongation. They do, however, require the presence of auxin to provide their effects. The cytokinins regulate growth at least in part by stimulating cell division, and their principal effects are clearly separate from those of the auxins and the gibberellins. The inhibitors include a wide array of chemical substances which may inhibit growth directly or give a secondary effect by inhibiting some component reaction relating to the growth regulators.

AUXINS

The study of auxins arose accidentally in the 1870s with the experiments of Charles Darwin. Darwin, better known for his ideas on evolution, did much thinking about plant growth. One recurring question was why plants turn toward the light. He found that a unilateral light source shining on the coleoptile of a grass seedling caused it to bend toward the light. The coleoptile did not bend if the tip was covered with tinfoil or an opaque glass shield, but it did bend if only the base was covered. Darwin hypothesized in his book *The Power of Movement in Plants,* published in 1880, that some stimulus passed from the tip down to the growing zone and there exerted a specific growth effect. About 1914, Paal found that if he cut off the coleoptile tip, no further growth occurred. If the tip were placed on an agar block and then the agar block placed on a decapitated coleoptile asymmetrically, curved growth occurred which Paal attributed to unequal distribution of the growth substance.

In 1926, F. W. Went (who has worked more recently in the United States with the California Institute of Technology Phytotron and the Missouri Botanical Gardens) began his classical *Avena* coleoptile studies as an evening project while completing his military service in Utrecht, Holland. He demonstrated that the active substance which caused the coleoptile to grow and to bend when exposed to unilateral light could be extracted into agar blocks. He perfected a test to more quantitatively measure the amount of auxin in the agar block based on the curvature resulting from the asymmetric placement of agar on a decapitated coleoptile. Went's technique for assaying growth hormone activity has been replaced by the *Avena* straight growth test, but he contributed greatly to the idea of a sensitive bioassay as a basis for quantifying a response to a growth regulator.

The challenge of identifying the chemical nature of the auxins was met by Kögl, Haagen-Smit, and Erxleben in 1931. These Dutch workers initially extracted a growth hormone from human urine and identified it as auxin *a*. They subsequently identified auxin *b* from corn oil. Finally, heteroauxin (now called IAA or indole-3-acetic acid) was extracted from urine and fungi (Fig. 7.1). Other workers have failed to reisolate auxins *a* and *b* and question their natural occurrence, but it is generally accepted that IAA is widespread in plants. It is especially concentrated in actively growing areas like the stem apex, young fruits, and germinating cotyledons.

Both endogenous and exogenous growth regulators cause plant movements or tropisms such as those described for the coleoptile by Darwin. A movement in response to light is a phototropism; movement toward light would be a positive phototropism, movement away from light a negative phototropism. Growth is symmetrical if the growth regulator is equally distributed in the plant; if it is not, curvature results. A stem reacts to gravity by growing upward, demon-

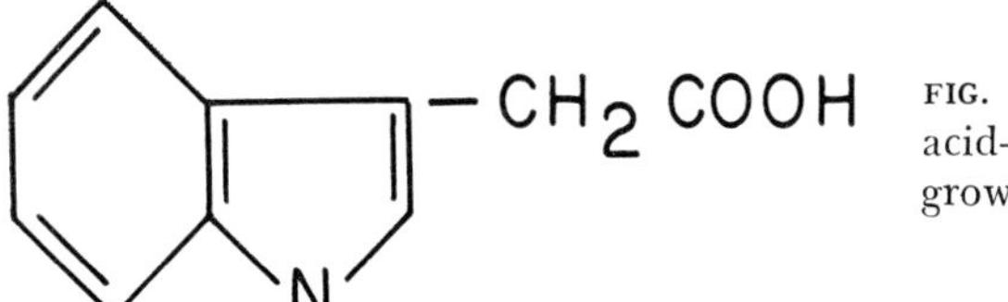

FIG. 7.1 ☙ Indoleacetic acid—the primary plant growth regulator.

strating a negative geotropism, probably caused by a gravitational redistribution of the growth regulator to the lower side of the stem and stimulating that side to grow more rapidly.

It has been known for many years that phototropic movement in plants may result from the unequal distribution of auxin in the lighted and shaded sides of the organs, but debate continues as to the exact nature of these differential levels and how they arise naturally. The mechanism is still considered unknown. The two most popular hypotheses suggest (1) that the action of ultraviolet light destroys the auxin on one side of the stem, or (2) that unilateral light causes a redistribution of the auxin. Lam and Leopold (1966) suggest a third concept, a light reduction of auxin production. They found more diffusable auxin from below a darkened cotyledon of sunflower than from beneath a lighted one. The mechanism remains in question, but the basis for the curvature seems well founded.

Usually the application of auxin to stems in an intact plant produces no extra increment of growth (Fig. 7.2). It thus appears the stem is saturated by auxin produced in its own tip. Roots, however, must operate normally under conditions of more than adequate auxin, because extra auxin inhibits their growth.

AUXIN ACTIONS

Auxin activity is characterized by numerous actions (Fig. 7.3):

1. They regulate growth.
2. They play a central role in apical dominance and cause an

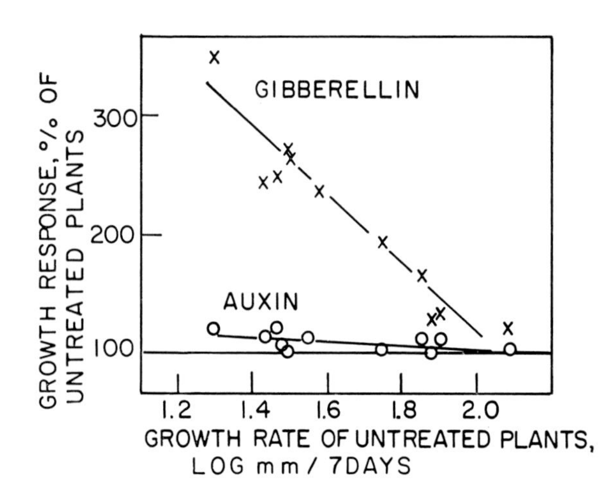

FIG. 7.2. ❧ The response of 11 pea varieties to auxin and gibberellin applications as compared to untreated checks (Brian and Hemming, 1955).

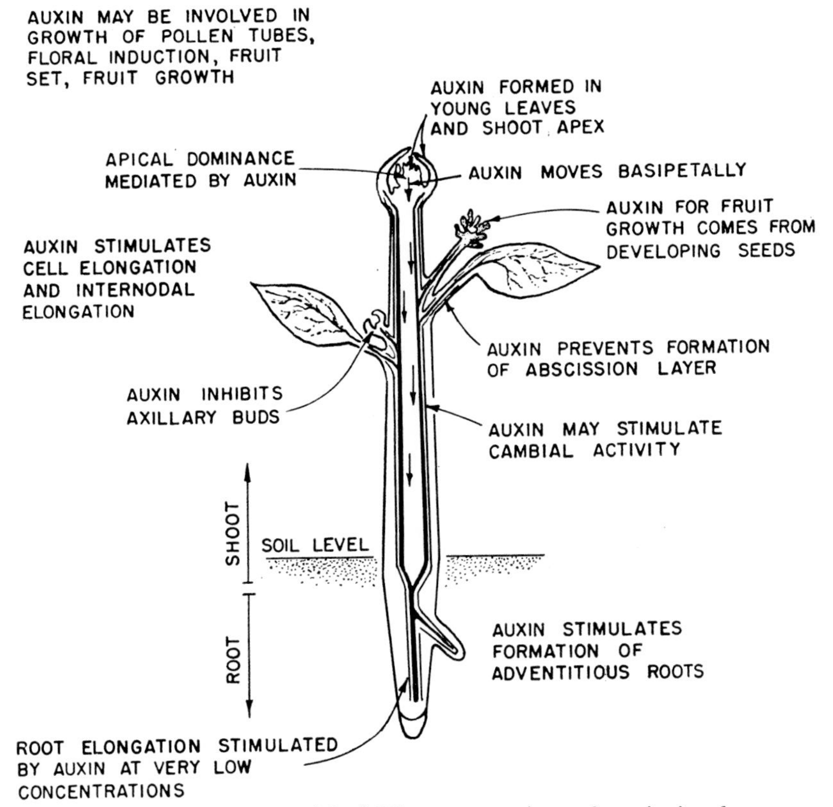

FIG. 7.3 ❧ The many actions of auxin in plant growth (Steward, 1964).

inhibition of lateral bud development by stimulating ethylene production.

3. They cause induction of root initiation, differentiation of the xylem, and influence the formation of the abscission layer.

4. They play a central role in the tropistic movements.

5. Their action appears to result from a softening of the cell walls.

6. They are formed in meristems and expanding tissues (the grass coleoptile, the dicotyledon apex, the embryo). In addition, parasitic or symbiotic organisms (mycorrhizae, nodule-forming bacteria) frequently provide auxin. The mycorrhizae cause root growth inhibition, whereas the nodule bacteria cause root hair curling.

7. Their transport is distinctly polar in the stem, moving from morphological tip to base at a velocity of 0.5–1.5 cm per hour.

8. Auxins function in cell elongation in conjunction with gib-

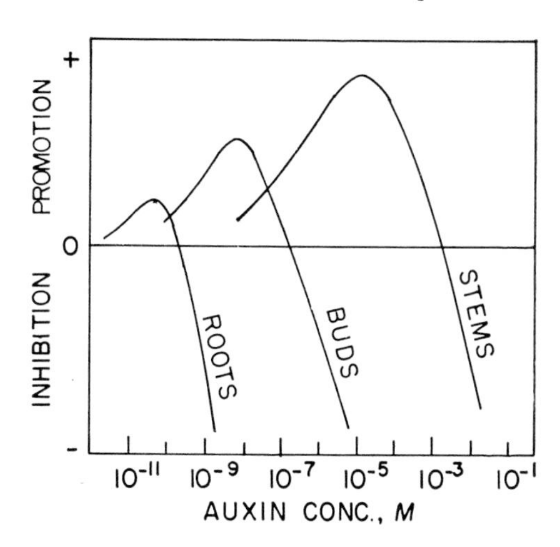

FIG. 7.4 ☙ The growth response of roots, buds, and stems to a range of auxin concentrations. Each organ may respond positively or negatively to the increasing supply (Thimann, 1937).

berellin and in cell division in conjunction with cytokinins.

9. Auxin transport is active, not passive, and is dependent on metabolic activity and oxygen.

Stems, buds, and roots all show positive and negative responses to auxins (Fig. 7.4). Stems have the highest optimum auxin response level, roots the lowest. Recognition of these reactions helps to explain many plant growth responses. Adequate auxin prevents the formation of an abscission layer through promotion of RNA and protein synthesis. Although not understood as yet, auxin has been shown to promote abscission in older leaf tissue. Gibberellin and abscisic acid (see Inhibitor section) have the opposite effect. Auxin appears to function by cell enlargement and simultaneous water uptake. It appears this effect of auxin is due to a softening of the cell wall. Once the wall is softened, osmotic pressure causes the cell to expand, until stopped by cell wall resistance at some new expanded size.

As researchers began to gain knowledge about auxins, there was a period of intensive effort to classify the characteristics of a molecule which were necessary to give growth activity. Among the characteristics suggested were: (1) an aromatic ring, (2) an acidic side chain, and (3) some spatial relationship between the two. However, there are such frequent exceptions to these and other classifications that current attempts for generalizations are few.

SYNTHETIC AUXINS

The synthetic auxins constitute a large and rather wide ranging group of compounds. Some of the most active in relation to toxicity and persistence (in ascending order) are: indoles, naphthyls, phenoxys, and benzoics. The indoles and naphthyls are useful herbicides and modifiers of fruit growth and ripening. The benzoics are used as herbicides. While this wide range of chemical materials can evoke the auxin effects on growth, only a much smaller group can evoke the correlative effects of natural auxin, apparently because of poor movement in the polar transport system.

AUXIN GROWTH INHIBITION

High concentrations of auxin may inhibit growth (Fig. 7.4). In order to consider the basis for this observation, it is important first to note that the growth promotive effect of auxin is postulated to be due to the connection of two attachments of the auxin molecule to the reaction sites (Fig. 7.5). The inhibitory act is a result of the attachment at the two sites by different auxin molecules, hence blockage of growth. This scheme fits the observation that growth reactions are generally stimulated by low concentrations of auxin and inhibited by higher ones. Antiauxins were once considered to be chemicals which closely resembled auxins in structure but which lacked some of the requirements for auxin activity. However, this concept has been

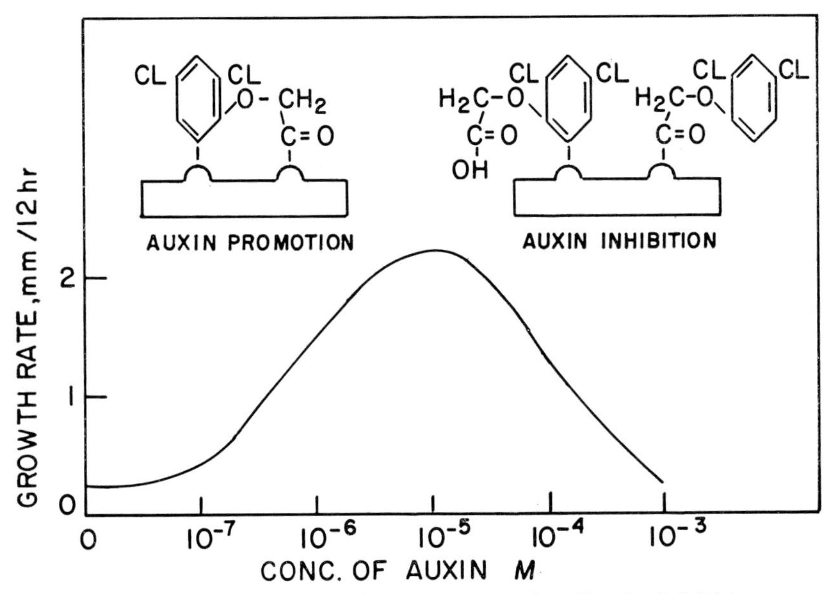

FIG. 7.5 A diagram to describe the inhibition caused by high auxin concentrations (Leopold, 1964, after Foster et al., 1952).

weakened by observations that several chemicals which should be expected to have antiauxin properties do not.

In summary, Leopold (1964) suggests that three major features of auxins stand out.

1. The auxins have a wide diversity of effects. In addition to influencing cell elongation, the tropisms, xylem and root differentiation, apical dominance, and the abscission processes, there are other effects on flower initiation and development, pollen-tube growth, fruit set and fruit growth, and seed germination. Almost every dynamic part of plant growth and development seems to be affected by auxin.

2. A wide diversity of other chemical controls may be interwoven with the auxin effects. The gibberellins, cytokinins, chelating agents. organic acids, and fatty acids may have growth-regulating power. Many of these growth regulators have a much wider range of systems which they stimulate.

3. There are systematic patterns of the auxin effects. The action of auxins as chemical messengers influencing patterns of plant development may be represented by the apical dominance induced by auxin suppressing lateral bud growth down the stem or by the tendency of auxin to stimulate vegetative and inhibit floral development, as discussed under growth correlations in Chapter 8.

GIBBERELLINS

The gibberellins represent a second group of natural growth regulators, but through a type of action which is distinctive from auxin (Gould, 1961; Phinney and West, 1961). The most pronounced effect of gibberellin is on stem internode elongation. It seems likely that gibberellin growth effects require the presence of auxins.

Japanese farmers were long aware of the presence of uncommonly tall rice seedlings in their fields which only rarely flowered and never lived to maturity. They concluded that the plants which grew in this manner were diseased and named the disease "Bakanae" or "foolish seedling disease," since the plant which looked so promising early in the season did not set seed. In 1926 the fungus *Gibberella fujikoroi (Fusarium moniliforme)* was isolated and identified as the pathogen. From this fungus the Japanese workers extracted gibberellin and were able to demonstrate the same height increase effect with the cell-free extract as had been observed for the diseased plant.

The Japanese studied the gibberellins intensively in the 1930s, but Western workers did not become actively engaged in gibberellic acid

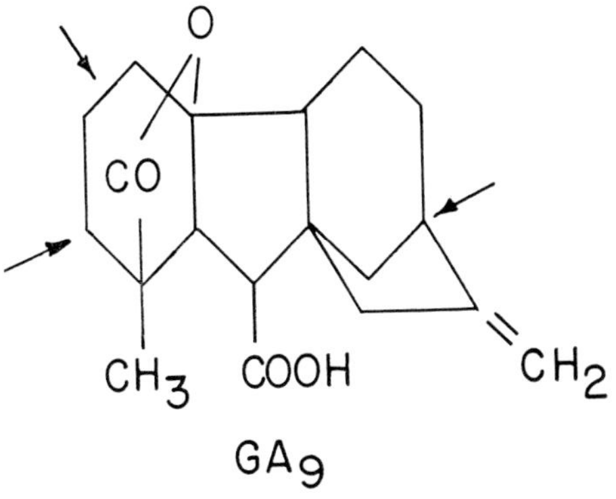

FIG. 7.6 ☙ The structure of gibberellic acid 9.

research until the 1950s. The following decade saw a flurry of research activity on these growth regulators, and gibberellins were shown to be widely distributed in higher plants. The structure in Figure 7.6 represents the primary gibberellin (GA_9) in a group of twenty or more.

GIBBERELLIN ACTIONS

In addition to the initial effects on stem elongation, further studies have shown that gibberellin may:

1. result in a diminution of leaf area.
2. stimulate flowering in long-day plants and replace the requirement for cold or long days in some biennials.
3. prevent the formation of root initials, but not inhibit root elongation.
4. break seed and bud dormancy.
5. elongate the internodes of genetically brachytic dwarf plants (Fig. 7.7).
6. interact with IAA in apical dominance.

Jacobs and Case (1965) found that gibberellin and IAA restored apical dominance more effectively than IAA alone in decapitated *Pisum* shoot apices. Translocation studies with C^{14} IAA revealed that gibberellin caused more IAA to be present and effective far from the site of application, but gibberellin added alone stimulated the growth of side shoots rather than inhibiting them. These authors conclude that the effect of gibberellin is an auxin-saving one rather than directly affecting auxin transport.

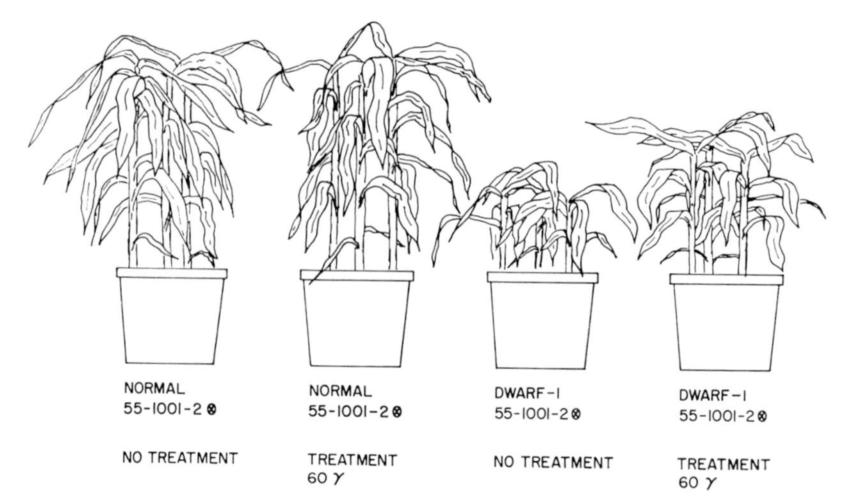

FIG. 7.7 ☙ Gibberellin applied to dwarf corn (D_1) causes it to grow as tall as normal corn. Left to right: normal control, normal plus gibberellin (60μg/plant), dwarf control, dwarf plus gibberellin (60μg/plant) (Phinney, 1956).

GIBBERELLIN VERSUS AUXIN

The contrasting actions of gibberellins and auxins are as follows. Both stimulate growth, but only gibberellin restores growth in dwarfs (Fig. 7.7). It appears that the gibberellins have their most distinct effect on the brachytic dwarfing genes, that is, normal plants except with shortened internodes. Only auxins seem to be generally able to inhibit growth at superoptimal concentrations. Both may cause fruit set, but the stimulations of flowering and dormancy control are almost unique to the gibberellins. While direct field applications are few, interest in gibberellin study continues as a means of understanding plant growth phenomena.

Not all genetic dwarfs may be returned to normal plant size by gibberellin. Gibberellin-treated dwarf alfalfa plants did not attain heights comparable to untreated normal plants (Pauli and Sorenson, 1961). Forage dry matter yield increases were obtained from gibberellin applications, and the effects of gibberellin and nitrogen were additive. However, the practical conclusion was that gibberellin had no advantage over a fertilizer application for increasing the growth of grass. The height of summer annual forages was increased, but the pounds of dry forage did not change significantly and the protein content was lowered slightly due to an increase in stemminess. These two factors reduced forage quality.

The effects of gibberellin on male sterility have been studied by numerous plant breeders. Gibberellin was used to induce a usable degree of male sterility in nonvernalized Wong barley in the greenhouse. Under field conditions, induced male sterility in vernalized winter and spring barley was not great enough to be of practical interest in a crossing program. Gibberellin caused abnormally early anther development, and male sterility was brought about by loss of nuclei at late stages of microsporogenesis or early stages of pollen maturation. Interaction of gibberellin with naphthaleneacetic acid and triiodobenzoic acid induced a higher degree of male sterility when gibberellin treatments were repeated from 10 days after planting to heading. The technique would appear to work only on winter types that produce spikes without cold treatment. Male sterility was induced in two inbred lines of corn with gibberellin, but in wheat gibberellin did not cause male sterility.

Several workers have studied the effect of gibberellin on seedling vigor. Allan et al. (1961) found that soaking seeds in solutions of gibberellin stimulated slow-emerging wheat varieties to emerge at rates comparable to rapidly emerging varieties under both greenhouse and field conditions. Gibberellin treatment did not stimulate coleoptile elongation of any of the varieties tested and adversely affected the seedling survival of three varieties. Gibberellin depressed wheat seedling emergence by causing premature unfolding of leaves from the coleoptile while the plant was still beneath the soil surface. The rates of laboratory germination were increased for three varieties and field emergence increased for one variety by an application of gibberellin. No other character was consistently increased, nor were yields of sorghum consistently increased by foliar sprays. Cotton varieties differed in seedling response to gibberellin. Rapid emergence and seedling height were positively correlated. The faster emerging, taller seedling types gave the greatest response to gibberellin.

The preceding discussion presents a variable picture on crop plant responses to gibberellic acid. A few interesting positive responses are noted, and such observations stimulate a continuing interest in gibberellin research.

In summary, the gibberellins are natural growth-regulating substances in plants and are also involved in the development of pathological symptoms. The lack of natural gibberellins appears to be involved in dwarfism. The gibberellins can trigger several developmental events which are ordinarily under environmental control including

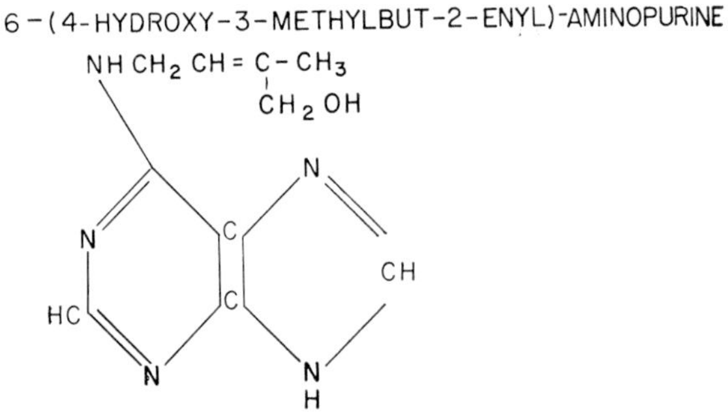

6-(4-HYDROXY-3-METHYLBUT-2-ENYL)-AMINOPURINE

$NH\,CH_2\,CH = C(CH_2OH) - CH_3$

FIG. 7.8. ☙ Zeatin, a naturally occurring cytokinin.

flowering, breaking of dormancy, and other temperature and photoperiod-controlled steps. The implication is that the gibberellins are involved in the regulatory systems which are responsible for the developmental responses of the plant to environmental cues.

CYTOKININS

The cytokinins regulate growth at least partly by stimulating cell division. This effect is not clearly separate from the auxins and gibberellins, for cytokinins may require auxin for their action, and either auxins or gibberellins may themselves stimulate cell division. However, the main effects of auxin and gibberellin are on cell enlargement, while the generic name cytokinin has been adopted for chemical substances which stimulate cell division or cytokinesis. A rather wide range of natural and synthetic materials have been reported to have cytokinin activities.

Cytokinins have been studied intensively for a shorter time than the auxins or the gibberellins. Lang (1967) reports work by C. O. Miller on zeatin, one of the identified cytokinins from higher plants (Fig. 7.8). In the young corn kernel zeatin occurs not only in the free form but also in the form of a nucleoside and a mono- and possibly a diphosphate nucleotide.

Early work with the synthetic cytokinins suggested that these chemicals were very immobile in the plant. However, further studies with natural cytokinin-type materials suggests they are mobile and that a cytokinin formed in the root may control some part of leaf senescence or grain development.

Cytokinin action seems to be extremely varied. In addition to affecting cell division, these growth regulators cause marked alterations in the protein and nucleic acid components in tissues, an effect which could be a basic part of the cytokinin effects on cell division. Protein, DNA, and RNA are all present in higher concentration when the plant has been treated with the cytokinin. These higher concentrations are considered to be due to less degradation of protein, DNA, and RNA rather than to increased synthesis rates. However, synthesis is increased in some cases. One of the most striking effects of the cytokinins is on differentiation of vegetative tissue into buds. They are also strong promoters of bud growth and thus modify apical dominance. Cytokinins have caused an increase in leaf enlargement, and synthetic cytokinins increased the cold tolerance of all plant species tested as well as increasing resistance to herbicides and pesticides. In summary, this brief survey of the most recently classified group of growth regulators indicates that cytokinins occur widely in plants and participate in many aspects of growth and development.

GROWTH INHIBITORS

NATURAL

The most common growth inhibitors are aromatic organic materials, although there are such diverse inhibitors as ascorbic acid, some fatty acids, and some metallic ions. The inhibitors are at present a poorly defined group of growth regulators. Simple phenyl compounds commonly cause growth inhibition. These include phenols, benzoic acids, cinnamic acid, and coumaric acids (Leopold, 1964). Some natural growth inhibitors are an integral part of the mechanism of dormancy in seeds and buds, and many natural inhibitors in plant extracts have a synergistic effect on growth, promoting auxin-induced growth at low concentrations and inhibiting this growth at higher concentrations.

Two special cases of natural inhibitors merit individual discussion: abscisic acid and ethylene.

Abscisic Acid. Abscisic acid has been isolated from young cotton fruit (Ohkuma et al., 1963). It inhibited the IAA-induced straight growth test of *Avena* coleoptiles but not the effect of gibberellin in increasing internode length in dwarf corn. Like gibberellin it promotes abscission of leaves and in fact is even more effective for inducing abscission at a given concentration than is gibberellin.

Ethylene. Ethylene at low concentrations inhibits the light-induced opening of the bean hypocotyl hook. This observation is suggested to be one of multiple steps in which auxin inhibits the hook opening by inducing the production of ethylene. Hook opening appears to be a response in which ethylene serves as a natural growth regulator. Carbon dioxide in turn may be involved as a growth regulator through its antagonism to the action of ethylene. Light causes a decrease in ethylene production and increases the production of carbon dioxide. At high concentrations IAA causes an inhibition of elongation and a swelling of stems, a response also attributed to the stimulation of ethylene production (Burg and Burg, 1966). Amchem 66–329 and 2,4-D have shown effects similar to those described for IAA.

SYNTHETIC

Some synthetic growth regulators currently or potentially of use in agricultural practice are considered here: Amo 1618, CCC, phosphon, maleic hydrazide, TIBA, NAA, and 2,4-D.

Amo 1618, CCC, and Phosphon. These appear to function as antigibberellins, acting to inhibit ring structure formation and gibberellin biosynthesis in peas and fungi (Baldev et al., 1965; Mertz and Henson, 1967). These were some of the first researches to pinpoint the action of synthetic plant growth regulators to one very definite physiologic mechanism. Lang (1967) reviewed work by several researchers which shows that by the use of these inhibitors the endogenous regulation of shoot elongation, flower initiation, and sex expression (determination of male and female) are governed by gibberellin. In a study on alfalfa, 200–250 mg of phosphon per pound of soil reduced plant height 38%, the number of stems 60%, and the racemes per plant 49% (Norwood et al., 1963).

Maleic Hydrazide. Maleic hydrazide (MH) depresses tiller formation and exerts direct effects on cell division but not on cell expansion during seedling growth. There is no specific interaction between MH and gibberellin either on lettuce seed germination or on the growth of wheat seedlings. Currently MH is used to inhibit suckering in tobacco and tillering in rice, and has been studied as a turf growth retardant.

TIBA. Triiodobenzoic acid (TIBA) is considered an auxin synergist; that is, it may enhance or inhibit auxin, depending on the concentration, but does not have a hormone effect of its own. TIBA causes reduced auxin transport (Galston, 1947). Applied as a spray to soybeans in early blossom, TIBA inhibits apical dominance; increases branching; enhances flower formation; modifies leaf structure (interveinal puckering), leaf color (deeper green), and leaf orientation (more upright); causes an overall change in leaf canopy shape (Fig. 7.9); increases the percentage of dry matter going into the seed and pod (Table 7.1); and reduces lodging (Greer and Anderson, 1965). The lodging reduction is the result of reduced plant height; for optimum yield benefit, this height reduction should not exceed 15–20%. TIBA apparently assists the young fruit in achieving a massed meristem status more rapidly and efficiently. Flower drop is reduced and more of the plant's photosynthate is directed toward the seed. TIBA had no effect on nodulation (Burton and Curley, 1966).

NAA. Naphthaleneacetic acid (NAA) has been used as a floral thinner, to inhibit apple abscission, and to promote uniform flowering of pineapple. A single application of NAA and 2,4-D will cause flower initiation in pineapple. NAA application on soybeans increased stem diameter, reduced lodging, and caused the pods to form higher on the plant (James et al., 1965). However, there was no change in soybean seed yield, and if applied late NAA caused a decrease in the rate of leaf abscission.

2,4-D. 2,4-D effects are primarily growth promotive. Corn sprayed with 2,4-D had an increased nitrate reductase activity (Beevers et al., 1963) in the leaves and thereby a reduced nitrate content. Smith and Harrison (1962) suggest that 2,4-D affects the dry weight and the content of ash, calcium, and phosphorus in barley by the direct effect of a reduction in the root absorbing area. 2,4-D altered the distribution

TABLE 7.1. Dry weight of plant parts of soybeans treated with TIBA at beginning bloom (Greer and Anderson, 1965)

	TIBA (ppm/application)			
Plant Part	0	10	20	40
	Cwt/A			
Stem	17.6	18.2	18.1	13.5
Petiole	5.5	5.8	6.0	4.7
Leaf	13.8	15.7	16.1	14.8
Pod	8.0	10.4	10.7	10.2
Seed	6.6	8.6	9.5	11.2
Total	51.5	58.7	60.4	54.4
	Pods and seeds as % of total dry matter			
	28.3	32.4	33.4	39.3

FIG. 7.9 ☘ Hawkeye soybeans. Above, the umbrella canopy of untreated plants. Below, the conical-shaped canopy of soybeans treated with TIBA (courtesy I. C. Anderson, Iowa State Univ.).

of Ca^{45} and P^{32} within the plant and stimulated the initiation of lateral root primordia from the pericycle.

In minute amounts 2,4-D has been shown to increase yields in field beans, soybeans, corn, barley, wheat, sugar beets, and potatoes (Wort, 1966; Huffaker et al., 1967). There is the danger that 2,4-D may become phytotoxic under certain conditions even at very low concentrations, but the addition of chelated iron compounds (FeEDTA, etc.) has a safening effect which eliminates any phytotoxic effects while growth promotion and yield increases are maintained. Huffaker et al. (1962) observed that 2,4-D sprays had a stimulatory effect on the dark fixation of CO_2 which was attributed in part to a secondary effect of 2,4-D, such as an increase in enzyme production (i.e., PEP carboxylase). Wheat and barley responded best to 2,4-D applications at the 5- to 7-leaf stage. Therefore, stage of growth, the formulation used (different esters reacted differently), and the type of iron chelate "safening agent" all were important in achieving a maximum response. One key quality effect was a greatly increased protein level in wheat. As a brief note here the herbicides Atrazine and Simazine also have been reported to increase the nitrogen content of susceptible plants when applied at low concentrations.

Defoliants and Desiccants. These terms are used to describe chemicals utilized to speed the drying rate of certain crops. A defoliant causes the early formation of an abscission layer between the stem and the leaf; a desiccant is essentially a contact herbicide. Both types of materials have value in certain cases but must be applied with precise timing in order not to kill the plant before physiologic maturity. Under the limited growing season of the north central states these chemicals have not been used successfully, but they are of value in the southern and western states, especially on cotton.

Pentachlorophenol (PCP), a contact herbicide, and NH_3 as a gas are often used as desiccants on cotton. Addition of 2,4-D to defoliants used on cotton retarded the defoliants' action (Texas Agr. Bull. MP 597, 1962), but when added to a desiccant like PCP it would enhance the action of PCP and also prevent undesirable regrowth which otherwise interferes with cotton stripper harvester. In seed cotton, however, 2,4-D increased the number of malformed seedlings and reduced germination. Defoliants have been used, but seldom with success, in other crops.

As knowledge increases about the role of growth regulators in plant development, several possibilities appear. First, it may be possible to directly manipulate plant growth more favorably by applications of certain chemicals, as is now done with TIBA and MH. Secondly, a fuller knowledge of how the growth regulator functions may provide the plant breeder a model to develop new varieties. Finally, this increased knowledge of growth regulators can improve the cultural management given to plants. There are broad vistas for development in this area of plant agriculture in the decades ahead. ☙

LITERATURE CITED

Allan, R. E., O. A. Vogel, and J. C. Craddock, Jr. 1961. Effect of gibberellic acid upon seedling emergence of slow and fast emerging wheat varieties. *Agron. J.* 53:30–32.

Baldev, B., Anton Lang, and A. O. Agatep. 1965. Gibberellin production in pea seeds developing in excised pods. Effect of growth retardant Amo 1618. *Science* 147:155–56.

Beevers, L., D. M. Peterson, J. C. Shannon, and R. H. Hageman. 1963. Comparative effects of 2,4-dichlorophenoxyacetic acid on nitrate metabolism in corn and cucumber. *Plant Physiol.* 38:675–79.

Bird, L. S., and D. R. Ergle. 1961. Seedling growth differences of several cotton varieties and the influence of gibberellin. *Agron. J.* 53:171–72.

Brian, P. W., and H. G. Hemming. 1957. The effect of maleic hydrazide on the growth response of plants to gibberellic acid. *Ann. Appl. Biol.* 45: 489–97.

Burg, S. P., and E. A. Burg. 1966. The interaction between auxin and ethylene and its role in plant growth. *Proc. Natl. Acad. Sci.* 55:262–69.

Burton, J. C., and R. L. Curley. 1966. Influence of triiodobenzoic acid on growth, nodulation and yields of inoculated soybeans. *Agron. J.* 58:406–8.

Christie, A. E., and A. C. Leopold. 1965. On the manner of triiodobenzoic acid inhibition of auxin transport. *Plant Cell Physiol.* 6:337–45.

Craigmiles, J. P., and J. P. Newton. 1962. The effect of gibberellin on forage crops. *Crop Sci.* 2:467–68.

Foster, R. J., D. H. McRae, and J. Bonner. 1952. Auxin induced growth inhibition: A natural consequence of two point attachment. *Proc. Natl. Acad. Sci.* 38:1014–22.

Freeman, J. E., and H. H. Hadley. 1963. Response of 1-dwarf, 2-dwarf, and 3-dwarf strains of milo to repeated foliar applications of gibberellic acid. *Crop Sci.* 3:332–34.

Galston, A. W. 1947. The effect of 2,3,5-triibdobenzoic acid on the growth and flowering of soybeans. *Am. J. Botany* 34:356–60.

Galston, A. W., and P. J. Davies. 1969. Hormonal regulation in higher plants. *Science* 163:1288–97.

Gould, R. F. (ed.) 1961. Gibberellins. *Advan. Chem. Ser.* 28, p. 167.

Greer, H. A. L., and I. C. Anderson. 1965. Response of soybeans to triiodobenzoic acid under field conditions. *Crop Sci.* 5:229–32.

Haber, A. H., and J. D. White. 1960. Action of maleic hydrazide on dormancy, cell division and cell expansion. *Plant Physiol.* 35:495–99.

Hemberg, T. 1955. Studies on the balance between free and bound auxin in germinating maize. *Physiol. Plantarum* 8:418–32.

Huffaker, R. C., M. D. Miller, and D. S. Mikkelsen. 1962. Effects of 2,4-D, iron and chelate supplements on dark Co_2 fixation in cell-free homogenates of field beans. *Crop Sci.* 2:127–29.

Huffaker, R. C., M. D. Miller, K. G. Baghott, F. L. Smith, and C. W. Schaller. 1967. Effects of field application of 2,4-D and iron supplements on yield and protein content of wheat and barley and yield of beans. *Crop Sci.* 7:17–19.

Jacobs, W. P., and D. B. Case. 1965. Auxin transport, gibberellin and apical dominance. *Science* 148:1729–31.

James, A. L., I. C. Anderson, and H. A. L. Greer. 1965. Effects of naphthaleneacetic acid on field-grown soybeans. *Crop Sci.* 5:472–74.

James, N. I., and S. Lund. 1965. Induction of male sterility in barley (*Hordeum vulgare* L. Emend. Lan.) with potassium, gibberellate and other plant growth regulators. *Agron. J.* 57:269–72.

Lam, S., and A. C. Leopold. 1966. Role of leaves in phototropism. *Plant Physiol.* 41:847–51.

Lang, A. 1967. Plant growth regulation. *Science* 157:589–91.

Leopold, A. C. 1964. *Plant Growth and Development.* McGraw-Hill, Inc., New York.

Mertz, D., and W. Henson. 1967. The effect of the plant growth retardants Amo 1618 and CCC on gibberellin production in *Fusarium monoliforme*: Light stimulated biosynthesis of gibberellin. *Physiol. Plantarum* 20: 187–99.

Miller, M. D., D. S. Mikkelsen, and R. C. Huffaker. 1962a. Effects of stimulatory and inhibitory levels of 2,4-D, iron and chelate supplements on juvenile growth of field beans. *Crop Sci.* 2:111–13.

———. 1962b. Effects of stimulatory and inhibitory levels of 2,4-D and iron on growth and yield of field beans. *Crop Sci.* 2:114–16.

Misra, G., and G. Sahu. 1963. Effect of maleic hydrazide on rice. *Phyton* 20(1): 35–39.

Morgan, D. G., and G. C. Mees. 1958. Gibberellic acid in the growth of crop plants. *J. Agr. Sci.* 50:49–59.

Nelson, P. M., and E. C. Rossman. 1958. Chemical induction of male sterility in inbred maize by the use of gibberellate. *Science* 127:1500–1501.

Norwood, B. L., P. C. Marth, and C. H. Hanson. 1963. Effects of phosphon on growth of alfalfa. *Crop Sci.* 3:241–42.

Ohkuma, K., J. L. Lyon, and F. T. Addicott. 1963. Abscisin II. An abscission-accelerating substance from young cotton fruit. *Science* 142:1592–93.

Partheir, R., and R. Wollgiehn. 1961. Über den Einfluss des Kinetins auf den Eiweiss—und Nukleinsäure Stoffwecksel in isolierten Tabakblattern. *Ber. Deut. Botan. Ges.* 74:47–51.

Pauli, A. W., and E. L. Sorensen. 1961. Vegetative and reproductive responses of dwarf alfalfa plants to gibberellic acid. *Crop Sci.* 1:269–71.

Pauli, A. W., and F. C. Stickler. 1961. Effects of seed treatment and foliar spray applications of gibberellic acid on grain sorghum. *Agron. J.* 53:137–39.

Phinney, B. O. 1956. Growth response of single-gene dwarf mutants in maize to gibberellic acid. *Proc. Natl. Acad. Sci.* 42:185–89.

Phinney, B. O., and C. A. West. 1961. Gibberellins and plant growth. *Handbuch Pflanzenphys.* 14:1185–1227.

Porter, K. B., and A. F. Wiese. 1961. Evaluation of certain chemicals as selective gametocides for wheat. *Crop Sci.* 1:381–82.

Smith, L. H., and C. M. Harrison. 1962. Effect of 2,4-dichlorophenoxyacetic acid on seedling development and uptake and distribution of calcium and phosphorus in barley. *Crop Sci.* 2:31–34.

Stickler, F. C., and A. W. Pauli. 1961. Gibberellic acid as a factor affecting seedling vigor and yield of varying seed sizes of winter wheat. *Crop Sci.* 1:287–90.

Steward, F. C. 1964. *Plants at Work.* Addison-Wesley Publishing Co., Inc. Reading, Mass.

Stowe, B. B., and T. Yamaki. 1959. Gibberellins: Stimulants of plant growth. *Science* 129:807–16.

Thimann, K. 1937. On the nature of inhibitions caused by auxin. *Am. J. Botany* 24:407–12.

Texas Agricultural Experiment Station Bulletin, MP597. 1962. Possible disadvantages of using 2,4-D with pentachlorophenol as a desiccant for cotton.

USDA. 1961. *Dwarfing Plants with Chemicals—A Promising Agricultural Technique.* ARS 22–65.

Van Overbeek, J. 1966. Plant hormones and regulators. *Science* 152:721–31.

Wittwer, S. H., and M. J. Bukovac. 1958. The effects of gibberellin on economic crops. *Econ. Botany* 12:213–55.

Wort, D. J. 1966. Effects of 2,4-D nutrient dusts on the growth and yield of beans and sugar beets. *Agron. J.* 58:27–29.

❧ CHAPTER EIGHT ❧ **GROWTH AND DIFFERENTIATION:** WITH SPECIAL EMPHASIS ON SHOOT AND BUD DEVELOPMENT

❧ THE DEVELOPMENT of the plant body is a complex process which involves many internal and external factors. The development process can be divided into growth and differentiation. Growth is considered to include both cell division and cell enlargement and results in an increase in plant size as well as in fresh weight. Differentiation is relatively more difficult to define but may include lignification of cell walls, protoplasmic hardening, formation of unique cellular inclusions like the alkaloids, or formation of flowering substances. Dry weight increases reflect the accumulation of photosynthate; however, these weight increases are properly thought of as parallel to, and not the same as, growth.

The processes of growth have a Q_{10} (activity rate increase with 10° C increase in temperature) of 2 to 3, indicating that they are chemical in nature. Three primary steps are energized by the energy consumed: cell division, cell enlargement, and anatomical differentiation. Division and enlargement are universal phenomena of plant cells. The uniqueness of a particular plant is developed via the rate at which these two steps occur and especially by the nature of differentiation. Internally, differentiation produces a great variety of cells organized into tissues and organs. External differentiation results from greater growth in one direction than another, producing the shape and form which distinguishes one kind of plant from another.

Growth is obviously a phenomenon of the entire plant, but the discussion to follow will center on the shoot-bud development (for a discussion of leaf and root growth see Chapters 1 and 9). The shoot serves as the central axis of the plant (Fig. 8.1). From the apical meristem, buds (undeveloped shoots) are initiated. These buds may remain dormant or they may expand into leaves, flowers, or branches. The genotypic and environmental factors determining that further development are discussed later in the chapter.

FACTORS ESSENTIAL FOR PLANT GROWTH

Cardinal temperatures, light, water, energy, the hormones, enzymes, nucleic acids, and amino acids are essential for growth to occur.

CARDINAL TEMPERATURE

Temperature, as it affects growth, is defined over a desirable range of cardinal points from minimum through optimum to maximum. For corn the cardinal temperatures are:

	°C	°F
Maximum	45	113
Minimum	10	50
Optimum	30–35	89–95

This means the corn plant's growth will be stopped by temperatures outside the stated maximum and minimum and will grow most rapidly at the optimum temperature. Cardinal temperatures differ among crops as well as for different organs of a crop plant and for the various stages of development. The maximum and minimum values are most important in determining the climatic ranges over which a crop may be expected to survive. The optimum values suggest the desirable temperatures for maximum productivity.

LIGHT

An increase in light intensity may increase photosynthesis and thus more food is provided for growth. The increased light intensity also increases transpiration and, if the rate of transpiration is rapid enough, the plant may wilt. However, photosynthesis is not slowed as quickly by the transpirational stress; therefore, the photosynthate may be provided at a high rate, organ expansion is slowed, and the photosynthate is left over for differentiation of either an anatomical or chemical nature.

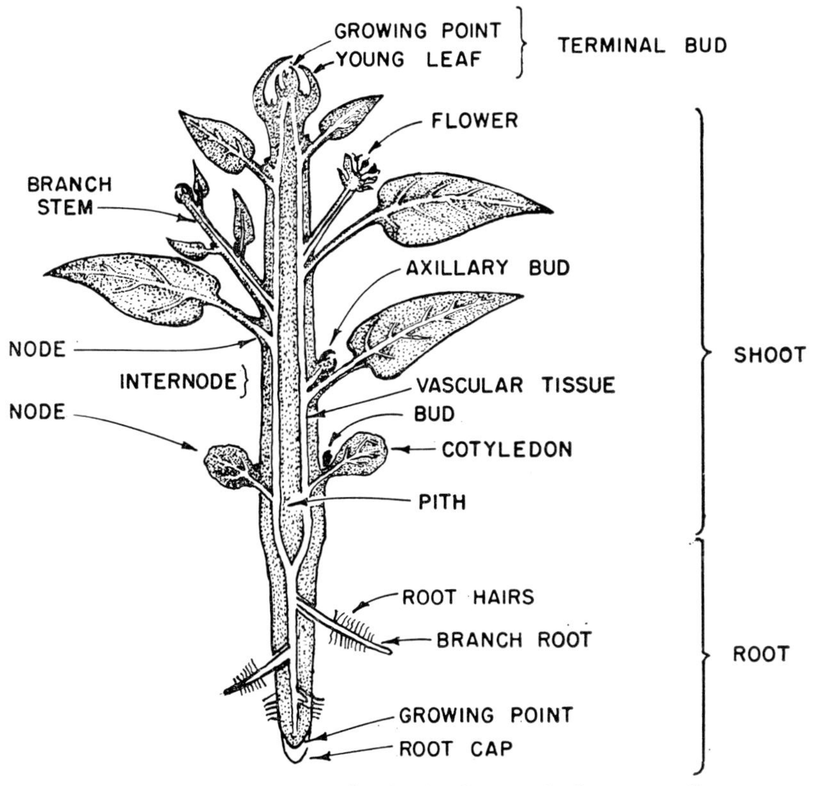

FIG. 8.1 ☙ The fundamental plant parts (from *Horticultural Science* by Jules Janick. W. H. Freeman and Company, © 1963).

Low light intensities have indirect effects on growth as follows. In total darkness, the plant grows tall and spindly and develops small, minimally expanded leaves. In normal light the same plants would be shorter (because more growth hormone is destroyed) and develop large leaves. Under moderate shade transpiration is reduced more than photosynthesis. The shaded plants may be taller and have larger leaves because the water supply within the growing tissues is better. With heavier shading photosynthesis is reduced to an even greater degree and small, weak plants result. The greatest response to light and shade is shown by the leaf. Leaves grown in the sun tend to contain more sugar and less water, while those grown in moderate shade have less sugar and more water and are relatively more expanded and less differentiated.

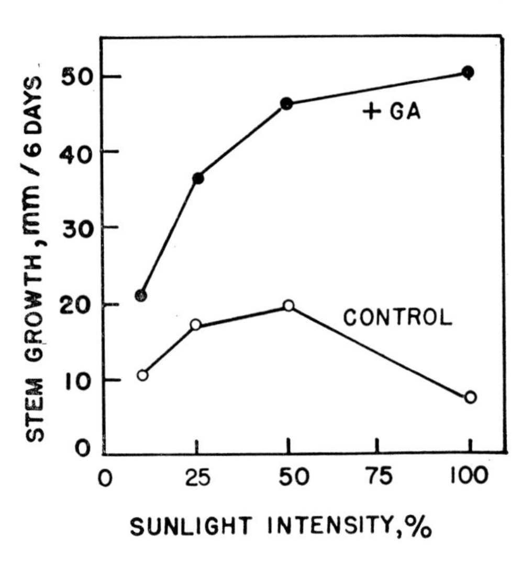

FIG. 8.2 ❧ Gibberellic acid (GA) prevents the inhibition of stem elongation by sunlight (Lockhart, 1961).

In addition to the influence of light intensity on photosynthesis and transpiration, an additional growth response is controlled through the effect of light on growth hormone destruction. As shown in Figure 8.2, full sunlight reduced stem elongation as compared to 50% sunlight. This effect could be offset by additional supplies of gibberellic acid and suggests that light is destroying the natural supply. It is not only an effect of light intensity but can also be shown to be tied to light quality. Over the range of visible light, different wavelengths cause growth effects, as described in Figure 8.3. The blue wavelengths affect phototropistic responses whereas leaf expansion and hypocotyl unhooking are favored by red light wavelengths. The pigment phytochrome, which controls many plant responses, is known to play a role in some of these reactions. The pigment changes from a form absorbing light around 660 mμ (red light) to a form absorbing 730 mμ (far red light). When left in the dark or exposed to far red light, it changes to the 660 absorbing form. In natural light it changes to the 730 absorbing form. Experimentally these changes can be induced by use of single wavelength 660 or 730 mμ light. Thus for example, 660 mμ light provides maximum leaf expansion and hypocotyl unhooking.

CARBOHYDRATE AND OXYGEN

Carbohydrate and oxygen are jointly necessary for growth. As discussed in Chapter 1, the primary source of energy for growth comes from ATP and $NADPH_2$ generated in photosynthesis, but this energy

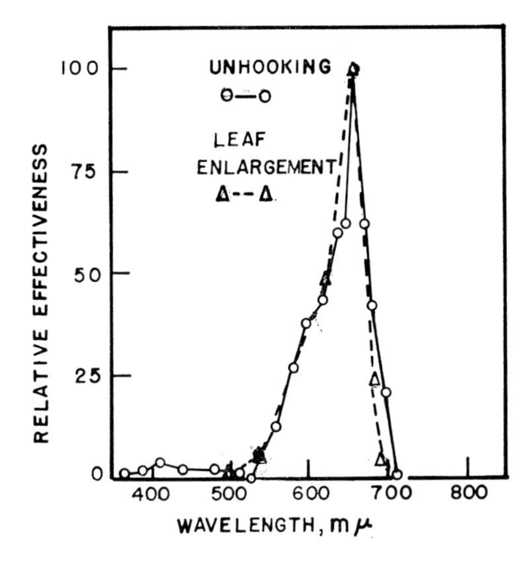

FIG. 8.3 Leaf enlargement and hypocotyl unhooking are stimulated by red light (625 mμ) (Downs, 1955; Withrow et al., 1957).

may be stored in carbohydrate or other plant foods and used later on. Thus respiration of the carbohydrates in the presence of oxygen does provide energy for growth in the dark.

NITROGEN AND MINERALS

Nitrogen and minerals serve as building blocks in numerous growth constituents, as discussed in the chapter on mineral nutrition. Nitrogen plays a key role in ring structures of ATP and the growth hormones. The minerals serve as cofactors in stimulating enzymatic action.

WATER

Water functions in providing a reaction solution for cellular activity and plays a key role in providing the cell turgidity necessary for cell enlargement. In addition, the effect of water in maintaining turgid cells and fully expanded leaves is important to enhance photosynthetic processes.

PLANT GROWTH REGULATORS

Growth of plant parts is preferentially stimulated by differing concentrations of plant growth regulators. Indoleacetic acid serves to stimulate cell division in conjunction with cytokinin, and with gibberellin functions to stimulate cell elongation.

CELL DIVISION

Cell division is concentrated in the meristems and is limited in location and time. It is characterized by rapid protein synthesis. Further, cell division requires IAA (indoleacetic acid) and cytokinin (Chapter 7). The IAA is looked upon as a by-product of protein synthesis because it is thought to be formed from the amino acid tryptophan. Throughout the plant, meristematic regions occur which function to expand the associated plant part. A meristem may be defined as formative tissue capable of cell division and containing one to many cells. Meristems throughout the plant commonly compete with one another; in fact, much of the art of growing crops for seed depends upon the management of this competition.

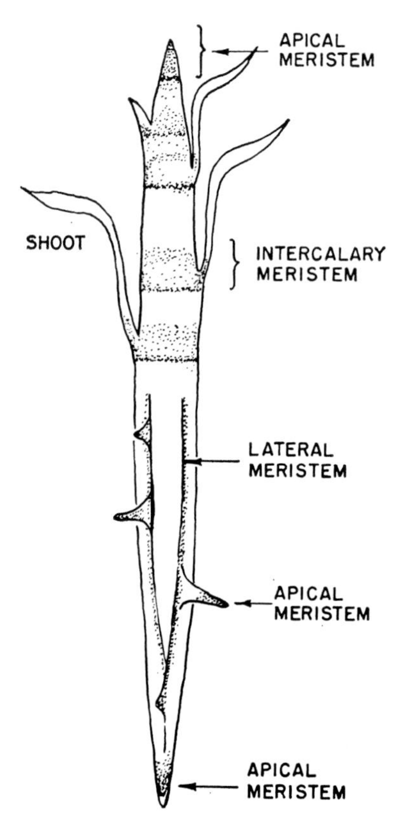

FIG. 8.4 ☙ The meristems of the plant (from *Horticultural Science* by Jules Janick. W. H. Freeman and Company, © 1963).

Apical meristems occur at the terminal point of a root, shoot, or branch (Fig. 8.4). The role of the apical meristem as it relates to the action of several growth regulators will be discussed at length later. *Lateral* meristems expand the breadth or diameter of an organ and are parallel to the axis of that organ. The vascular cambium is a specialized lateral meristem from which xylem and phloem originate. Other lateral meristems may be active in the expansion of leaf width. The *intercalary* meristem occurs at a point removed from the apex between previously differentiated tissue, but it is actively growing primary tissue. The pulvinus in a grass stem contains such an intercalary meristem. This particular meristem occurs at the base of the internode. Its growth occurs at differential rates around the stem circumference if a stem lodges, allowing the plant to grow erect again. Leaf blades and sheaths often have intercalary meristems at their base which function to extend leaf length.

A useful concept when discussing meristems is to differentiate between those classed as diffuse and those classed as massed. The diffuse meristem has a low number and/or activity of cells; hence an external source of growth hormone is required to keep it going. This would be true for the cambium and for a recently fertilized egg. If a rudimentary organ does not develop into a massed meristem over a given time period, it is commonly starved or aborted. Very often crop management practices are designed to favor rapid formation of a massed meristem. For example, corn populations are kept at a moderate level partly to favor the development of the ear shoot and to minimize the likelihood that the plant will abort it. Wheat seeding rates are moderate to allow first the formation of some tillers and further to allow these tillers to form heads. The massed meristem, then, is one containing cells of sufficient number and/or activity to perpetuate cell division and to control the flow of carbohydrate to the meristem.

CELL ENLARGEMENT

Cell enlargement has a dual requirement of a pliable cell wall and cell turgidity. Indoleacetic acid acts to retard secondary cell wall thickening and allow the wall fibrils to slip. The turgor pressure of the cell interior then pushes out the walls and enlargement results. A combination of division and enlargement provides the basis for meristematic activity described in the next section.

A combination of buds or meristems may be found at one site on the plant. In grasses, a unit complex of such meristems which perform different functions may be referred to as a *phytomer.* The phytomer, or unit of structure, of a grass in the vegetative stage consists of a node, a lateral bud, an internode, and the leaf which is attached around the circumference of the next higher node. The phytomer is therefore a telescoped version of one full node-internode potential in the plant. The oat plant has seven phytomers which expand above the ground and four less expanded ones at ground level. Corn commonly has fourteen phytomers above the ground and ten near the surface.

GROWTH CORRELATION

APICAL VERSUS LATERAL GROWTH

Meristematic regions have a distinct influence on overall plant development. Apical dominance connotes that a terminal apex, as a massed meristem, controls growth, often repressing the growth of the auxiliary buds in favor of the central axis. The apex may be capable of channeling carbohydrate to itself in a unidirectional movement and very often against a concentration gradient. Partial dominance is reflected by weak or sporadic axillary bud development.

Langer (1956) emphasizes that the onset of the reproductive growth stage in the meristem of the shoot apex imposes an inhibitory influence on the beginning of growth of lateral buds. However, those buds which had started to develop prior to the differentiation of the floral parts continue to grow (Campbell, 1961). During the latter stages of reproductive development of the shoot apex (i.e. at flower emergence), the inhibitory influence apparently disappears in some grasses. Langer found this to coincide with the production of a large number of new tillers in timothy. Removal of the young reproductive apex also ends the inhibition of growth in the axillary buds and so encourages the production of new tillers (Whyte et al., 1959). Aspinall (1963) observed a similar reaction in barley.

Compact corn, a variety with shortened internodes, is better able to produce grain under conditions of population stress because it terminates growth earlier than normal corn (Sowell et al., 1961). Presumably the compact corn loses its apical dominance sooner and thus is better able to support reproductive growth. This corn does not readily form basal tillers, but essentially the ear shoot represents a successful side branch, a specialized axillary bud formation. This same phenomenon is often observed in soybeans. Both determinate and indeterminate growth habits occur in this species. The determinate type completes its vegetative growth and then the stem apex is converted to a floral condition. Pods are set at the stem apex as well as at many nodes down the stem. The indeterminate type does not have this distinct change from the vegetative to the reproductive stage but rather continues to form leaves at the stem apex at the same time flowers are forming and pods are being set at lower nodes on the stem. The determinate types lose their apical dominance and allow more lateral buds (in this case pods) to form successfully.

Branching or tillering results from the expansion of axillary buds. These buds may be found as a crown, that is, a mass of buds at the base of the main stem or stems as in alfalfa or bromegrass. Or the buds may be found at all or most of the nodes up the plant stem. It is commonly observed that the new crop varieties give higher yields not because of more tillers but rather because more of those which are expressed initially succeed in survival and develop to maturity. Tillering and branching are of great agronomic importance, and the agriculturist depends upon this reaction to multiply the number of heads, pods, or ears per plant or per original seed planted. The importance lies in the contribution made by the yield components, as discussed at the end of the chapter. For example, the cereals often form two to ten tillers per plant under field conditions. Tiller growth and leaf growth are closely related, and the forage crops can survive a multiharvest or grazing because they regenerate a supply of tillers several times each season. In established perennial forage stands, with fairly stable tiller densities, the rate of growth per tiller largely determines productivity, and the effect of environmental conditions on growth per tiller is far more important than effects on the number of tillers. During stand establishment the opposite condition holds true. Therefore, it is often difficult to separate precisely the effect of various treatments on number versus growth of tillers. Tillering is usually measured as the rate at which tillers emerge from the leaf sheath, but by then considerable growth has occurred. Tiller primordia appear to be initiated at the shoot apex almost as soon as the primordia of the subtending leaves.

Factors Affecting Tiller or Branch Growth and Expansion.

GENOTYPE. The basic factor in determining the plant's tillering response is its genetic makeup. Oat varieties have distinctly different tillering habits (Frey and Wiggans, 1957a). Low-tillering spring varieties in 12-inch spacings developed 5.7 to 7.1 tillers per plant. High-tillering spring varieties ranged from 9.7 to 13.9 tillers and winter varieties from 9.7 to 14.2 tillers per plant in the same spacing. Similar results for wheat are reported (Milthorpe and Ivins, 1966, pp. 25–27) but the range in numbers of tillers is less. Pasture species also differ greatly both in amount and duration of tiller production (Mitchell, 1956).

TEMPERATURE. Temperature interacts with photoperiod, and discussing either singly is of limited value. For oats, a cool-season annual, increasing temperature decreases tillering. However, longer days favor more tillering. Thus Wiggans and Frey (1957a) found that for oats planted at weekly intervals in Iowa, beginning in early April, the number of head-producing tillers per plant increased for the first five planting dates and then decreased. Plants grown at 58° F required 4–9 days longer to produce heads than those grown at 70° F. The number of head-producing tillers was highest and about equal for the 18- and 24-hour photoperiods at 70° F and for the 15-, 18-, and 24-hour photoperiods at 58° F. As shown in Figure 8.5, the total number of tiller buds declined, then rose again as daylength shifted from short to long under the 70° F temperature, whereas head-producing tillers steadily increased as the photoperiod lengthened. The 58° F temperature gave a higher total number of potential tillers on a short photoperiod, but the number of head-producing tillers showed a similar increase as daylength increased.

For rice an increasing temperature, particularly in the early stages of growth, had the effect of inducing apical dominance. After heading of the main culm, the hitherto suppressed primary and secondary tillers began to develop. This effect occurred earlier and at higher nodes when the temperature was high at the early stage (Oda and Honda, 1963) and demonstrates how apical dominance and lateral bud suppression come and go through the plant's life cycle. Beinhart (1963) observed a decreased degree of branching in white clover under increased temperatures. He concluded that summer-hardiness of this species was a function of the number of active meristems in a particular strain. These observations emphasize a strong interaction of species and temperature in relation to tillering responses.

PHOTOPERIOD. These effects are commonly confounded with temperature, as shown in Figure 8.5 for oats, and may be confounded with the changes in light energy attendant with changes in daylength. Nevertheless, there does appear to be a genotype-photoperiod interaction. For example, Kentucky bluegrass tillers more under short days and is favored further by cool temperature (Peterson and Loomis, 1949)

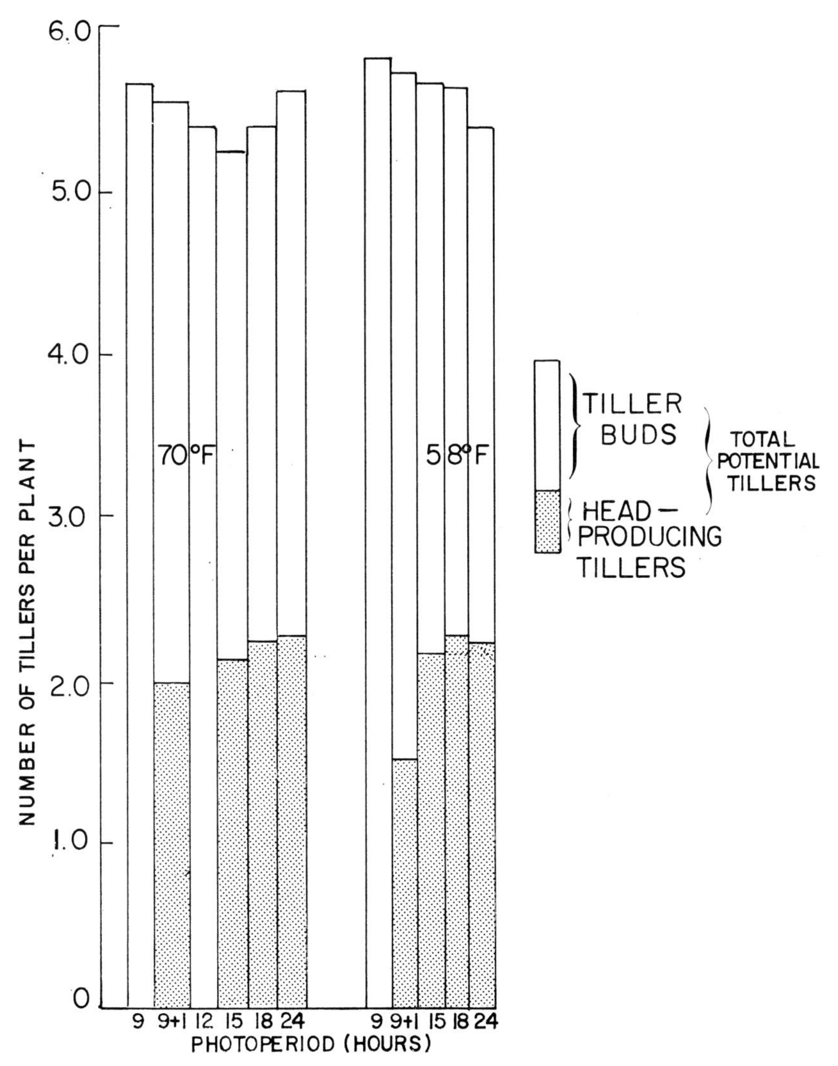

FIG. 8.5 ☙ Number of tiller buds, head-producing tillers, and potential tillers per plant for five varieties of oats planted at different photoperiods and temperatures (Wiggans and Frey, 1957a).

TABLE 8.1. Effect of daylength and temperature on the number of tillers per 100 Kentucky bluegrass plants (Peterson and Loomis, 1949)

Temperature	Hours of Light		
	11	15	19
Cool	160	114	116
Warm	142	116	106

(Table 8.1). Tillering of established alfalfa plants was increased by short photoperiods whereas tillering of seedlings was increased by long photoperiods (Cowett and Sprague, 1962), thus demonstrating a shift in response related to plant age. For other species long photoperiods favor increased tillering, and it is for these species that difficulty arises in separating the light quantity and photoperiod effects.

PLANT COMPETITION AND LIGHT INTENSITY. Tillering is strongly influenced by these two factors, and their individual effects are very difficult to separate. Under field conditions they must be considered as analogous. Lower planting rates result in more light for the individual plant and thus more tillers are formed. Increasing the density of Gaines wheat from 8 to 25 plants per square foot reduced tillers per plant from 3.72 to 3.37 (Nelson, 1964). Mitchell and Coles (1955) studied the influence of light intensity on ryegrass tillering. They suggest that the number of tillers in shaded plants was determined by the amount of light energy available for photosynthesis to the plant as a whole and not by light intensity just at the base. Hertz (1962) found that more tillers were produced by bromegrass kept free of weeds than when weeds were clipped or a companion crop was used, again underlining the light effect as one of improving photosynthate supply.

Recently plant populations have been significantly increased in corn production. Many hybrids available were not selected under these intense conditions of plant competition. In some such hybrids the ear shoot frequently failed to develop (barrenness). Sass and Loeffel (1959) suggest the barrenness to be a result of no silk formation by the ear shoot. Zieserl et al. (1963) have laid heavy emphasis on the role of nitrate reductase in relation to this reaction. They suggest that low light conditions do not allow nitrate reduction in certain hybrids and as a result protein synthesis is slowed and poor ear formation occurs (see Chapter 6).

For forage grass seed production, upright species are grown in rows to improve panicle number (orchardgrass and bromegrass). Alfalfa seed producers plan for one plant per square foot to obtain maximum tillering and vigor. In contrast, 5 to 20 alfalfa plants per square foot are used for forage production. Hence the fact that reduced plant competition increases tillers per plant is regularly used to maximize seed produced by a new variety.

In many fiber crops a thick stand is used to reduce light intensity per plant and thus to reduce branching. This reduces the number of branches breaking through the long fibers of the main stem phloem and sclerenchyma. For soybeans relatively thick planting is used to reduce branching. If large branches are allowed to develop they often lodge or break, whereas a single central stem is less likely to lodge.

NITROGEN. Tillering is highly dependent on the nitrogen supply. At later stages of growth, high nitrogen supplies are not necessary for optimal production in a closed-canopy crop, but a freely tillering plant always reacts to a nitrogen shortage by a drop in dry matter production caused by reduced tillering. When a plant is purposely kept from tillering, as for example when closely spaced to other plants, a limited nitrogen supply should be provided. The management practice of relatively low nitrogen is designed to keep growth from being too extended under the low light conditions of close spacing. In a sparse crop, however, high nitrogen is required for optimal production, since tillering is highly dependent on the nitrogen supply. The nitrogen-plant density-tillering interaction is described in Table 8.2 (Frey and Wiggans, 1957b).

WATER SUPPLY. Diminished water supplies reduce tillering for many economic crops, but the most pronounced effect comes from a water-nitrogen interaction. In a dry season little tillering occurs and heads per unit are determined primarily by the number of seeds planted. In a moist season tillering is stimulated and yield is more related to tillers formed than basic number of seeds planted. In contrast, quackgrass is stimulated to form more rhizomes (modifed tillers) as the soil dries down (Gardner, 1942). However, this quackgrass response appears to be a survival response of a species evolved under dryland conditions.

GROWTH REGULATORS. Tillering was early shown to be under hormonal control in barley (Leopold, 1949). Mechanical destruction of the apical meristem caused tiller formation to increase, but tillering was inhibited by an injection of the growth hormone naphthaleneacetic acid into the apex. From other workers come reports of less tillering

TABLE 8.2. Effect of nitrogen fertilizer and seeding rate on the tillers per oat plant (Frey and Wiggans, 1957)

Pounds N/Acre	Seeding Rate (bushels/acre)		
	1	3	5
0	1.02	1.00	1.02
20	1.18	1.12	1.03
40	1.40	1.35	1.16
80	1.68	1.35	1.19

when gibberellin was applied externally to plants. However, tiller numbers do not appear to be invariably influenced by gibberellin. Cowett and Sprague (1962) increased tillering in alfalfa by an application of antiauxin TIBA, and Greer and Anderson (1965) report more axillary branches in beans treated with TIBA.

DEFOLIATION. It appears that unless flowering stems and their apical meristems are removed, defoliation is unlikely to increase tillering in single plants. As indicated by work like that of Singh and Colville (1962), clipping sorghum plants and removing the stem apex increases the number of side branches developing from aboveground buds, but the loss of leaves does not appear to be a factor. Cowett and Sprague (1962) found that soil moisture, light intensity, temperature, mineral nutrition, and cutting treatment altered bud and stem development concurrently as they affected vigor and overall growth. Clipping or grazing, lodging, and growth regulators may remove or override apical dominance and favor tillering.

In summary, anything that increases growth commonly increases tillering unless the factor so favors the central apex that side branches are suppressed.

TOP VERSUS ROOT GROWTH

This second growth correlation will be discussed at length in the chapter on roots. It is sufficient to emphasize here that in general the organ (root or top) nearest the factor limiting in supply is favored in relative growth. Therefore, the root is favored when water or nitrogen is limiting; the top is favored under limitations resulting in less carbon dioxide fixation such as defoliation or a reduced supply of light. In the field there is a shifting balance between root and top.

GROWTH AND DIFFERENTIATION

If conditions favor rapid vegetative growth, photosynthate may be used so rapidly as to produce a succulent, relatively undifferentiated plant. If growth conditions are not favorable (nitrogen, temperature, water limiting), carbohydrates accumulate and become the stimulus for whatever differentiation process the plant in question is capable of carrying out. This may be the development of protein and other protoplasmic inclusions that favor winterhardiness in alfalfa or other perennials. It may mean the production of inclusions such as alkaloids (tobacco), essential oils, gums, resins, terpenes, saponins, and various substances used as drugs.

Certain crop management practices are necessary to guard against too great a differentiation tendency. Even though cell wall differentiation contributes to the strength of cotton, flax, hemp, and other fibers, extreme differentiation may be associated with reduced length in cotton and with woodiness and brittleness in hemp. Shade-grown tobacco is used for premium cigar wrappers because the higher humidity and somewhat lower light stimulate leaf growth, reduce differentiation, and result in a large, thin, nonbrittle leaf.

In summary, the growth-differentiation balance is based upon the tendency of carbohydrate supplies to be preferentially utilized in growth, particularly in top growth, whenever conditions are sufficiently favorable. Any factor, such as a deficient supply of water or nitrogen, which checks growth without correspondingly reducing photosynthesis will tend to increase any differentiation response of which the plant is capable.

VEGETATIVE VERSUS REPRODUCTIVE GROWTH

Light and temperature commonly function to trigger the flowering response in plants, shifting the apex from a vegetative to a floral condition. While many earlier workers had touched on the role of light in this phenomenon, the work of W. W. Garner and H. A. Allard in the 1920s with Maryland Mammoth tobacco at Beltsville, Maryland, is looked upon as the classic definition of daylength response. They recognized three daylength classes: short-day, long-day, and day-neutral plants.

The inductive effect of light in shifting the plant to the reproductive condition is now known to be one step in a chain that may

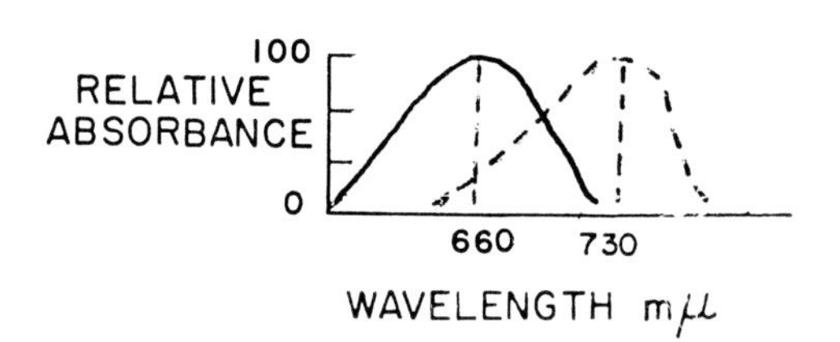

FIG. 8.6 ❧ Absorption spectra for phytochrome in red (–) and far-red (– – –) absorbing forms.

include: (1) a change in length of night (this being the real control as shown experimentally, although in nature night and day lengths are obviously directly related), (2) night temperature, or (3) exposure to low temperature. Fall flowering species respond to lengthening nights; summer flowering species apparently have a short-night response.

In recent years brilliant work by S. B. Hendricks, H. A. Borthwick, and their associates at Beltsville has shown that the induction of the photoperiodic response depends on a reversible photoreaction involving the pigment *phytochrome.* This pigment is known to assume two forms, a red light-absorbing and a far red light-absorbing form (Fig. 8.6). The same kind of light which inhibits the flowering of short-day plants will promote the flowering of long-day plants. The red light wavelengths are effective in interrupting the dark period, but this red light effect may be completely negated by the subsequent application of far red light. The control of flowering thus appears to be a result of the ratio of the two forms of phytochrome in the plant. This complexity in triggering the developmental process involved in the formation of flowers, fruits, and seeds is evidence of the evolutionary values of sexual reproduction. Flowering may be delineated in three phases: the *induction* of the reproductive state in leaves, the *initiation* of floral meristems, and the morphological *development* of the flower (Fig. 8.7).

Induction may occur when the plant has reached a state of puberty or "ripeness to flower." It may be stimulated by light, temperature, or both. Induction is essentially a chemical change. No gross morphological changes are discernible. However, with more sophisticated techniques of cell study, visible changes can be shown to occur rather early. For example, the apical meristem of the vegetative shoot of *Chenopodium album* (lambsquarters) exhibits alterations in cytoplasmic structure as early as three hours after the plant has been subjected to one photoinductive cycle which promotes flowering (Gifford and Stewart, 1965).

Flower initiation is often a reaction to light duration and results

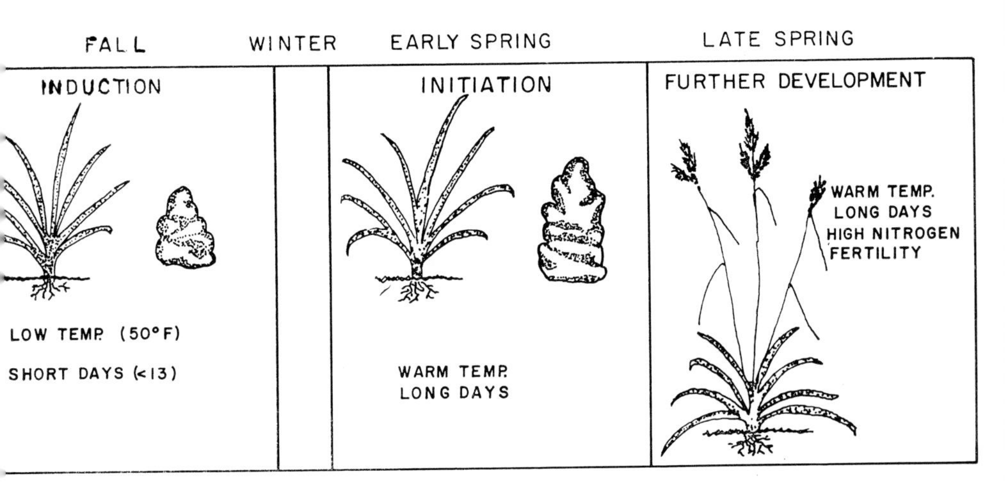

FIG. 8.7 ☙ Diagrammatic representation of flowering in orchardgrass (Gardner and Loomis, 1953).

in a visible formation of floral primordia. For many perennials this step may occur in the early spring after a fall induction period. For annuals it would occur as a continuation of reproductive development as the plant shifts away from the vegetative phase. *Bud development* is a macroscopic process and ends with the bud opening into a visible flower.

As indicated previously, both temperature (thermoperiod) and light (photoperiod) affect these changes. In addition, the puberty of the plant apparently is controlled by a balance of its mature to immature leaves. The mature leaves seem to favor the flowering response, the immature leaves favor vegetative growth. Hail-damaged corn and soybeans tend to mature their seed later than undamaged plants. This is because hail removes the relatively more mature leaves and then, as the growing point forms new leaves, the relative number of immature leaves is greater than on a plant not receiving hail damage (Weber and Camery, 1953; McAlister and Krober, 1958). The plant requires an additional period of time to develop the mass of mature leaves needed to favor reproduction. Similar effects occur in wheat and flax (Hella and Stoa, 1964).

Returning to the growth correlations which result from the factors of control mentioned above, it is useful to consider the outline developed by Loomis (1953, p. 209). He divided the developmental process of the fruiting plant into those which are predominantly growth and those classed as differentiation.

1. Growth of the vegetative plant—accompanied by the usual differentiations of such development but not by flowering or fruiting.
2. Differentiation of the ready-to-flower condition (puberty or the ability to be inducted).
3. Growth of the essentially vegetative parts of the flower.
4. Differentiation of the spore mother cells.
5. Growth of the gametophytes.
6. Differentiation of gamete cells—followed by pollination and fertilization.
7. Growth of the ovules and the fruit.
8. Differentiation of special flavors, oils, protein.

Growth correlation in plants seems rather clearly to be associated with the action of growth hormones, either in stimulating growth or in mobilizing and channeling the supplies of growth materials. In correlations between growth and differentiation, growth may be thought of as producing its own hormones and monopolizing the food supply. Differentiation is dependent on the leftover supplies of carbohydrate and hence is negatively correlated with growth.

Evidence for a balance between vegetative hormones and flowering hormones which are active in floral induction is available from many sources. Under this hypothesis, the apparent antagonism between vegetative growth and fruiting would be due to the stimulation of auxin production by rapid growth and nitrogen fertilization. Slowly growing plants would tend to flower because of low auxin production. Rapidly growing plants should have a high auxin production which would tend to prevent floral induction. Such plants would soon develop many mature leaves, however, which should bring them into balance for induction. Finally, the competitive ability of the various organs varies with the species and with environmental conditions, but for many plants the order seems to be: rapidly growing young fruit → vegetative buds → flowers → freshly pollinated fruits.

ACCUMULATION AND STORAGE

With the preceding consideration of growth correlations summarized, a logical next step is to consider the "levels of reversibility" which occur in various plant reactions as carbohydrate moves through the plant. Four "levels of reversibility" are cited below.

1. Sugar and starch occur in a highly reversible condition in the chloroplast. During periods of rapid photosynthesis, not all of the

sucrose, glucose, and fructose is rapidly removed from the chloroplast region; instead, some is temporarily stored there as starch. This starch may be readily reconverted to sucrose and translocated during periods of reduced photosynthesis or during the dark period.

2. Sucrose may be stored in the stem or stalk. This would be the condition briefly for corn before the sucrose moves into the grain. In fact, with lack of an ear (barrenness) the corn stalk is a relatively good sink for carbohydrate accumulation. The stalk is the normal site of storage for sugarcane.

3. Carbohydrate accumulation may occur in a modified stem area or in the root. For most crops this represents an accumulation that carries the plant through a dormancy period and serves to generate perennial growth. Such accumulation then is a relatively fixed time span of storage. It may occur in the crown and root of alfalfa, in the rhizome of bromegrass, quackgrass, Kentucky bluegrass, or in the lower leaf sheath of orchardgrass.

4. Carbohydrate accumulation in the seed is the least reversible type on the parent plant. Once sugar moves to the seed it is lost to the parent plant and can be put to use only by the embryo of the new generation. Evolutionarily, the plant has developed an irrevocable commitment to the survival of the species when it accumulates starch or other reserve materials in a seed structure.

YIELD COMPONENTS

In an effort to more carefully evaluate the several factors contributing to the final yield of a crop and especially to a crop harvested for seed, agronomists have developed the concept of yield components. The final grain yield is usually considered to be a product of the following equation:

$$\text{Yield} = \text{f}\left(\frac{\text{plants}}{\text{area}} \times \frac{\text{heads}}{\text{plant}} \times \frac{\text{seeds}}{\text{head}} \times \frac{\text{weight}}{\text{seed}}\right)$$

Tillering and branching have a distinct effect on the yield components by affecting heads per plant. As plants per acre change, the number of tillers or branches changes. However, tiller number per plant is often inversely related to number of plants and on a curvilinear basis. For example, as the plants per acre in wheat increases, the number of potential heads per plant declines, and very likely the seeds per head and weight per seed declines as well. However, since these relationships are curvilinear there is the opportunity to select a combination that will provide maximum yield. In addition, it may be that one of the factors, possibly seeds per head, changes relatively little. The other factors can be increased, seeds per head held relatively constant, and thus increase yield.

Grafius (1965) presents an analysis of these ideas in geometric terms. He suggests that yield be thought of as a parallelepiped (rectangular box) with the three edges representing number of heads per unit area, average number of kernels per panicle, and average kernel weight. He then emphasizes that a unit change in one edge may contribute vastly more than any other edge depending on the shape of the parallelepiped. A comparison of several economic grain crops shows the following yield components. Most of these crops have several

Oats	Sorghum	Soybeans	Corn
heads/plant	heads/plant	pods/plant	ears/plant
seeds/head	seeds/head	seeds/pod	seeds/ear
weight/seed	weight/seed	weight/seed	weight/seed

ways in which they may adjust to environmental shifts. However, current Corn Belt hybrids do not significantly adjust the heads/plant or heads/area factor in response to the environment. This component was apparently present in corn in the mid-1800s but was bred out for hand-harvesting convenience. Breeders have found it very difficult to reincorporate a multieared character. They have attempted to do so by using teosinte, popcorn, and southern prolific parentage and are beginning to succeed in adding this component of yield to corn again.

In studying the anatomy of multieared development, Collins (1963) points up that the critical period occurs during the three weeks before the plant has visible flowers. During this time the fate of the second ear is decided. This is the period after tassel elongation is nearly complete, and following this the cob grows rapidly. It appears abortion is initiated when the second ear fails to compete with the top ear during the three days before silking (Collins et al., 1965).

Corn has a visible potential ear at every leaf axil below the top six or seven. Before tassel initiation, visible axillary buds develop from bottom to top. Then the tassel expresses apical dominance for a time and suppresses axillary buds. After the tassel formation is complete, the ear shoots develop from the top to the bottom. In two-eared types, the first ear does not seem to suppress the second. Because many corn hybrids do not adjust to improved environmental conditions by forming more ears per plant, much emphasis has been placed on plant population increases.

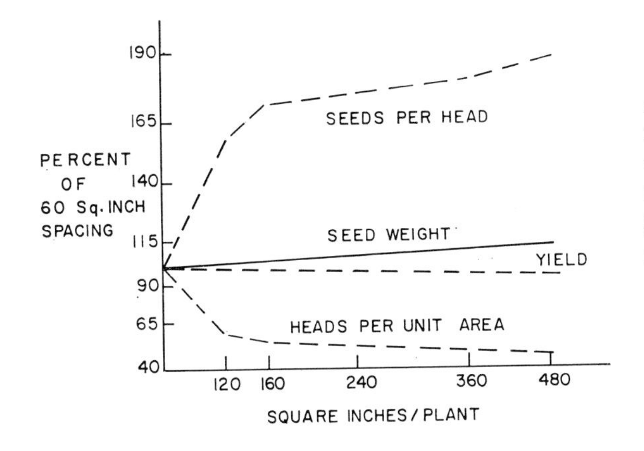

FIG. 8.8 Mean relative yield and components of yield for grain sorghum grown in eastern Kansas (Stickler and Wearden, 1965).

The interaction of yield components and population and the contribution of the individual components to yield is described for sorghum in Figure 8.8. As each plant has more space to use, the seed weight increases slightly, but the number of seeds per head gives the most dramatic increase. Sorghum has a tillering capacity which prevents a straight line decline in heads per unit area as wider spacing occurs, but tillering cannot fully compensate for the fewer plants.

Increased plant population is commonly used in other crops if they are planted at a time when tillering or branching is not favored. Late-planted oats may be seeded 50% heavier. Kentucky bluegrass seeded in the spring, a less favorable time, is seeded two to three times as heavy as in the more favorable fall months.

LODGING

A key crop production problem associated with the stem is lodging. As the plant reaches the heading stage an environmental stress or a weak-stemmed variety may allow the plant to fall partly or completely over. Lodged plants yield less and are usually much more difficult to harvest. If plants are subject to lodging, yield increases can often be obtained by split applications of nitrogen. Alternatively, some of the most dramatic yield increases in the world's agriculture have been achieved by shortening the lower internodes of the stem and thus reducing the lodging of the crop. This was achieved in the Pacific Northwest with Gaines wheat, which withstands high nitrogen fertility and irrigation to give 200-bushel yields. Wheats developed in Mexico and rice developed in Japan and the Philippines yield very well partly because they stand up even when well fertilized.

The breaking strength of oats and wheat stems was much more closely associated with wall thickness than with stem diameter (Hancock and Smith, 1963). Stem breakage was found to occur in the third or fourth internode. To reduce lodging, small-grain breeders have selected for more solid stems in the commonly hollow-stemmed cereals, and corn breeders have selected for rind thickness. However, McNeal et al. (1965) found highly significant negative correlations between bushels of wheat per acre and stem solidness. A possible explanation is that development of a solid pith uses food reserves that otherwise would have gone into grain development. Nevertheless, there is most likely interaction between stem strength, lodging resistance, and harvestable yield. For each increase of 1,000 corn plants per acre, the crushing strength decreased 34.8 pounds, rind thickness decreased 0.02 mm, and stalk lodging increased 1.56% (Thompson, 1964). These observations suggest that some minimum quantity of dry matter must be laid down in the stem to prevent lodging.

Stem strength is important not only in a generalized way, but it is generally assumed that very high yields can be obtained only with a cereal which can tolerate a high nitrogen fertilization rate without lodging. The two varieties compared in Figure 8.9 describe this relationship. Variety I gave similar yields to variety II at low nitrogen levels, but as nitrogen levels were increased it lodged and did not rise in yield (Milthorpe and Ivins, 1966). Because of this, it is common for the breeder selecting for maximum yield ability to place his genetic lines under the pressure of a high nitrogen fertility rate.

SHOOT AND BUD CHARACTERISTICS OF SPECIFIC CROPS

RICE

Direct-sown rice may start tillering from the first node above the cotyledonary node. In transplanted rice the first tiller usually appears at a node above the fourth. Leaf number on a tiller declines progressively with the rise in tiller order. That is, the initial group of tillers off the main stem are referred to as first-order tillers. They have more leaves than the tillers which in turn arise from them—the second-order tillers.

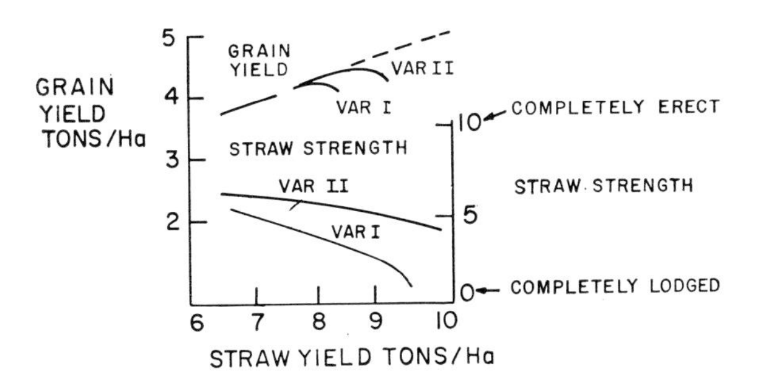

FIG. 8.9. ☙ Grain yield-straw strength interactions for two varieties of wheat (after van Dobben, in Milthorpe and Ivins, 1965).

SUGARCANE

The original crop is planted by preparing stem sections that sprout new plants. The generations which arise as tiller crops from the base of the old stem after harvest are referred to as ratoon crops. Cane growers used to go for several years without replanting and harvested the ratoon crops, but because ratoon crops give lower yields and also because of smut invasions, it is becoming more common to take only one or two ratoon harvests and then to replant.

SORGHUM

Sorghum is a very good example of a perennial grown in environmental conditions that result in its use as an annual. Nevertheless, it tillers profusely after setting its first crop of heads on main stems and tillers. These green shoots may cause harvest problems. There is also the potential problem of prussic acid poisoning if livestock is turned into the harvested grain field before frost has killed these new tillers.

Southern growers are attempting to get two harvests per season from one planting by harvesting the first crop of heads at 90–120 days at high moisture. Then a new crop of heads forms from tillers during the next 90–120 days (Hogg and Collins, 1964). Sorghum and rice are examples of species commonly grown as annuals but which may provide ratoon crops; most other annuals provide only one harvest. The perennial forages may all be considered as forming ratoons.

ALFALFA

The tendency to tiller, set seed, and tiller again is made use of in alfalfa seed production. The grower allows two to four seed-setting periods, then chemically dries the crop and harvests the several seed sets at one time.

LADINO CLOVER

Moderate shading as compared to full sunlight resulted in an increase in (1) growth as measured by number of nodes produced, (2) frequency of axillary buds developing into branch stolons, and (3) number of nodes that rooted (Trautner and Gibson, 1966). The results support the concept that slight shading by a companion crop favors the persistence of a white clover stand. This benefit of shading is more likely the result of modifying soil surface temperatures and moisture favoring branching rather than the result of reducing light directly.

FORAGE GRASSES

The stage of tiller development at the time of defoliation by grazing or clipping is critical for many grasses. New tillers must be developed in bromegrass before old ones are removed lest the plant be thrown into dormancy. To accomplish this, the apical meristem must have lost sufficient dominance to allow the crown buds to begin expansion.

Reed canarygrass stems and rhizomes branch readily and make this grass useful for establishment by sprigging of its plant material where its seeds and seedings might not survive. In addition, this grass can survive silting around its base. If silting occurs, buds on stems protruding above the silt generate new shoots readily.

TURF

In discussing branching habits, the tendency of the shoot to grow inside the sheath (intravaginal branching) or through the sheath (extravaginal branching) affects the nature of the turf formed. Extravaginal branching causes the development of the most dense and continuous sod or turf.

WEEDS

Certain weeds are particularly troublesome to control because their buds develop so effectively into new plants if the main stem is

TABLE 8.3. Effect of topping and suckering on yield, value, and composition of tobacco (Chaplin et al., 1964)

Treatment	Lb/A	$/A	$/Cwt	% Alkaloids	% Sugar
Top and sucker	1,806	1,184	65.65	2.80	18.20
Top	1,487	964	64.99	2.36	17.30
No topping or sucker	1,390	889	63.99	1.76	13.30
LSD 5%	62	44	1.31		

removed. Quackgrass exhibits a strong ability to form new shoots and roots from rhizomatous material. Pigweed and lambsquarters are examples of annual broad-leaf weeds capable of branching very effectively from buds low on the stem. Thus they are difficult to control by mowing.

SWEET CORN

Growers believe that plant performance is better if suckers (tillers) are removed. This has not been found to be true in several different studies (Snell, 1965). A number of earlier studies have shown the sucker to be capable of filling an ear on the main stem if the leaves are stripped from the main stem after the ear is well started. However, the effect of the sucker is most likely neutral under standard field conditions.

TOBACCO

Tobacco is topped (stem apex removed) to prevent flowering. With apical dominance gone, branches (suckers) form, and these must be removed manually or suppressed chemically. As discussed in Chapter 7, this can be done with maleic hydrazide. Using both practices (Table 8.3) gives a significant dollar return per acre (Chaplin et al., 1964). ☙

LITERATURE CITED

Aspinall, D. 1963. The control of tillering in the barley plant. 2. The control of tiller-bud growth during ear development. *Australian J. Biol. Sci.* 16:285–304.

Beinhart, G. 1963. Effects of environment on meristematic development, leaf area and growth of white clover. *Crop Sci.* 3:209–13.

Campbell, A. C. 1961. A theoretical basis for grazing management. *Proc. New Zealand Soc. Anim. Prod.* 21:18–32.
Chandraratna, M. F. 1964. *Genetics and Breeding of Rice.* Longmans, Green & Co., Ltd., London. Chap. 2.
Chaplin, J. F., Z. T. Ford, and R. E. Currin. 1964. Some effects of topping heights and suckering flue-cured tobacco. *S. Carolina Agr. Exp. Sta., Bull.* 510.
Collins, W. K. 1963. Development of potential ears in Corn Belt *Zea mays. Iowa State J. Sci.* 38:187–99.
Collins, W. K., W. A. Russell, and S. A. Eberhart. 1965. Performance of 2-ear type of Corn Belt maize. *Crop Sci.* 5:113–17.
Cowett, E. R., and M. A. Sprague. 1962. Factors affecting tillering in alfalfa. *Agron. J.* 54:294–97.
Downs, R. J. 1955. Photoreversibility of leaf and hypocotyl elongation of dark grown red kidney bean seedlings. *Plant Physiol.* 30:468–73.
Frey, K. J., and S. C. Wiggans. 1957a. Tillering studies in oats. I. Tillering characteristics of oat varieties. *Agron. J.* 49:48–50.
———. 1957b. Tillering studies in oats. IV. Effect of rate and date of nitrogen fertilizer application. *Proc. Iowa Acad. Sci.* 64:160–67.
Gardner, F. P., and W. E. Loomis. 1953. Floral induction and development in orchardgrass. *Plant Physiol.* 28:201–17.
Gardner, J. L. 1942. Studies in tillering. *Ecology* 23:162–74.
Gifford, E. M., Jr., and K. D. Stewart. 1965. Ultrastructure of vegetative and reproductive apices of *Chenopodium album. Science* 149:75–77.
Grafius, John E. 1965. A geometry of plant breeding. *Mich. State Univ. Res. Bull.* 7.
Greer, H. A. L., and I. C. Anderson. 1965. Response of soybeans to triiodobenzoic acid under field conditions. *Crop Sci.* 5:229–32.
Hancock, N. I., and E. L. Smith. 1963. Lodging in small grains. *Tenn. Agr. Exp. Sta., Bull.* 361.
Hector, J. M. 1936. *Introduction to the Botany of Field Crops,* Vols. I & II. Johannesburg, South Africa, Control News Agency, Ltd.
Hella, A. N., and T. E. Stoa. 1964. Simulated hail injury to wheat and flax. *N. Dakota Res. Rept.* 12.
Hertz, L. B. 1962. Effect of certain fertility and management treatments on the growth and early development of tillers of two varieties of smooth bromegrass. *Agron. J.* 54:139–41.
Hogg, P. G., and J. C. Collins. 1964. High-moisture grain sorghum fits feed program. *Crops and Soils* 16:13–14.
Janick, J. 1963. *Horticultural Science.* W. H. Freeman and Co., San Francisco, Calif.
Langer, R. H. M. 1956. Growth and nutrition of timothy *(Phleum pratense).* I. The life history of individual tillers. *Ann. Appl. Biol.* 44:166–87.
———. 1963. Tillering in herbage grasses. *Herbage Abstr.* 33:141–48.
Leonard, E. R. 1962. Interrelations of vegetative and reproductive growth with special reference to indeterminate plants. *Botan. Rev.* 38:353–410.
Leopold, A. C. 1949. The control of tillering in grasses by auxin. *Am. J. Botany* 36:437–40.
Lockhart, J. A. 1961. Photoinhibition of stem elongation by full solar radiation. *Am. J. Botany* 48:387–91.

Loomis, W. E. 1953. *Growth and Differentiation in Plants.* Iowa State University Press, Ames.

McAlister, D. F., and O. A. Krober. 1958. Response of soybeans to leaf and pod removal. *Agron. J.* 50:674–77.

McNeal, F. H., C. A. Watson, M. A. Berg, and L. E. Wallace. 1965. Relationship of stem solidness to yield and lignin content in wheat selections. *Agron. J.* 57:20–21.

Milthorpe, F. L., and J. Ivins. 1966. The growth of cereals and grasses. *12th Easter School in Agricultural Sciences.* Butterworths, London.

Mitchell, K. J. 1956. Growth of pasture species under controlled environment. I. Growth at various levels of constant temperature. *New Zealand J. Sci. Tech.* 38A:203–16.

Mitchell, K. J., and S. T. J. Coles. 1955. Effects of defoliation and shading on short-rotation ryegrass. *New Zealand J. Sci. Tech.* 36A:586–604.

Nelson, C. E. 1964. A management experiment with Gaines wheat under irrigation. *Wash. Agr. Exp. Sta., Circ.* 435

Oda, Y., and T. Honda. 1963. Environmental control of tillering in rice plants. *Sci. Rept. Res. Inst., Tohoku Ser. D* 14:15–36.

Pauli, A. W., F. C. Stickler, and J. R. Lawless. 1964. Development phases of grain sorghum (*Sorghum vulgare* pers.) as influenced by variety, location, and planting date. *Crop Sci.* 4:10–13.

Peterson, Maurice L., and W. E. Loomis. 1949. Effects of photoperiod and temperature on growth and flowering of Kentucky bluegrass. *Plant Physiol.* 24:31–43.

Sass, J. E., and F. A. Loeffel. 1959. Development of axillary buds in maize in relation to barrenness. *Agron. J.* 51:984–86.

Singh, S. S., and W. L. Colville. 1962. Effect of clipping on yield and certain agronomic characters of irrigated grain sorghum. *Agron. J.* 54:484–86.

Snell, R. S. 1965. Manual removal of suckers from fresh market sweet corn in New Jersey. *Agron. J.* 57:338.

Sowell, W. F., A. J. Ohlrogge, and O. E. Nelson, Jr. 1961. Growth and fruiting of compact and Hynormal corn types under a high population stress. *Agron. J.* 53:25–28.

Stickler, F. C., and A. W. Pauli. 1963. Yield and yield components of grain sorghum as influenced by date of planting. *Kans. Tech. Bull.* 130.

Stickler, F. C., and S. Wearden. 1965. Yield and yield components of grain sorghum as affected by row width and stand density. *Agron. J.* 57:564–67.

Thompson, D. L. 1964. Comparative strength of corn stalk internodes. *Crop Sci.* 4:384–86.

Trautner, T. L., and P. B. Gibson. 1966. Fate of white clover axillary buds at five intensities of sunlight. *Agron. J.* 58:557–58.

V. Booysen, P. de, N. M. Tainton, and J. D. Scott. 1963. Shoot-apex development in grasses and its importance in grassland management. *Herbage Abstr.* 33:209–13.

Weber, C. R., and M. P. Camery. 1953. Effects of certain components of simulated hail injury on soybeans and corn. *Iowa Agr. Exp. Sta., Res. Bull.* 400.

Wiggans, S. C., and K. J. Frey. 1957a. Tillering studies in oats. II. Effect of photoperiod and date of planting. *Agron. J.* 49:215–17.

Wiggans, S. C., and K. J. Frey. 1957b. Tillering studies in oats. III. Effect of rate of planting and test weight. *Agron. J.* 49:549–51.

Whyte, R. O., T. R. G. Moir, and J. P. Cooper. 1959. *Grasses in Agriculture.* Plant Production and Protection Division. Food and Agriculture Organization of the United Nations. FAO Agricultural Studies No. 42.

Withrow, R. B., W. H. Klein, and V. Elstad. 1957. Action spectra of photomorphogenic induction and its photoinactivation. *Plant Physiol.* 32:453–62.

Zieserl, J. F., W. L. Rivenbark, and R. H. Hageman. 1963. Nitrate reductase activity, protein content and yield of four maize hybrids at varying plant populations. *Crop Sci.* 3:27–32.

❧ CHAPTER NINE ❧ ROOT GROWTH AND DEVELOPMENT

❧ THE ROOT represents one of the three main organs of the plant body, but much less attention has been paid to its physiology than to those of aboveground parts. Roots are indeed analogous to the submerged part of icebergs in that the part beneath the surface is virtually disregarded until some problem arises. This neglect is due partly to the laborious nature of root studies. Yet, as Weaver and Clements (1938) suggest, that part of the plant environment beneath the surface of the soil could be under the control of the agriculturist to a greater extent than the aboveground microclimate. While little can be done toward changing the composition, temperature, or humidity of the air or the quantity of light available, much may be done by proper cultivation, fertilization, drainage, and irrigation to influence the structure, fertility, aeration, and temperature of the soil and hence its suitability for root growth.

The root plays a primary role in the plant's anchorage, water uptake, and mineral nutrition. The distribution of roots through the soil, the effect of the soil environment on root proliferation, and the effect of growth of the plant top on the flow of food to the roots determine how well the root will perform the anchorage and absorption functions. An understanding of the root's anatomy, patterns of growth for different species, and factors affecting growth will provide a foundation for improved crop management.

ROOT STUDY TECHNIQUES

The distribution of roots within the soil may be studied most accurately by digging a trench near the plant, then either drawing or photographing the root distribution which is outlined on the face of the trench. As an alternative a cross section of the root system may be impaled on a grid of metal rods, cut from the ground, and washed free of soil. The grid retains the roots in their approximate position and provides a qualitative estimate of their distribution (Weaver, 1926). More recent root distribution studies have been undertaken by placing radioactive nutrient sources at varying depths in the soil profile and then measuring the radioactivity present in an aboveground part (Hammes and Bartz, 1963) or in the root proper. A related technique has been to estimate root activity in a given zone of the soil by the changing moisture content at a particular site (Linscott et al., 1962).

The physiological functions of roots have been very thoroughly studied with water or sand culture methods. The classic reference describing water culture techniques was prepared by Hoagland and Arnon (1950). The plants were grown in distilled water with or without inert sand present. To this system the desired nutrients were added, and by plant analysis and visual observation of deficiency symptoms, nutrient requirements were determined.

ROOT FUNCTIONS AND STRUCTURAL DIFFERENTIATION

The root performs four key functions for the plant. First, it is the site of absorption of water and nutrients. To enhance the root's ability to perform this function soil management is heavily concerned with good drainage and aeration, a pH optimal for the species, and an ample supply of nutrients. Secondly, the root anchors the plant and gives it stability. Then after the nutrients are absorbed the root serves as a translocation system. Finally, in all plants to some degree and in certain plants exclusively (e.g. alfalfa) the root serves as a site of storage for accumulated carbohydrates (Troughton, 1957).

To perform these functions, the root structure develops as follows. The root tip, one of the plant's apical meristems, is surrounded by a root cap. New root cap cells are formed constantly in the inner portion of the cap. Meanwhile the walls of the outer root cap cells become gelatinous and slough off into the soil. This loss of cells lubricates the root tip and protects it from mechanical injury as the root is forced through the soil.

The root grows by division in the root tip and then by cell elongation in the region immediately behind the root meristem. Still further back from the tip, cells differentiate into many diverse structural types (Fig. 9.1). One important differentiation process occurs in the

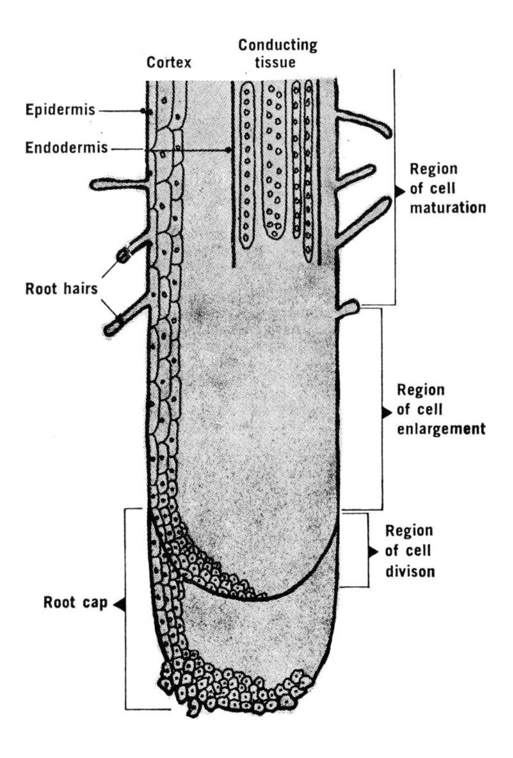

FIG. 9.1 ✤ Longitudinal section of a young barley root showing zones of division, enlargement, and maturation (Tanner, 1967).

epidermal cells. Some of these cells form long, tubular outgrowths called root hairs which begin as small swellings in the epidermal cell walls but may reach a final length of several millimeters. The number of root hairs is vast; there may be more than 200 root hairs on each square millimeter of root surface. They provide a very efficient contact with the soil by a mucigel on the root hair surface which is rich in iron oxides and microbes. This mucigel provides a continuous contact between the root cells and soil particles and enhances nutrient uptake.

In the differentiation zone, cells begin to take on the characteristics of their mature condition. As the root tissue ages, periderm (cork cambium) activity occurs, most commonly in the pericycle but sometimes in the endodermis or the cortex. As suberized (corky) tissue is formed from this cork cambium, the outer zone of the root is sloughed off, removing the epidermal cells and root hairs. Therefore, new root hairs must be constantly developed in the forward zone of differentiation. Water movement to the root in unsaturated soil by diffusion and vapor transport is too slow to be an important factor in supplying the requirements of rapidly transpiring plants. Therefore, roots must continue to proliferate unexploited zones to utilize soil-stored moisture effectively. In general the roots advance into new areas of the soil profile at random and then branch more profusely where soil conditions are especially favorable. If soil moisture and oxygen supply are optimal for root growth, the major development will be positively geotropic and under the control of growth hormones, as discussed in Chapter 7.

The outermost layer of cells of the vascular cylinder is the pericycle, the point of formation for the apical meristem of lateral roots which form opposite the points of the xylem star or arch. Thus sugar beets have two vertical rows of lateral roots and soybeans have four. As the lateral roots continue to develop, they force their way outward through the cortex. The vascular bundle of the root interior to the pericycle functions to translocate nutrients and water upward and to nourish the root with food synthesized by leaf and stem.

PATTERNS OF ROOT GROWTH

A thorough knowledge of root structure and growth patterns is useful to agriculture in many ways. Where banding of fertilizer is desirable to reduce nutrient fixation by the soil, it is important to know where the roots are growing so that the fertilizer band may be placed in the most advantageous point. While some workers have shown that one rootlet can supply a plant's needs (Duncan and Ohlrogge, 1958), practical agriculture more commonly depends upon full exploration of the available profile by the root system. Knowledge of root growth patterns provides a basis for judging the desirable depth of cultivation, the adaptation of a crop to a particular soil, tile spacing, cropping systems, and the depth of irrigation desired under certain soil conditions.

MONOCOTYLEDONS

For agronomic crops the fibrous and taproot patterns are the most common. The monocots, especially the grasses, form a primary and a nodal root system. The primary or seminal roots develop at the level of the seed (Fig. 9.2). Upon germination, the first organ to emerge is the radicle, which breaks through the coleorhiza (cell layers surrounding the root tip) and then through the tip of the seed. Very soon seminal roots appear at the scutellar node, the junction in the embryo between the radicle and the plumule. These roots may function throughout the life of the plant in several of the annual range grasses

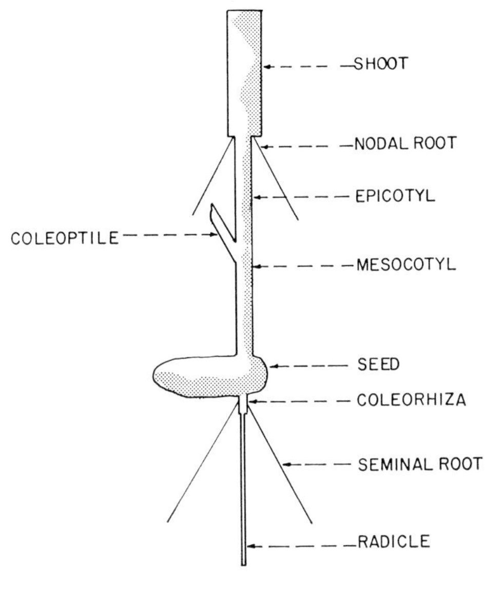

FIG. 9.2 Schematic drawing of a young grass seedling showing the location of radicle, seminal, and nodal root origins (Troughton, 1957).

and small grains. In corn it is commonly observed that the mesocotyl disintegrates after a few weeks of growth, causing the primary root system to be detached from the developed plant. For practical purposes the primary root system can be considered important to corn for three to six weeks.

In corn the number of seminal roots varies widely among types adapted to different environmental conditions. There are commonly fewer seminal roots in southern corns which were often planted at considerable depth in earlier years. Flint corns, adapted to wet soils, develop many seminal roots. Corn Belt hybrids most often form three seminal roots.

Initially the seminal roots and the radicle of corn grow on a line parallel to the long axis of the seed. They respond very little to geotropic and thermotropic stimuli until they are several inches long (Hawkins, 1963).

If it is assumed that the corn kernel falls in the seeding furrow broadside, this would leave the long axis parallel to the ground surface and would suggest that a fertilizer band would best be placed level with the seed for early interception. Lyon and Yokoyama (1966) observed that the primary root of wheat was not precise in its positive geotropism. They suggest that when curvature does occur it may result from more growth hormone being supplied to the side of the root on the embryo's axis (see Chapter 7). Thus the observed root epinasty and geotropic response may come from an auxin imbalance.

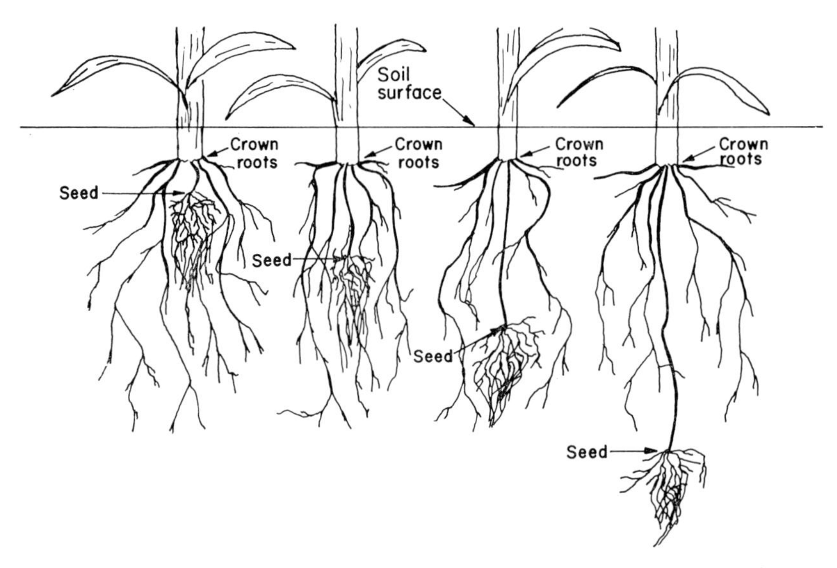

FIG. 9.3 ❧ Corn planted at 2, 4, 6, and 10 inches deep. The crown was formed at nearly the same depth regardless of the planting depth.

The nodal root system of monocots is developed from the basal nodes of the stem. Their depth of initiation in the soil profile is determined by a hormone reaction triggered by light. As the coleoptile emerges through the soil surface, both the coleoptile and the mesocotyl are stopped in growth. The node at the base of the coleoptile is commonly ½–¾ inch below the soil surface at this time (Fig. 9.3). Subsequently the secondary root system, which will become the dominant root system of the monocot, begins to form from a group of densely compacted nodes at or near the surface. Early whorls are fine and densely branched and develop from the bottom of the stem; the later ones are formed higher on the stem and become thick and sparsely branched, performing a stem supporting function. Early formed nodal roots grow about 25° from the horizontal, later nodal roots grow at 45° angles, and the last formed grow nearly straight downward. Brace or coronal roots develop at nodes above the ground surface where they break through the point of attachment of lower leaf sheaths and cause the destruction of those leaves.

DICOTYLEDONS

The dicots, especially the group of economic legumes, form a distinct taproot, the growth response of which is strongly geotropic.

The radicle emerges in the region of the hilum but very rapidly starts a perpendicular downward growth. The taproot may develop into a multibranched condition as represented by birdsfoot trefoil (Fig. 9.4). These branch roots arise at localized points along the vascular strand. Some species (e.g. alfalfa) most commonly have an unbranched taproot. However, modifications of rooting habit and stem growth are represented by broad-crowned types of alfalfa which form lateral rhizomes and are capable of withstanding grazing under arid conditions. The root spreading type develops shoot buds on the roots from which new plants develop (Smith, 1951; Heinrichs, 1963; Avendano and Davis, 1966).

On shallow soils that may cause heaving problems in Alaska, the taproot has been amputated by pulling a knife horizontally through the soil at a depth of six inches. This results in alfalfa forming a multibranched fibrous-like root system that is less subject to heaving damage.

In summary, the rooting patterns of monocots may be described as a two-phase system. The primary roots developing from the seed are weakly geotropic and more commonly develop in a horizontal plane. The nodal or coronal root system serves the plant a greater part of its life and develops in an increasingly vertical direction as new whorls of nodal roots are formed.

In contrast, the dicotyledons have a taproot that is very positive

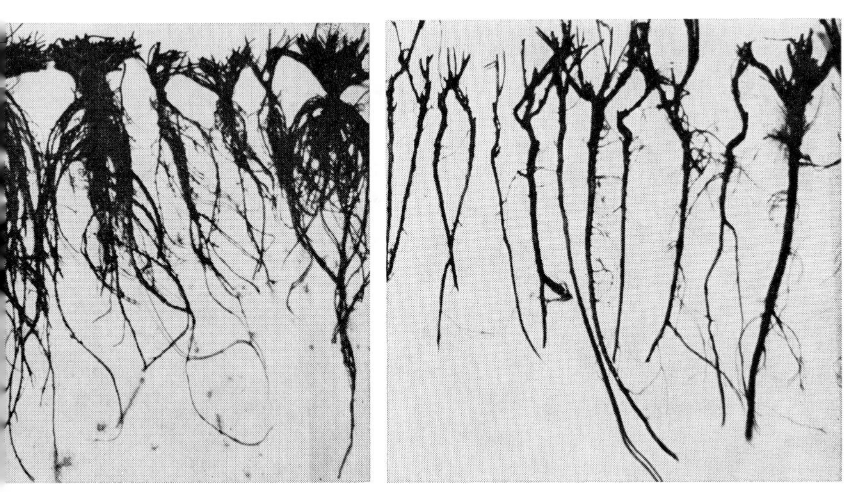

FIG. 9.4 ☙ A comparison of birdsfoot trefoil (left) and alfalfa (right). The trefoil represents a branched taproot, the alfalfa an unbranched taproot (Photo courtesy Louis Greub).

in its geotropic response. From this taproot laterals develop which may grow horizontally for a time and then turn downward.

FACTORS AFFECTING ROOT GROWTH AND DISTRIBUTION

Brouwer (1966) emphasizes that "an understanding of the effect of external conditions on root growth and development is possible only if at the same time attention is paid to shoot growth and development. In uniform media, the rate of growth in root length is relatively constant over long periods and up to root lengths of two meters, but is very sensitive to changes in the carbohydrate supply." Still further, as Pearson (1966) reminds us, it is important to distinguish between those conditions that alter root development by limiting overall plant development and those that directly affect a part of the root. Any root study, then, must carefully evaluate whether the effect considered is a direct or an indirect one. First the factors commonly thought of as primarily related to the soil environment will be considered, and then the root-shoot relationships will summarize the interactions as emphasized by Brouwer.

HEREDITARY OR GENETIC POTENTIAL

Environmental factors may modify root growth but the basic difference between a fibrous root system and a taproot system will not be changed. For example, a marked difference was observed in the ratio of branch roots to main roots in four inbred lines of corn. These differences were inherited and not subject to alteration by the medium in which the plants were grown (Weaver, 1926, p. 50). The rooting habit of two lines of corn is depicted in Figure 9.5. Several genes are presumed to be involved in determining root type. Plant breeders have consciously selected for certain root types as they sought lines that would better withstand corn rootworm attack. Those varieties selected for greater drouth tolerance or reduced root lodging display an increased mass of roots. The variance of root growth is as striking as for any aboveground character

FERTILITY

Fertilization enhances the inherent tendencies of the root. There commonly is an increased proliferation of corn roots in a zone con-

FIG. 9.5 ☙ The root systems of two inbred lines of corn demonstrate the wide variance in root systems that occurs within a species (Photos courtesy F. F. Dicke).

taining more organic matter (Weaver, 1926) or in a fertilizer band (Duncan and Ohlrogge, 1958), although this does not appear to occur for all species. While these authors have shown that branch root formation is affected by nitrogen or phosphorus, Brouwer (1966) suggests that the determining factor was not the presence of these elements in the root environment but the nutrient status of the plant overall. Isensee et al. (1966) found that roots which contacted fertilizer exhibited deformities and were shorter than untreated roots. Apparently seminal and branch roots (first order) are deformed or killed, but higher-order roots proliferate in a fertilizer band, at least for some species, if the concentration in the band is relatively low.

Nitrogen. An increased supply of nitrogen, within limits of practical interest, causes more top growth in relation to root growth (i.e. it increases the shoot-root ratio). Thus an adequate supply of nitrogen favors the use of available carbohydrate for growth of the top. In addition, a greater nitrogen supply tends to increase the quantity of growth hormone present in the plant. This favors top growth and inhibits root growth. However, nitrogen fertilizer does increase the total dry weight of roots. And nitrogen-fertilized roots were consistently higher in growth hormone activity than unfertilized roots (Wilkinson and Ohlrogge, 1962). The best root growth in cereal crops is obtained by applying a high level of nitrogen early and then allowing its concentration to diminish over the season. This gives a greater leaf area early and then more photosynthate later on to nourish the roots. Linscott et al. (1962) found that nitrogen-fertilized corn plants used considerably more water during the period 40 to 65 days after planting than did unfertilized corn. Root development was also greater during this time. This result points up the benefit that nitrogen fertility seems to give by promoting deeper and more profuse rootings early in the season when the plant has a heavy nutrient uptake. In addition, such a plant is better equipped to withstand drouth later on. A reduced supply of nitrogen results in more root elongation and less branching (Bosemark, 1954).

Phosphorus. Phosphorus-fertilized plants develop more roots than plants not fertilized with phosphorus. However, this is not a direct effect. First, the phosphorus fertilization increases top growth, more carbohydrate is photosynthesized, and this in turn increases root growth. Wilkinson and Ohlrogge (1962) showed that extracts from phosphorus-fertilized roots had less hormonal activity than extracts from nitrogen-fertilized roots and about the same activity as unfertilized roots. However, these workers found that phosphorus did cause a direct increase in root hair proliferation. Pearson (1966) reported that phosphorus did not have to be at the site of growth to provide normal root development; this suggests that subsoil phosphorus may have no advantage over phosphorus in the surface layer in terms of adequate root development.

Considerable work has been done to evaluate the best ratio of nitrogen to phosphorus in fertilizer mixtures, especially for applications to be made at planting. A 1:5 ratio of nitrogen to phosphorus (Wiersma, 1959) seems to favor the most profuse root system in a fertilizer band for corn.

Potassium. Potassium seems to have no direct effect on root exploration of the soil. It does not change the extent of root growth or the profuseness of branching. However, it is very important to the internal functioning of the root. If potassium is inadequate, the translocation system is poorly formed or may degenerate, cell organization is poor, and the cells lose their differential permeability.

In summary, it appears that many of the effects of fertilizers are indirect and give increased root development only after top growth increases. Weaver (1926) was very interested in the potential benefit of deep fertilization to stimulate root growth. Experimental work over the years since then has produced both positive and negative results with this practice. However, deep fertilization experiments have most commonly been done in conjunction with deep plowing, and the effects of these two have not been clearly separated. It may be that the variability results from the depth of root development in relation to the plant's nutrient demand. Perhaps roots do not penetrate deeply enough at the time the plant exhibits its heavy nutrient demand. A mere increased availability of nutrients would do little to increase that rate of penetration.

SOIL TYPE, STRUCTURE, AND BULK DENSITY

Corn is commonly observed to extend roots five feet deep in a medium- to coarse-textured glacial till but seldom below three feet in fine-textured subsoils (Pearson, 1966). Pore size is undoubtedly a factor in mechanical restraint of roots in compacted soil. If the tip is larger than the pore, it may not force its way in, although this supposition is under question (Barley and Greacen, 1967). Phillips and Kirkham (1962) concluded that soil compaction reduced corn seedling root growth by mechanical impedance and not through lack of aeration when corn seedlings were grown at a constant temperature for one week in clay. They further state that this does not imply "that aeration or other physical properties will not limit corn growth later on." Bertrand and Kohnke (1957) concluded that the reduction of root growth in dense soils was not entirely mechanical but also caused by lack of oxygen. Increasing bulk density has commonly been checked in relation to root growth. It must be related to soil moisture and texture to be useful in predicting or interpreting root growth, because the texture of the soil has a distinct effect on the relative portion of coarse to fine pores and this in turn affects the oxygen space available at a given moisture tension.

In summary, roots must respire to grow. If there is too much water or too little pore space containing air, root growth is slowed or stopped.

While both factors—mechanical impedance and lack of oxygen—play a role, oxygen supply appears to be more often limiting over a wider range of conditions (Rickman et al., 1966).

OXYGEN

A nitrogen atmosphere (N_2 gas) will inhibit root growth. At least a limited supply of the necessary oxygen must be available in the soil or soil water for root growth to occur. Letey et al. (1965), in a very sensitive test on barley roots, found that roots receiving adequate oxygen removed more water from the soil than roots receiving inadequate oxygen. While this is the case in general, it is important to note that certain plants are capable of transporting oxygen internally from the leaf to the root. For example, rice transports oxygen through its vascular system very efficiently and grows well under flooded conditions. Jensen et al. (1964) have demonstrated that oxygen can also be transported through the vascular system of corn and released by the roots. Oxygen released from roots may be a common factor in sustaining the rhizosphere effect in which abundant microbial activity is observed on the surface of plant roots.

CARBON DIOXIDE

Carbon dioxide may build up to a toxic level in compacted or wet soils. However, under natural conditions it appears the more important criterion is the relative levels of carbon dioxide and oxygen that are present. Geisler (1967) describes some very instructive carbon dioxide-oxygen interactions. A mixture of oxygen and carbon dioxide gave greater growth than oxygen alone when various systems were evaluated for their influence on radicle extension in corn. Concentrations of carbon dioxide up to 2% promoted root growth when oxygen was absent or in low concentration. A combination of 8% carbon dioxide and 7% oxygen had a very deleterious effect on root growth, but if the oxygen content was below 7%, there was a reduced susceptibility to carbon dioxide.

SOIL pH

A pH outside the range of 5.0–8.0 has a potential direct effect in limiting root growth, but within that range and under most field conditions the effect is indirect. For Corn Belt soils a pH in the 5.0–

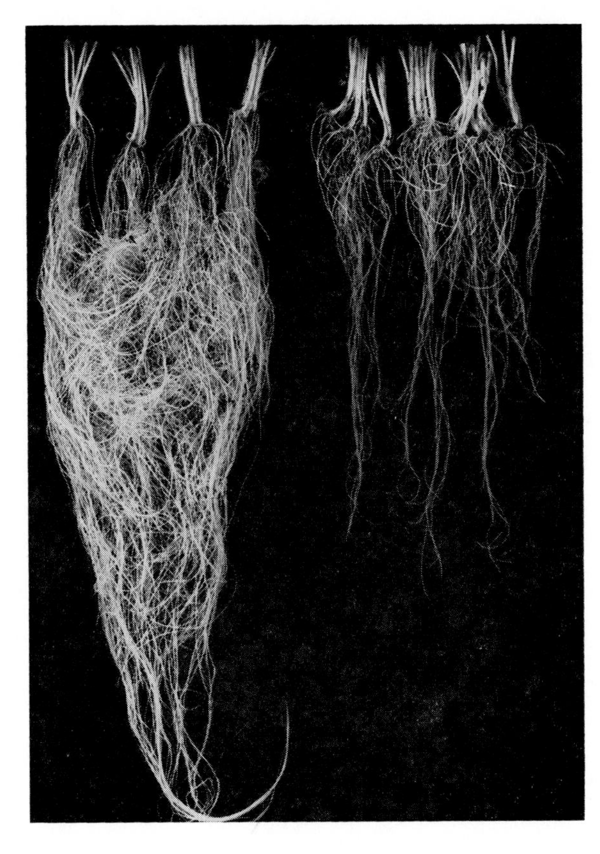

FIG. 9.6 ☙ The variation in aluminum tolerance in two lines of wheat (Photo courtesy C. D. Foy).

6.0 range increases the quantity of aluminum and manganese present in the soil solution; both have a direct toxic effect limiting root growth. Plant breeders can select for lines that are less sensitive to the presence of aluminum and manganese, as has been done with aluminum-tolerant wheat and barley lines (Fig. 9.6). These lines raise the pH in the immediate vicinity of the root. This tendency to alter pH is genetically controlled (Agr. Res., 1965).

TEMPERATURE

The shoot to root ratio of several species has been shown to increase as temperature increases (Anderson and Kemper, 1964). Moreover, low root zone temperatures adversely affect the dry weight of corn at all stages of development, and a high root zone temperature stimulates total growth (Knoll et al., 1964). So while the shoot growth is relatively more favored by an increase in temperature, root growth

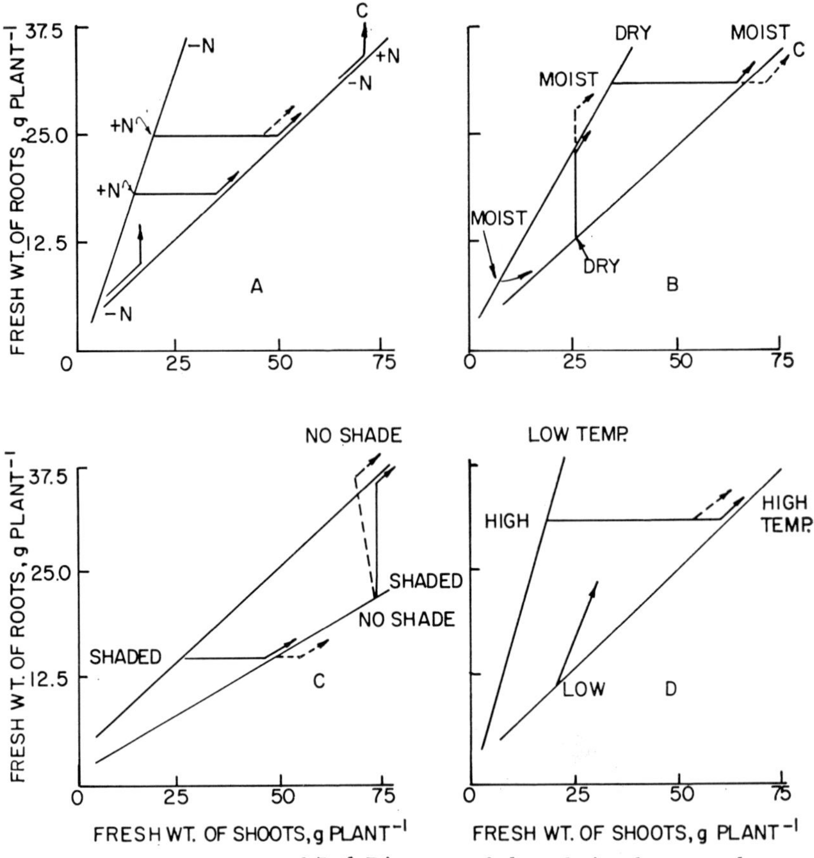

FIG. 9.7 ❧ Diagrams of the relation between shoot and root growth under various external conditions. Continuous lines: relationship under constant conditions indicated; arrows: weights at which conditions have been changed to those shown. Connecting lines show changes in growth pattern if no morphological adaptations have resulted; broken lines show the most likely changes (after Brouwer, 1966).

is also favored on an absolute scale, and the higher temperature favors the entire plant's development (Fig. 9.7D).

WATER

Water is essential for root growth as roots do not grow in or through a dry soil layer. In addition, lateral movement of water in soils is very slight. Therefore, continuous water uptake is secured only because the root system is continuously growing a thick network of roots and root hairs into moist soil. It is estimated that a rye plant develops an average of three miles of new roots with 55 miles of root hairs each day over a four-month growing period. The effects of moisture stress on plant development were summarized by Loomis and Worker (1963), who found that moisture stress reduced vegetative growth and increased the sucrose concentration on a fresh weight basis in the roots of sugar beets (Fig. 9.7B). Beet purity and sucrose yield were increased by nitrogen deficiency but not by moisture stress.

Root weight in blue panicgrass was significantly and linearly lowered as the degree of soil-moisture stress was increased (Wright, 1962). Wheat varieties demonstrated root pattern differences at different moisture levels that helped explain their yield differences (Hurd, 1968).

Transportation of water takes place partly through the soil and partly through the plant. As water becomes depleted at a particular point, root development there is slowed down. Growth occurs more rapidly where water availability is higher since cell elongation is limited less in such areas. Therefore, as the soil dries down from the surface, the zone with adequate water present is paralleled by a zone of actively growing roots.

Energy is required to move water through the roots. This energy requirement increases with decreasing root diameter; therefore, at a distance from the main stem smaller roots absorb relatively little water from the soil (Weaver, 1926, p. 189).

PLANT COMPETITION AND DEFOLIATION

Both factors reduce root growth (Norden, 1964). Increased plant population causes increased aboveground competition and has a much more pronounced effect on dry weights of roots than on volume of soil occupied. In Norden's study the dry weight of roots per plant decreased 72% as the population was increased from 5,000 to 25,000 plants per acre. However, the total root weight per acre did increase up to a 20,000-plant population level.

Defoliation significantly reduces root weight (Wright, 1962). Both competition and defoliation cause less per plant root development due to the reduced carbohydrate synthesized per plant and hence the reduced quantity of food available for root growth. Shading favors shoot over root growth (Fig. 9.7C).

SHOOT-ROOT RATIO

The ratio of shoot to root development is often of as much interest in understanding crop management effects as is the absolute yield of either the shoot or the root. Environmental factors will be discussed here as they influence this ratio.

As a general principle the organ (or organ complex, i.e. shoot or root) nearest the factor that is limited in supply is favored in relative growth. Relative root growth is most favored under limited supplies of water or mineral nutrients. The shoot is favored relatively by limitations of carbohydrate synthesis or limited light. An organ that does not photosynthesize or store food must depend upon a food-producing organ for its growth. Therefore, under low light intensity the shoot is favored and the shoot-root ratio increases. The observations in Figure 9.7 (Brouwer, 1966) support these generalizations.

Conversely the shoot depends on the root for its mineral and water supply. The shoot-root ratio declines with a diminishing nitrogen supply (Fig. 9.7A). This occurs partly because of less shoot growth and is also related to greater root extension. Under low nitrogen conditions, an increase in supply of potassium has very little effect (less than 10% change) on the absolute weight of the shoot and root, and the shoot-root ratio is unchanged by potassium. Under a high nitrogen level, an increased level of potassium gives a 60% increase in absolute weight and favors the shoot sufficiently to increase the shoot-root ratio (Brouwer, 1966). A limited water supply decreases root growth slightly but causes an even more striking reduction in shoot growth (Fig. 9.7B). Under certain management conditions, this response is used to favor the development of a more extensive root system, thus providing a better root system to support increased shoot growth later.

It is generally recognized that the optimum temperature for root growth is lower than for shoot growth. For perennial plants this means that in the early spring, with high carbohydrate reserve in the plant, root growth gets ahead of shoot growth (Fig. 9.7D).

Under field conditions there is a shifting ratio between shoot and root. After a rain more top growth occurs. But as water becomes somewhat deficient and is evaporated from the leaves, the free energy of water is greater in the roots. As water stress increases, there is too little water available for cell elongation in the top, carbohydrates accumulate (since photosynthesis is not reduced as quickly by water stress as is growth), and a greater proportion of the photosynthate goes to the root (Loomis, 1953).

Conversely, after defoliation, and assuming an adequate water supply, those carbohydrates produced by the leaf are used to produce relatively more top growth, and root growth is slowed until either the top is well formed again or a water stress overrides the "light stress" and the root becomes favored once more. Thus the plant under water stress does not seek water, nor does the plant under limited nitrogen grow deeper to find nitrogen. Rather, when these factors are limiting, the top is slowed in growth and relatively more carbohydrates are available to the root.

ROOT GROWTH IN CORN

Weaver (1926) and Foth (1962) superbly describe the several phases of root development in corn. Their descriptions relate primarily to the secondary root development. During the early weeks growth is angular and proliferates the upper layer of soil (15 inches in Foth's study). During phase 2, vegetative top growth continues while the roots "fill" this upper zone and there is limited root expansion (leader roots) down to 36 inches. Phase 3 is characterized by rapid stem elongation and deep root penetration. Foth suggests that this deep penetration occurs very close to the time of tasseling. This observation underlines the need for adequate water in the upper soil zone during the critical time of flowering. Eighty percent of the total roots developed in the 15–36-inch zone were formed in phase 3. The appearance of the successive whorls of nodal roots of corn changes considerably from the fine, closely branched roots of the first whorl to thick, sparsely branched roots of the highest whorls. It is generally recognized that the effectiveness per gram of root weight of the fine seminal roots is much greater than that of the nodal roots.

Brace roots, which constitute a large part of the final root mass, are formed by the time the corn kernels are in the "early milk" stage. During the 5th phase—seed maturation—there is no significant root development. Total root weight does not change significantly after 80 days of growth. Figure 9.8 summarizes the data presented by Foth (1962) and emphasizes the heavy reliance of corn on relatively shallow roots. While the mass of roots, then, are in the upper 6 inches of the soil, it appears that during periods of water stress the deeper roots can provide the plant's water needs relatively well. Since the roots deeper in the profile have fewer root hairs, less total mass, and fewer branches, their importance in nutrient absorption is presumed to be low. However, the fertility of lower horizons is commonly much lower than that of the surface layers, and it is not well understood how important the

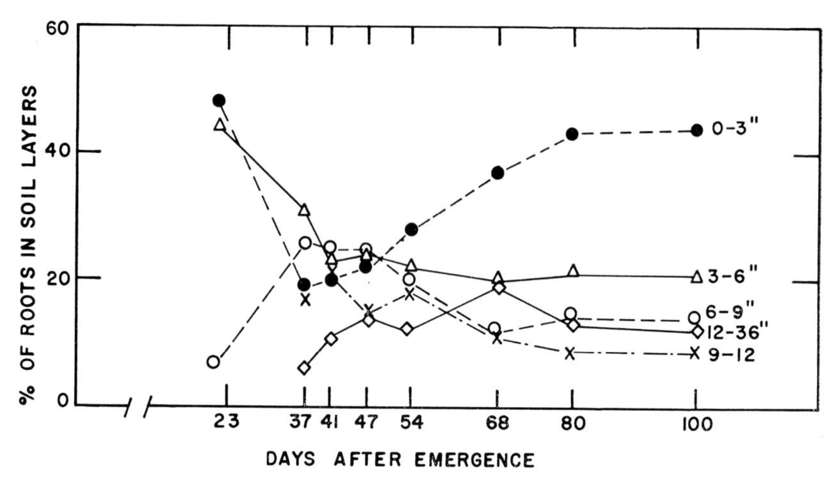

FIG. 9.8 Percentage of corn roots in several layers of the soil profile 23 to 100 days after emergence (Foth, 1962).

contribution of the deeper roots might be if they were supplied with adequate amounts of nutrients and moisture.

ROOT GROWTH IN SORGHUM

Sorghum appears to have a root system which contributes strongly to its drouth tolerance. Miller (1916b) made the following comparison between sorghum and corn grown in Kansas (Table 9.1). He observed that the volume of soil proliferated by the two species was similar, but the sorghum proliferated that volume with twice as many roots, thus providing a more complete extraction of the water. Sorghum hybrids have been noted by several researchers to make significant root growth after flowering occurs (McClure and Harvey, 1962). While this presumably contributes to drouth tolerance, it also limits the amount of photosynthate going into the grain.

ROOT GROWTH IN ALFALFA

The common alfalfa root type is a taproot that may penetrate deeply. Roots two months old have been found to a depth of 3 feet. Later penetration may depend on soil type and water level, but 10–15-foot depths are not uncommon, and depths over 100 feet have been

TABLE 9.1. Comparison of root development in sorghum and corn grown in Kansas (Miller, 1916b)

	Sorghum (dwarf milo)	Corn
Height (feet)	3.0	6.0
Root depth (feet)	6.0	6.0
Root spread (feet)	3.5	3.0
Secondary roots per cm	10.0	5.0

recorded. Irrigated alfalfa, in contrast, only roots to a depth of 6 feet, with one-third of the root system developed in the top 6 inches and over two-thirds in the top 24 inches. Weaver (1926) studied alfalfa roots in lowland and upland soils and found marked differences in penetration and type. The lowland soil was deep, fertile, and well aerated and the roots penetrated as far as 12 feet. Where their descent was terminated by the water table, there was little branching. In the upland soil, with a hard, chalky, clay subsoil, the roots rarely extended beyond 7.5 feet. Here root branches in the first 8 inches were two to three times as frequent as in the lowland soil and the branches were much larger.

While the common rooting habit for alfalfa is the taproot, both creeping-rooted and rhizomatous types have been developed. These types are particularly desirable where alfalfa is to be used for grazing. Carlson et al. (1964) report that such adventitious roots originate in the primordial dome and are the result of unusual meristematic activity in some cells of the phellogen. Warm short days favor this root site development.

CATION EXCHANGE CAPACITY OF ROOTS

One of the theories proposed to explain ion uptake by roots is cation exchange. Whether or not this theory is in fact a major basis for ion uptake is not fully agreed upon, but it is a characteristic of the root worthy of discussion (Drake et al., 1951). To provide the H^+ for exchange with cations from the soil, CO_2 may combine with water to form carbonic acid (H_2CO_3), or organic acids of the root may release H^+ to the soil. The H^+ from these acids is adsorbed on the root surfaces. Once the new cations (i.e. Ca^{++}, K^+) are on the root surface, they may exchange with interior H^+ across the cell wall.

Plants vary considerably in the magnitude of their root cation exchange capacities (CEC). For example, the CEC of dicotyledons (legumes) is generally higher than that of monocotyledons (grasses).

The legumes tend to preferentially absorb divalent cations over monovalent cations, and for the grasses monovalent cations are favored. Under certain field conditions where a grass-legume mixture is grown, the grass may persist and the legume die out, possibly because the grass is more effective in absorbing potassium.

In conclusion, it is to be reemphasized that an understanding of root development is exceedingly important in crop management. Soil factors may affect that development and are changed by tiling, fertilization, and tillage practices to favor it. However, it is equally important to recognize the strong influence which the shoot growth exerts over the root development and to consider this shoot-root relationship in evaluating crop growth questions in the field. ❧

LITERATURE CITED

Agricultural Research. April 1965. Aluminum tolerance—How is it inherited? 13–15:3–4.

Anderson, W. G., and W. D. Kemper. 1964. Corn growth as affected by aggregate stability, soil temperature, and soil moisture. *Agron. J.* 56:453–56.

Asher, C. J., and P. G. Ozanne. 1961. The cation exchange capacity of plant roots and its relationship to the uptake of insoluble nutrients. *Australian J. Agr. Res.* 12:755–66.

Avendano, R. E., and R. L. Davis. 1966. Lateral root development in progenies of creeping and noncreeping-rooted *Medicago sativa* L. *Crop Sci.* 6:198–201.

Barley, K. P., and E. L. Greacen. 1967. Mechanical resistance as a soil factor influencing the growth of roots and underground shoots. *Advan. Agron.* 19:1–43.

Bertrand, A. R., and H. Kohnke. 1957. Subsoil conditions and their effects on oxygen supply and the growth of corn roots. *Soil Sci. Soc. Am. Proc.* 21:135–40.

Boehle, J., L. T. Kardos, and J. B. Washko. 1961. Effect of irrigation and deep fertilization on yields and root distribution of selected forage crops. *Agron. J.* 53:153.

Bosemark, N. 1964. The influence of nitrogen on root development. *Physiol. Plant.* 7:497–502.

Brouwer, R. 1966. Root growth of grasses and cereals. In F. L. Milthorpe and J. D. Ivins (eds.). *Growth of Cereals and Grasses.* Butterworth and Co. Ltd., London.

Carlson, G. E., V. G. Sprague, and J. B. Washko. 1964. Effects of temperature, daylength and defoliation on the creeping rooted habit of alfalfa. *Crop Sci.* 4:284–86.

Duncan, W. G., and A. J. Ohlrogge. 1958. Principles of nutrient uptake from fertilizer bands. II. Root development in the band. *Agron. J.* 50:605–8.

Drake, M., J. Vengris, and W. G. Colby. 1951. Cation—Exchange capacity of plant roots. *Soil Sci.* 72:139–47.

Foth, H. D. 1962. Root and top growth of corn. *Agron. J.* 54:49–52.

Foy, C. D., A. L. Fleming, G. R. Burns, and W. H. Armiger. 1967. Characterization of differential aluminum tolerance among varieties of wheat and barley. *Soil Sci. Soc. Am. Proc.* 31:513–21.

Geisler, G. 1967. Interactive effects of CO_2 and O_2 in soil on root and top growth of barley and peas. *Plant Physiol.* 42:305–7.

Hammes, J. K., and J. F. Bartz. 1963. Root distribution and development of vegetable crops as measured by radioactive phosphorus injection technique. *Agron. J.* 55:229–33.

Hawkins, George. 1963. Some factors affecting early root development of corn. Ph.D. thesis, Iowa State Univ.

Heinrichs, D. H. 1963. Creeping alfalfa. *Advan. Agron.* 15:317–37.

Hoagland, D. R., and D. I. Arnon. 1950. The water culture method for growing plants without soil. *Calif. Agr. Exp. Sta., Circ.* 347.

Hurd, E. A. 1968. Growth of roots of seven varieties of spring wheat at high and low moisture levels. *Agron. J.* 60:201–5.

Jensen, C. R., J. Letey, and L. H. Stolzy. 1964. Labeled oxygen: Transport through growing corn roots. *Science* 144:550–52.

Isensee, A. R., K. C. Berger, and B. E. Struckmeyer. 1966. Anatomical and growth responses of primary corn roots to several fertilizers. *Agron. J.* 58:94–97.

Klebesadel, L. J. 1964. Modification of alfalfa root system by severing the tap root. *Agron. J.* 56:359–61.

Knoll, H. A., D. J. Lathwell, and N. C. Brady. 1964. Effect of soil temperature and phosphorus fertilization on the growth and phosphorus content of corn. *Agron. J.* 56:145–47.

Letey, J., W. F. Richardson, and N. Valoras. 1965. Barley growth, water use and mineral composition as influenced by oxygen exclusion from specific regions of the root system. *Agron. J.* 57:629–31.

Linscott, D. L., R. L. Fox, and R. C. Lipps. 1962. Corn root distribution and moisture extraction in relation to nitrogen fertilization and soil properties. *Agron. J.* 54:185–89.

Loomis, R. S., and G. F. Worker, Jr. 1963. Responses of the sugar beet to low soil moisture at two levels of nitrogen nutrition. *Agron. J.* 55:509–15.

Loomis, W. E. 1953. *Growth and Differentiation in Plants.* Iowa State College Press, Ames.

Lyon, C. J., and K. Yokoyama. 1966. Orientation of wheat seedling organs in relation to gravity. *Plant Physiol.* 41:1065–73.

McClure, J. W., and C. Harvey. 1962. Use of radiophosphorus in measuring root growth of sorghums. *Agron. J.* 54:457–59.

Miller, E. C. 1916a. The root systems of agricultural plants. *Agron. J.* 8:129–54.

———. 1916b. Comparative study of the root systems and leaf areas of the corn and the sorghums. *J. Agr. Res.* 6:311–31.

Mitchell, W. H. 1962. Influence of nitrogen and irrigation on the root and top growth of forage crops. *Delaware Agr. Exp. Sta., Tech. Bull.* 341.

Norden, A. J. 1964. Response of corn (*Zea mays* L.) to population, bed height and genotype on poorly drained sandy soil. I. Root development. *Agron. J.* 56:269–73.

Pearson, R. W. 1966. Soil environment and root development. In W. H. Pierre, D. Kirkham, J. Pesek, R. Shaw (eds.). *Plant Environment and Efficient Water Use.* Am. Soc. Agron. and Soil Sci. Soc. Am.

Phillips, R. E., and D. Kirkham. 1962. Mechanical impedance and corn seedling root growth. *Soil Sci. Soc. Am. Proc.* 26:319–22.

Rickman, R. W., J. Letey, and L. H. Stolzy. 1966. Plant responses to oxygen supply and physical resistance in the root environment. *Soil Sci. Soc. Am. Proc.* 30:304–7.

Smith, D. 1951. Root branching of alfalfa varieties and strains. *Agron. J.* 43:573–75.

Smith, S. N. 1933. Genetic investigations of the physiological response of inbred lines and crosses of maize to variations of nitrogen and phosphorus supplied as nutrients. Ph.D. thesis, Iowa State Univ.

Tanner, J. W. 1967. Roots: Underground agents. *Plant Food Rev.* 13:11–13.

Troughton, A. 1957. The underground organs of herbage grasses. *Commonwealth Bur. of Pastures and Field Crops, Bull.* 44.

Weaver, J. E. 1926. *Root Development of Field Crops.* McGraw-Hill Book Co., New York.

Weaver, J. E., and F. E. Clements. 1938. *Plant Ecology,* 2nd ed. McGraw-Hill Book Co., New York.

Wiersma, D. 1959. The soil environment and root development. *Advan. Agron.* 11:43–51.

Wilkinson, S. R., and A. J. Ohlrogge. 1962. Principles of nutrient uptake of fertilizer bands. IV. Mechanisms responsible for intensive root development in fertilized zones. *Agron. J.* 54:288–91.

Worley, R. E., R. E. Blaser, and G. W. Thomas. 1963. Temperature effect on potassium uptake and respiration by warm and cool season grasses and legumes. *Crop Sci.* 3:13–16.

Wright, N. 1962. Root weight and distribution of blue panicgrass (*Panicum antidotale* Retz.) as affected by fertilization, cutting height and soil-moisture stress. *Agron. J.* 54:200–202.

❧ CHAPTER TEN ❧ SEED DEVELOPMENT, GERMINATION, AND PRODUCTION

❧ THE SEED is an important component of much of the world's agriculture and serves as the bridge from one generation of plants to the next. While it is possible and sometimes necessary to vegetatively propagate a crop plant, most large-scale farming enterprises are built on the foundation of using seed to start a new crop. The primary consideration in this chapter will be the cycle of events related to seed development, germination, and seed production. However, the uses of seeds far transcend their natural purpose of species perpetuation, and a large part of the world's food and feed supply consists of seed crops.

Generally grains are rich in carbohydrates but poor in lipids. In the United States, despite the relatively low percentage of lipid present in corn, there is such a large volume of the crop processed for starch that sizable quantities of corn oil extracted from the embryo are available for marketing.

Legume seeds are usually rich in protein; many also contain either starch or other polysaccharides as a principal energy-reserve substance. Even though on the average legume seeds, like cereal grains, are low in oil content, two of the best known and most widely used oilseeds—soybeans and peanuts—are legumes. The two supply about one-fourth of the world's edible fats and oils. Soybean oil also goes into many nonfeed uses, and the beans themselves are man's most important leguminous food.

SEED ANATOMY

The seed develops from an ovule and at maturity consists of the following parts: the young, partially developed plant—the embryo;

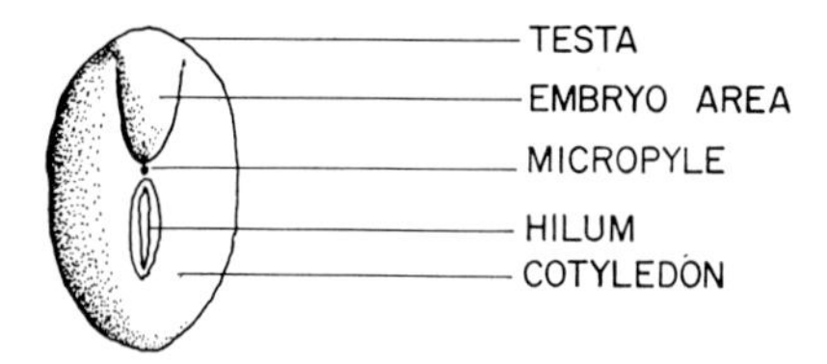

FIG. 10.1 ☙ Parts of a legume seed—the soybean.

variable amounts of endosperm (sometimes none) or another source of food supply; the protective layers on the surface—the seed coat or testa. The micropyle, site of entry of the sperm tube, may be obliterated or may remain in the form of a partially covered pore. A scar, the hilum of the soybean, which is highly permeable to water, occurs where the seed breaks away from the ovary wall (Fig. 10.1).

The embryo axis of legumes consists of the radicle, hypocotyl, and epicotyl. This embryo axis is partially imbedded in the cotyledons and receives germination and extension energy from the cotyledons. Legumes have little or no endosperm, but fats and oils stored in the cotyledons serve this role. A true seed coat, the testa, is present on the legume seed.

The embryo or embryo axis has the future vegetative organs initiated during its development, at least in the form of their apical meristems. The embryo axis in grasses consists of the plumule, radicle, and scutellar node, as shown in Figure 10.2. The attached scutellum acts to digest the food supply stored in the endosperm. The substances stored in the endosperm are varied, but the principal storage carbohydrate is starch in the form of starch grains. Hemicelluloses or related carbohydrates occurring in the thick cell walls also serve as storage materials. Proteins found in seed occur in two principal forms: (1) the glutens, amorphous in structure, and (2) the aleurone grains, composed of a proteinaceous substrate with a crystalloid body and a circular body. Glutens are common in the starch-

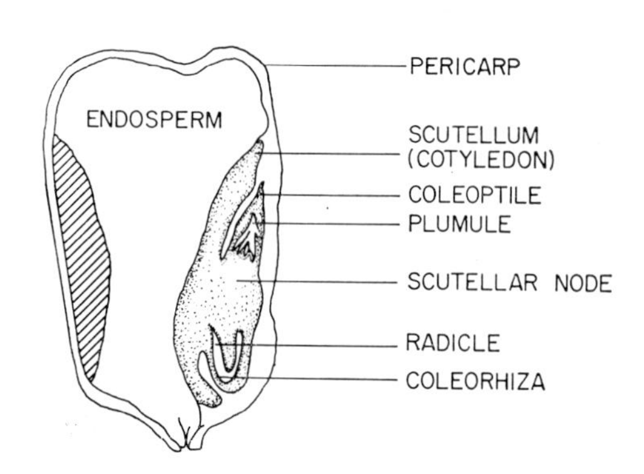

FIG. 10.2 ☙ Parts of a grass seed—the corn caryopsis.

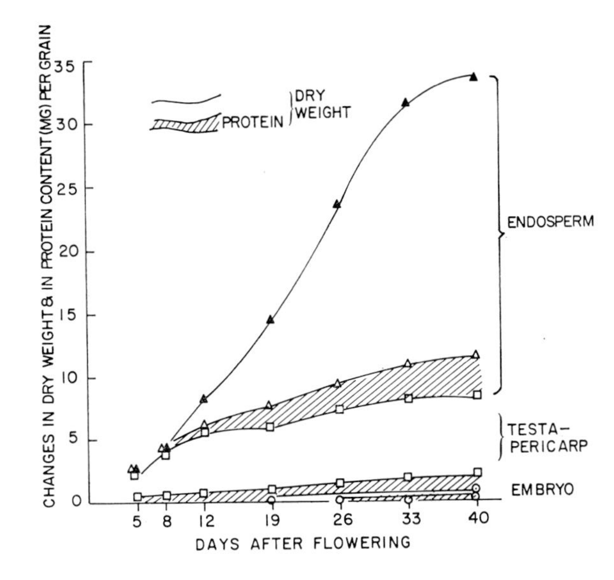

FIG. 10.3 Changes in the dry weight and protein content of wheat during maturation (Jennings and Morton, 1963).

containing cells of cereal grains. Starchless endosperm contains oil and fats as storage materials.

SEED DEVELOPMENT

Seed development represents the formation of a new sporophyte generation and is one of the cyclic events in a plant's life. Following fertilization of the embryo sac nuclei, seed development occurs in a sigmoid pattern of growth. The developing grain of wheat described in Figure 10.3 consists mainly of pericarp in the early stages and of endosperm in later stages. Both fresh weight and dry weight increase for about 35 days after flowering. At that point there is a marked decline in water content and little or no more increase in dry weight (Jennings and Morton, 1963). Figures 10.4 and 10.5 note the changes in sucrose, reducing sugars, starch, and nitrogen components. The shifting levels of these components appear to occur in like manner in many other species. The bulk of RNA increase and presumably cell division has occurred by the 14th day after fertilization. Nonprotein nitrogen is a high proportion of the total nitrogen early in the seed's development and remains high until grain hardening when protein nitrogen increases.

The development of corn may be divided into two phases (Ingle et al., 1965). The first 28 days following pollination are characterized

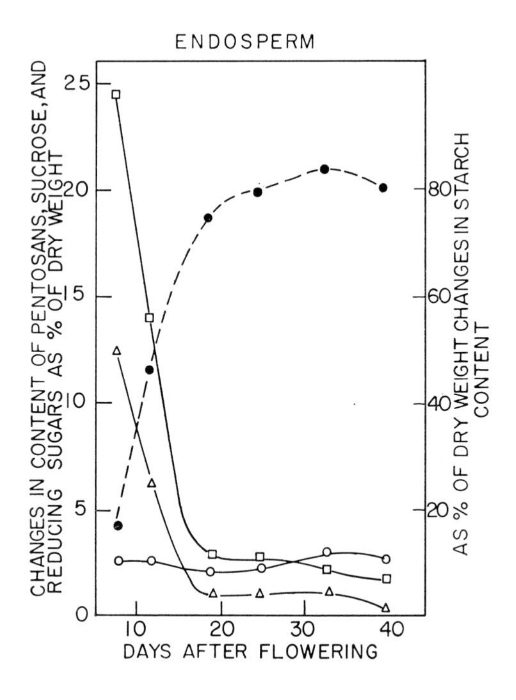

FIG. 10.4 ☙ Changes in content of pentosans (o), sucrose (□), reducing sugars (Δ), and starch (●) as a percent of the dry weight during time indicated after flowering (Jennings and Morton, 1963).

by an accumulation of soluble constituents (soluble nitrogen, amino acids, sugars, and nucleotides) and by the synthesis of protein, RNA, and DNA (Fig. 10.6). During the second phase (from 28 to 46 days) there is a utilization of the soluble constituents with further increases in the protein content, and the RNA content decreases rapidly. The embryo develops in a linear manner during the 46 days and at the end of 46 days is essentially complete. The components mentioned above account for approximately 20% of the total dry weight. Sixty to

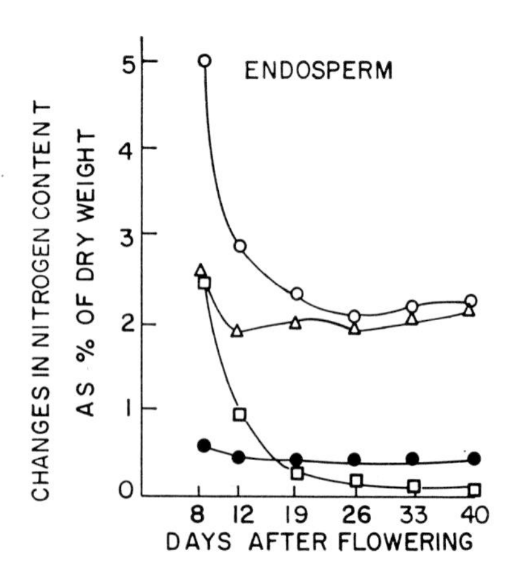

FIG. 10.5 ☙ Changes in total nitrogen (o), protein nitrogen (Δ), nonprotein nitrogen (□), and amide nitrogen (●) during time indicated after flowering (Jennings and Morton, 1963).

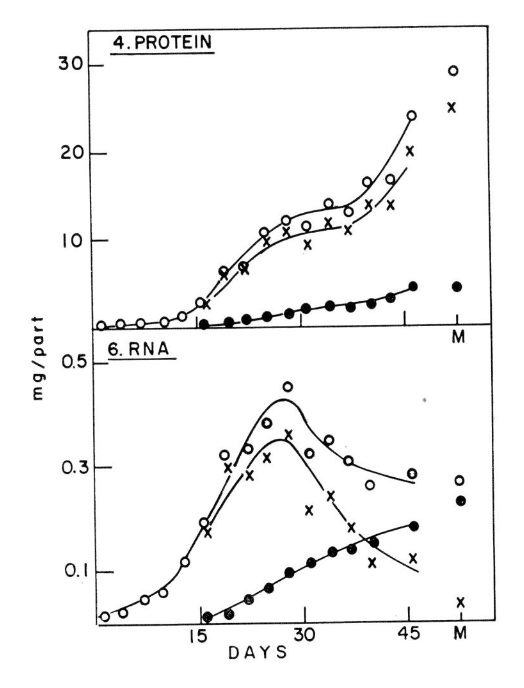

FIG. 10.6 Changes in protein and RNA content of whole grains (o), embryo (●), and endosperm (x) of corn over a 46-day developmental period after pollination (Ingle et al., 1965).

70% of the total dry weight would be starch and dextrins; this accumulates more rapidly in the last half of the 46-day period. Cellulose, waxes, and minerals would account for the remainder of the total dry weight.

The synthesis of oil in flax is shown in Figure 10.7 (Eyre, 1931). Little synthesis of oil occurs until the 10th day, then a sharp surge occurs between the 10th and 20th days, with a maximum at the 25th day. The iodine number begins to rise more sharply after the oil reaches a maximum percent (the iodine is absorbed by the C=C bonds). This indicates that development of unsaturation is a separate process from fat formation. The early oil formation represents a re-

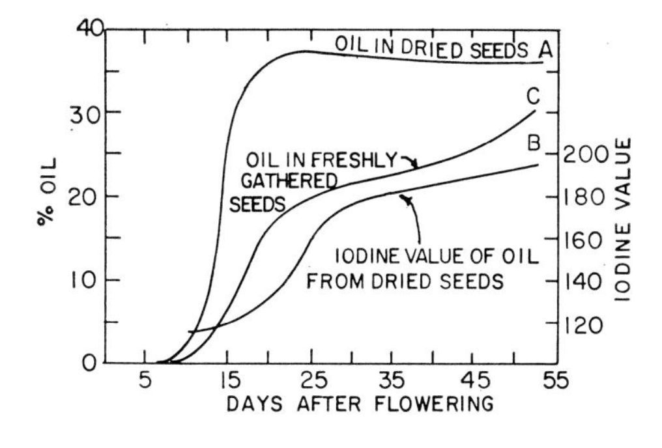

FIG. 10.7 Change in percent of oil and iodine value during flax maturation (Eyre, 1931).

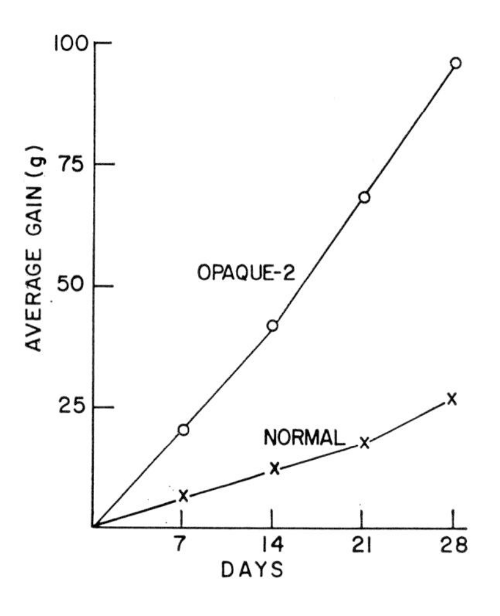

FIG. 10.8 ❧ Curves showing average weekly gains of rats fed on Opaque-2 corn (o) and normal Indiana Hybrid (453 x) (Mertz et al., 1965).

duction process from carbohydrate structures while the continued increase of unsaturation (greater iodine number) involves oxidation and development of more double bonds between carbons.

Plant breeders have made many modifications in seed composition by selection processes. A striking example is the case in which the strain Opaque-2 of corn has been shown to contain a significantly higher content of lysine, arginine, and aspartic acid (three essential amino acids). This results in a more rapid growth rate when fed to young rats, presumably because the amino acid composition of the grain is more satisfactory (see Fig. 10.8). This is a very significant step because the cereal grains previously have all been inadequate in their amino acid balance. The present challenge is to incorporate this high amino acid characteristic into higher-yielding lines.

Another example is a shift in the starch composition of corn. Standard varieties contain 28% amylose and 72% amylopectin. Breeders have modified this composition all the way from 100% amylopectin (waxy) to 80% amylose (high amylose). Such shifts can be very useful in providing a basic constituent for new foods or easier processing, but they may generate production problems, as noted by Crane (1964). He reports that high amylose corn had relatively poorer emergence and that higher kernel sugar was highly correlated with increasing seedling blight. Nevertheless, composition of the seed can be shifted, making it more useful as food or feed.

Seeds become capable of growing into a new plant long before they are classified as mature. Properly handled, immature seeds of corn will grow 20 days after fertilization, but they would not usually be considered capable of withstanding rapid dehydration or frost damage. The entire seed structure is not considered mature until about 55 days after fertilization. This suggests a delicate balance of conditions present in the seed as it matures. Freshly harvested immature seed may not germinate immediately. Sprague (1936) suggests that the mechanism inhibiting normal germination of freshly harvested immature seed operates in the scutellum of the embryo, since for germination corn kernels require only 35% moisture in the whole grain but 60% in the embryo. The size and development of the embryo is apparently controlled by natural inhibitors such as ascorbic acid (Howell, 1963).

SEED DORMANCY

Seeds which will not germinate after the mother plant dies but have a live embryo are classified as dormant. The causes of dormancy are varied and include: (1) rudimentary embryos, (2) physiologically immature embryos (inactive enzyme systems), (3) mechanically resistant seed coats, (4) impermeable seed coats, and (5) the presence of germination inhibitors (Amen, 1963). The first two categories may require a period of afterripening. The third and fourth categories resist the ready transfer of oxygen, water, and carbon dioxide. In legumes and some grasses this may be referred to as hard seed. The fifth category includes seed coats that can be leached with water to remove the inhibitor and thus allow germination. This feature adapts a species to not germinate until considerable moisture is present in its environment, thus favoring its likelihood of establishment.

In nature, then, hard-seededness and inhibitors aid in perpetuating the species. The hard seed of some legumes is a result of two factors: (1) the well-developed seed coat is composed of tightly packed, thick-walled, palisade (Malpighian) cells at right angles to the surface; (2) the hilum scar valve closes when the moisture level outside the seed is lower than that inside (Fig. 10.9). Thus hard seeds lose water in low humidity but do not gain it in high humidity. Hard seed and germination inhibitor dormancies are broken more rapidly at high temperatures and/or under moist conditions. The most common means of reducing this hard-seededness has been mechanical scarification. However, considerable embryo damage may result. Recent use of infrared light to disorganize the Malpighian layer has been successful (Works and Erickson, 1963). Infrared treatment provides a breakage of dormancy without damage to the embryo.

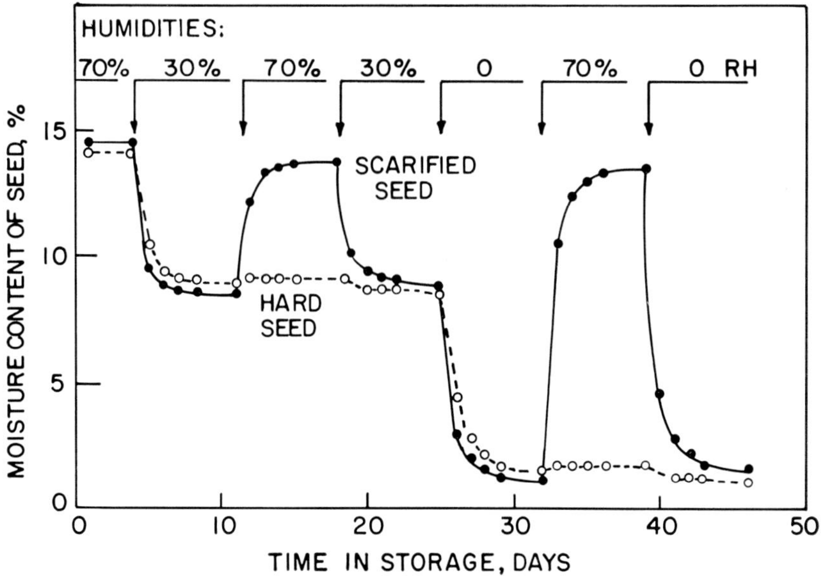

FIG. 10.9 ☙ When hard seeds of white clover are placed in alternating high and low relative humidities, the seeds lose moisture in lowered humidities but do not gain it back in elevated humidities because of the operation of the hilum valve. Scarified seeds can gain moisture readily (after Leopold, 1964).

Varying degrees of post-harvest dormancy were found in ten wheat varieties by Ching and Foote (1961). The dormancy remained for seeds stored at 3° C but disappeared within one month when stored at room temperature (38° C). This dormancy was shown to be caused by growth inhibitors which were water- and methanol-soluble materials. As another example of dormancy problems, the range grasses often fail to germinate promptly even under favorable environmental conditions. Germination inhibitors have been leached from the lemma, palea, and caryopses of *Avena fatua, Oryzopsis miliacea,* and *Bouteloua curipendula.* In *Stipa viridula* Trin. (green needlegrass) and *Oryzopsis hymenoides* Richer (Indian ricegrass) it has been shown that erratic germination is due to the lemma and palea acting as barriers to oxygen uptake. Peak germination of 72% occurred with green needlegrass seven years after harvest. Under laboratory conditions, a combination of chilling and KNO_3 treatment induced nearly complete germination immediately upon placement in an appropriate environment (Wiesner and Kinch, 1964).

Vernalization is a common requirement of temperate climate species before reproduction will occur. Most commonly the seedling plant or a well-developed one is vernalized. However, in some cases vernalization may occur before the developing embryo goes into a "resting state" dormancy on the parent plant. Such embryo vernalization reduces the growth requirements of the next generation, but it also reduces the degree of winterhardiness of the seedlings. Weibel (1958) succeeded in vernalizing winter wheat during the fruiting period, using the obligate winter variety Comanche. Low temperature seems to be effective only when the embryo is actively growing, and the low temperature and duration of the low temperature influence the effectiveness of the treatments.

SEED GERMINATION

Of all developmental phases of the plant, germination is probably the one which is most subject to control by the environment. Temperature stimulation for germination, for example, is restricted to a fairly narrow range in some species and delimits the time of year that species may grow. The requirements of the seed for ample water for rehydration often conflict with its requirement for efficient gaseous exchange. Dependence of germination on leaching of the seedcoat would appear to be a remarkable adaptation by plants in habitats where rainfall is the most variable factor.

The data on the effects of light on germination are voluminous and complex and do not seem to lend themselves to a simple evaluation in terms of survival (Evanari, 1961). Some crop seeds require light to germinate, others are inhibited by light. A light requirement is also common for germination of the seeds of many weeds.

Hoveland (1964) has shown that the root extracts of many grasses caused sharp reductions in the germination and seedling vigor of some legumes. Alfalfa seedlings were highly resistant; the clovers were very susceptible.

Prior treatment of seed has a distinct effect on the quality of germination. Rosenow et al. (1962) found that freezing immature sorghum seed with a moisture content of 34% or higher at 25° and 22° F resulted in greatly reduced germination. Seed with 23% moisture or less showed little injury at temperatures as low as 22° F for as long as eight hours. Seeds of grasses harvested by the windrow method germinated the same regardless of moisture levels at harvest and were consistently superior in germination to direct-combined seed. Seed

FIG. 10.10 ❧ Rapid water uptake by the embryo axis characterizes the first step in germination (after Ingle et al., 1965).

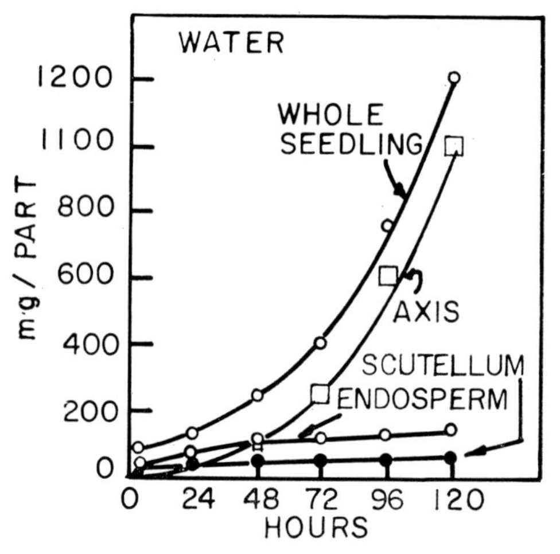

combined at less than 25% moisture did not heat appreciably in storage.

The rapid uptake of water is the most obvious change associated with the initiation of germination (Fig. 10.10). In the first 24 hours the water content in the seedling axis of corn increases 24%. By the fifth day over 80% of the water content is held by the axis (Ingle et al., 1965).

The dry weight pattern of the root and the shoot are different, as shown in Figure 10.11. The root begins to increase in dry weight in 20 hours, the shoot does not put on dry weight until 32 hours. Measurements of axis weight early thus indicate primarily what is occurring in the root.

At the start of germination, the initial water uptake causes the entire seed to swell. Wheat grains may increase in volume by 22% after 5 hours and 42% after 28 hours, which compares with the 46% reduction in volume which occurs during ripening (Percival, 1921). The swelling is due mainly to imbibition by the protein, which comprises about 13% of the grain. The starch, which comprises 60–75% of the grain, does not swell under normal conditions.

Respiration in barley apparently reaches a maximum a few hours after imbibition and then is constant until the tissue dies. For several oilseeds (soybean, castor bean, peanut) and in corn, there is an increase in the bulk and function of the mitochondrial fraction.

Metabolism during seed germination has been studied in considerable detail. It is not possible to precisely delineate the order in which metabolic reactions are initiated during germination, but the catabolism of several constituents and resultant energy release can be described.

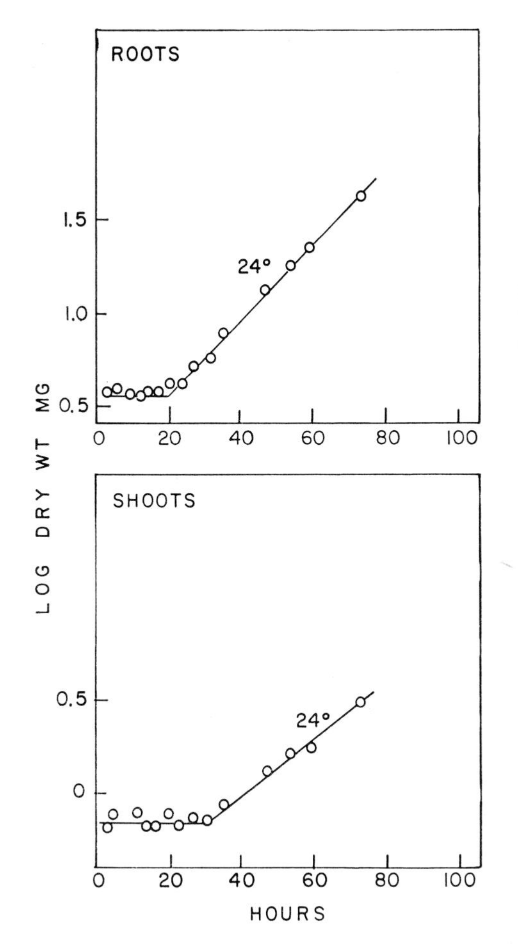

FIG. 10.11 ☙ Increase in dry weight of roots and shoots of *Phaseolus* seeds (Simon and Meany, 1965).

LIPIDS

There is a striking disappearance of reserve materials and a rapid conversion of fat to carbohydrate, especially in high-oil seeds. This occurs as a conversion of reserve fat to sucrose. For example, an ungerminated castor bean contains some 260 mg of fat (70% of its dry weight) and 15 mg of total carbohydrates. During germination in the dark over a period of 8 days, the total fat content of the seedling falls to about 50 mg while the carbohydrate rises to 230 mg, and this in spite of the fact that sugars are being used extensively in the growth and respiration of the seedling. Thus there is the production of more than one gram of sugar (principally sucrose) for each gram of fat consumed. This interconversion is strictly confined to the endosperm tissue (respiratory quotient about 0.35); sucrose moves into the growing seedling where it supports carbohydrate respiration (RQ = 1.0). This process is suggested to be one in which 2 carbon fragments (acetate) from β-oxidation of fatty acids are first synthesized into a dicarboxylic acid (via the glyoxylate cycle), followed by conversion of malate to hexose (sucrose).

On the basis of this information, Beevers (1961) summarized the process of sucrose formation from fatty acid as follows:

$$\text{C-16 fatty acid} \xrightarrow{\beta\text{-oxidation}} \text{8 acetyl CoA} + \text{28 H}$$

$$\text{8 acetyl CoA} \xrightarrow{\text{glyoxylate cycle}} \text{4 oxaloacetate} + \text{24 H}$$

$$\text{4 oxaloacetate} + \text{8 H} \xrightarrow[\text{reversed}]{\text{gylcolysis}} \text{sucrose} + 4\ CO_2$$

$$C_{16}H_{32}O_2 + 11\ O_2 \longrightarrow C_{12}H_{22}O_{11} + 4\ CO_2 + 5\ H_2O$$

Lipid + Oxygen ⟶ *Sucrose + Carbon Dioxide + Water*

CARBOHYDRATES

Barley has been studied intensively and is useful as a model of seed utilization of carbohydrate. The work of Paleg (1960) and others has shown that gibberellin released by the embryo triggers the activity of amylase in the aleurone or the scutellum; this α-amylase in turn breaks down the starchy endosperm and makes it available to the embryo. Simultaneously, β-amylase levels increase in the endosperm and speed the breakdown of starch while maltase functions to break short-chain saccharides to shorter chain lengths. All three—α-amylase, β-amylase, and maltase—thus mobilize the carbohydrate reserve into a form (sucrose, glucose, or fructose) which the plant embryo can use for energy release. The same basic reaction has been shown to occur in peas.

PROTEINS

Seed proteins are hydrolyzed to peptides and amino acids which can be translocated to the embryo for use in growth and synthesis of new protoplasm. There is a concurrent increase in RNA, probably associated with both an increase in mitochondria and an increase in enzyme synthesis.

Catabolism of carbohydrates and proteins in corn endosperm was stimulated by exogenous gibberellic acid (Ingle and Hageman, 1965b), as is the case in other species previously studied. In the intact plant the embryo axis produces a compound which has the same effect and is received by the endosperm 36–48 hours after the start of germination. It has further been noted that hybrids convert endosperm into seedling growth more rapidly but no more efficiently than does either of the parental lines, which suggests that the primary difference in the hybrid in seedling vigor may be its rate of gibberellin synthesis.

In aged seeds (caused either by a long period of storage or a short time of unfavorable storage) germination is not greatly impaired, but the vigor or growth rate of seedlings is often markedly reduced. Mitochondria from the dark-grown seedling axes of new and old soybeans were found to differ significantly in their respiratory activity (Abu-Shakra and Ching, 1967). The old seeds gave only 40–70% as much phosphorylation per unit of oxygen used as did the new seeds. Old seeds also had fewer mitochondria per unit seedling.

Seedling use of reserves is similar for the grasses (Brown, 1965; Fig. 10.12) and legumes (Simon and Meany, 1965). Each contains a rather large store of reserve materials and powerful hydrolytic enzymes. The same rapid uptake of water occurs as was shown earlier for corn, and it is enhanced by higher temperatures. At 24° C this expansion is smooth and uniform for beans (Fig. 10.13), but at 15° three stages become evident. There is an initial rapid expansion, a lag, and then a rapid expansion again, but one that is not equal to the 24° curve in its rate. At both temperature levels, most of the reserve material is consumed in dry weight increase rather than respiration (Fig. 10.14).

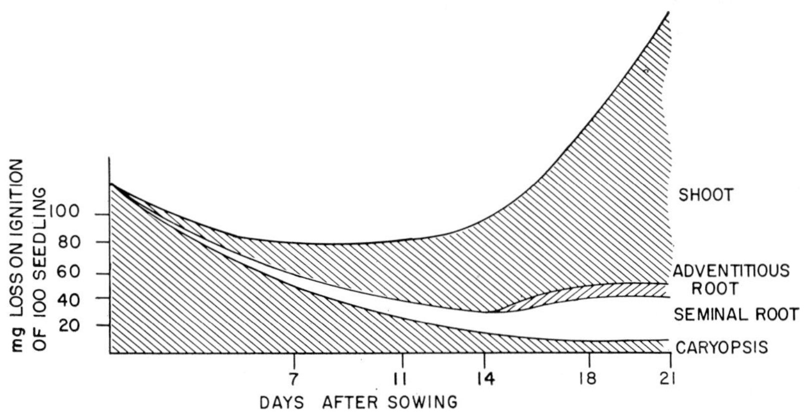

FIG. 10.12 ☙ Dry weight distribution in the seed, root, and shoot of barley for 21 days after sowing (Anslow, 1962).

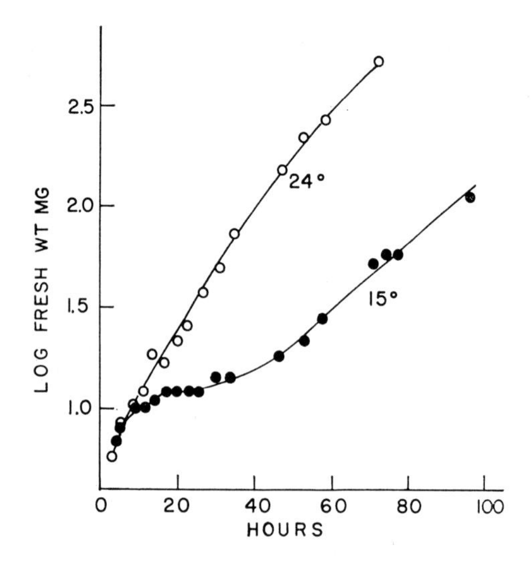

FIG. 10.13 ☙ Fresh weight increase of root-shoot axis in seeds of *Phaseolus* germinating at 24° and 15° C (Simon and Meany, 1965).

SEED QUALITY

The purchaser of seed wishes to obtain material which has the following characteristics: high germination percentages, freedom from weeds (especially primary and secondary noxious), and freedom from foreign material. To provide the seed purchaser these characteristics in his seed, federal legislation has been developed during this past half-century. The Federal Seed Act of 1912, completely revised in 1939 and amended in 1956 and 1958, has as its basic philosophy "truth in labeling." Seed laws do not decide what kind, variety, or quality of seed is good for the consumer (this decision is left up to the buyer); the law merely requires that the information be present on the tag. The law is meant to cover all seed transfers, but enforcement is difficult on the farmer-to-farmer level. It therefore becomes the farmer-purchaser's responsibility to insist that the farmer-seller comply with the seed law. Seed laws do vary from state to state, but the basic

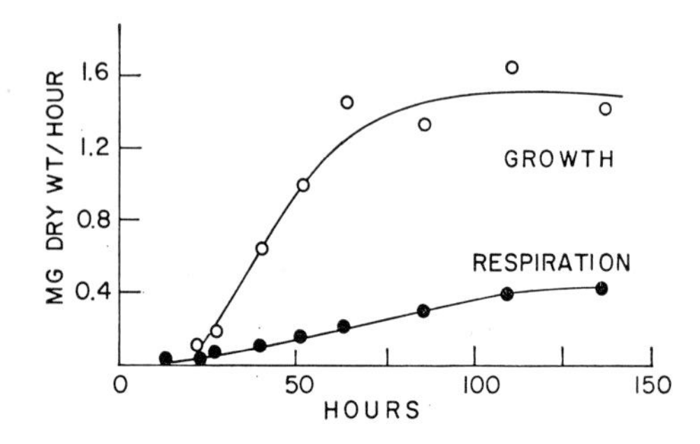

FIG. 10.14 ☙ Consumption of dry matter by *Phaseolus* root-shoot axis in growth and in respiration (Simon and Meany, 1965).

characteristics are similar. The minimum requirements in Iowa for 1969 are:

1. The germination percentage must be stated from a test conducted within the past nine months.
2. The purity provides a statement of the percentage of weight in the bag that is crop seed.
3. The seed lot must contain no primary noxious weeds.
4. The number and kind of secondary noxious weeds must be recorded.
5. Not more than 1.5% of the total weight can consist of weeds.

The purity test separates the following materials by percentages calculated on a weight basis: crop seed, weed seed, and noxious weed seed. Germination tests involve the pure seed fraction with standardized techniques conducted on one of several media: a sand bench, expanded mica, filter paper, or in a petri dish with a blotter paper insert. Photoperiod and temperature variations that may include alternating day and night temperatures are used on certain seeds. Counts are made of abnormal seedlings, dead seed, and hard seed (seed which remains solid but does not sprout). These germination tests require 8 to 28 days for completion depending on the species.

An innovation to seek more rapid estimates of viability has been the tetrazolium test. As a basis for this test, the chemical 2,3,5 triphenyl tetrazolium chloride (a reducing agent) turns actively respiring tissue a bright red. The tetrazolium is acted upon by dehydrogenase enzymes and upon the addition of hydrogen changes to formazon. Moistened seeds are cut, placed in the solution 2–6 hours in the dark, and then microscopically examined. The test is useful on grasses and legumes as a germination guide, but it is best used in conjunction with the more standardized tests. Additional laboratory tests may include bioassays to detect the presence of pesticides on seeds or cold or vigor tests to evaluate germination after the seed has been under biological stress.

The problem of low seed vigor as a result of physiological deterioration is discussed at length by Grabe (1967). He describes the following tests for vigor:

1. *The cold test,* used primarily for corn, determines how well seeds withstand rotting organisms under cold, wet soil conditions. Seeds are planted in a mixture of sand and unsterilized soil, held at 50° F for 7 days, then transferred to a warm temperature to germinate.

The cool temperature favors the activity of pathogens which enter the seed through cracks or breaks in the seed coat. Seed treated this way reflects the amount of mechanical damage in corn seed and the effectiveness of fungicide applications.

2. *Length of primary root or speed of germination* is a measure of seedling growth rate. The more rapidly the seedling expands, the greater vigor it is presumed to have.

3. *GADA* (glutamic acid decarboxylase activity) is an enzymatic assay which serves as an index of vigor under the assumption that it is associated with the rate of carbon dioxide evolution. This decarboxylase enzyme is active in seed and a high level should indicate high vigor.

4. *Permeability changes* can be tested by soaking the seed in distilled water and measuring the electrical resistances of the water. High electrical conductance means the seed coats have deteriorated and allowed materials to leach into the water which serve as electrolytes to conduct the electrical current.

Each test has merit for certain uses, best determined by a careful study of a particular problem area and correlating one of the above to the most meaningful, repeatable results as related to field performance.

Finally, varietal purity is an important component of seed quality. Seed laboratories do not commonly test for variety; but there are guidance tests available. For example, ultraviolet light may be used to separate seeds that do and do not fluoresce. The technique differentiates between groups of oat varieties and between annual and perennial ryegrass. A second varietal index, coleoptile color, is an identifying feature easily run in growth chambers. Phenol tests on wheat seed give identifying reactions from variety to variety. Nittler et al. (1964) describe many tests helpful in identifying alfalfa varieties, although these tests have the disadvantage of being somewhat slow, requiring 3–8 weeks to run for an acceptable degree of precision. The tests include winterhardiness category, degree of resistance or susceptibility to disease, flower color, and leaf shape and serration. However, it is necessary to reemphasize that none of these tests absolutely determines a particular variety, and therefore seed certification as discussed in the next section is a very important step in providing varietal purity.

SEED CERTIFICATION

Since varietal or genetic purity is so difficult to test, heavy reliance has come to be placed on a seed certification program to maintain seed

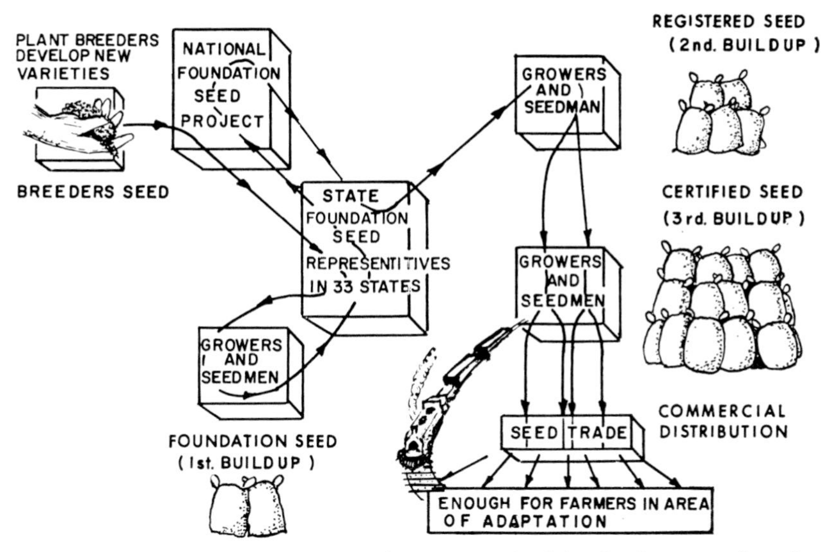

FIG. 10.15 ✤ Steps involved in the increase of seed supply from original breeder's seed to a large volume for farmer use (after Smith, 1962).

of known genetic background. Such a program is absolutely essential to preserve the positive attributes of any new variety.

Crop improvement associations have set up a four-step classification scheme for certified seed as follows (Fig. 10.15):

1. *Breeder's Seed*—seed directly controlled by the originating or sponsoring plant breeder or institution which provides the source for the initial and recurring increases in pedigreed stock.
2. *Foundation Seed*—the progeny of breeder's seed so managed as to maintain its specific genetic identity and purity. Its production is very closely controlled and often is handled directly by the staff of the seed certification agency. Foundation seed may be planted for the production of either registered or certified seed.
3. *Registered Seed*—the progeny of breeder's or foundation seed so managed as to maintain satisfactory genetic identity and purity. It is commonly planted by a select group of farmers, often men who have been certified seed producers for several years.
4. *Certified Seed*—the progeny of breeder's, foundation, or, most commonly, registered seed. Certified seed is purchased for production of commercial grain or forage crops. In some states certified seed may be used to produce more certified seed, but the trend is toward the use of registered seed only to produce certified seed. Where certified seed is used for seed production, it is often on a "limited generation" basis

where certified seed can be used to produce one more generation of certified seed but no more.

Crops grown for certification are isolated from other fields of the same species by a combination of distance and border rows. The distance is related to the expected distance pollen may blow or may be carried by insects. Border rows reduce this distance in corn by providing massive supplies of the desired pollen. Both field and bin inspections are made of the seed production procedure and careful records kept on the source of the seed stock. Thus when the seed receives a blue tag as certified seed, it is an indication of assured genetic purity.

"Drill box surveys" of seed planted, particularly for small-grain production, indicate there is much room for improvement in seed quality. A Pennsylvania survey found less than 16% of the oats planted to be of high value; 25% of the samples contained noxious weeds. An Oklahoma study on wheat showed 6% was mislabeled, 37% contained noxious weeds, and 38% had varietal mixtures. Grabe and Frey (1966) found farmers planting oats with germination percentages of 33, 27, 21, 16, and 1! Three-fourths of the farmers were sowing samples containing primary or secondary noxious varieties. These surveys indicate a strong need for use of better-quality seed.

SEED PRODUCTION

FORAGES

The United States and Canada formed an International Crop Improvement Association in 1919 to coordinate seed certification procedures. Prior to that time distribution of new varieties had no standardized pattern and their value was often quickly lost. As a further step, in 1949 the National Seed Foundation Program was established (Fig. 10.15). These two steps paved the way for the shift which has occurred in forage seed production, especially since World War II. Prior to 1940, much of the north central region of the United States produced its own grass and legume seed. With the development of Ranger alfalfa at the Nebraska station in 1943, the stage was set to seek the highest per acre yields of seed possible in order that this excellent variety might be widely available. The decision was made to move seed production to California where both higher yields (two to five times the yields obtained in the north central states) and better quality seed could be attained. The National Foundation Seed Act and the certification program provided a framework to make this a workable scheme. Guidelines were set up to be sure seed produced outside a region had the necessary characteristics (in this case winter-hardiness and disease resistance) when it was shipped back to the region in increased supply. States set up reciprocal tagging so, for example, seed grown in Oregon and blue-tagged as evidence of certification could be sold in Iowa. For both alfalfa and red clover, the practice continues whereby the variety is developed (or at least tested extensively) in the area of use, then taken to the West for large-scale seed increase.

Grass seed production is much more widely scattered than legume production, and significant quantities are produced in the area of use. There are approximately one billion acres of grassland (for nongrain use) in the United States. About 10% of this acreage is seeded every year, or about 1.3 times the U.S. corn acreage; as a result there is need for a large volume of seed annually.

Forage seed production is a very specialized industry, and frequently it is the grower's main source of income. Since grass and legume seeds have few alternate uses (compared to the cereal crops) there is constant concern about a balance between use and supply. Seed prices of these crops are likely to fluctuate more widely as a result. Yet while surplus seed production may occur for some seeds, the United States regularly imports approximately 800 million pounds of seed for species in short supply.

Alfalfa. Key production areas have developed in the Central Valley of California, the Willamette Valley in Oregon, the Columbia River Basin in Washington, and the Salt River of Arizona. For seed production, plants are spaced 12 inches apart in 24–36-inch rows. Pure stands are established with rates of 1–1½ pounds per acre. The field is trimmed in April and then the normal two or three flowering cycles are allowed to occur through the season. Only one harvest is made and this is done in late September after the field has been treated with a desiccant. If the weather is cool, diquat is used as a desiccant; if the weather is warm, a mixture of dinitro and oil is used. Following this treatment the crop can be combined directly. Good legume seed production areas are characterized by high light intensities, clear days, and low humidity conditions. To maximize seed set the cross-pollinated alfalfa crop is supplied with additional insect pollinators. Honeybees, alkali bees, and leaf-cutter bees are all useful pollinators, the latter two being particularly good because they are pollen seekers. Klostermeyer (1964) presents a very thorough discussion on the habits and management of these bees. The alkali bees nest in the ground and the leaf-cutter bees will use a 4 x 6-foot board drilled with 3/16 holes as their "hive." Careful management will give alfalfa seed yields of 1,000 to 2,000 pounds per acre.

Grasses. Grasses are also produced in pure stands and row culture for seed production. Most grasses produce a flowering stem just once each year, thus concentrating the time for seed harvest. One very important management practice is nitrogen fertilization. As described in Chapter 8, nitrogen enhances tiller formation and is credited with increasing seed production even more consistently than it does forage production. For cool-season grasses, nitrogen applied in the fall (at the time of tiller induction) or in the early spring (as the tiller bud is initiated) gives the best seed yield. For warm-season grasses, tillering initiation is later in the spring and this is the best time to apply nitrogen. For most tall-growing, upright species, nitrogen-fertilized solid stands do not yield as well as unfertilized row culture. For these species adequate light on each plant is apparently important to favor tiller growth, which in turn provides more seed heads. Carlson (1964) found that with Sterling orchardgrass, rows spaced 24 inches apart averaged higher in seed yield than 9- or 36-inch rows. The greatest response to the fertilizer treatment was obtained in the 9-inch rows. Phosphorus also gave a response, especially in the 9-inch rows. Spring fertilizer application resulted in more seed per panicle and higher weight per 100 seeds. Panicle size increased and panicle number decreased (i.e., less tillering) as width between the rows increased. Yields were similar for 9-inch rows fertilized in the fall and 24-inch rows fertilized in the spring.

Canode (1965) noted that perennial grasses produce little or no seed the year they are planted, suggesting the need for induction by cool, short fall days. These species reach a peak in seed yield the second or third year, then decline. The perennial sod-forming intermediate wheatgrass did not respond to higher nitrogen levels, but burning of the straw and stubble each September favored high seed yields. Mechanical thinning of the stand also improved yield. Thinning of the stand reduced lodging and the number of culms per yard of row, resulting in a negative correlation of these factors and seed yield. This suggests that seed weight is a more important component of yield in such species than is seed number—an exception to the more general rule that seed number is the primary component of yield. Seed production of creeping red fescue has been improved by limiting nitrogen and reducing the vegetative growth to 3 inches after each harvest. Therefore, it is important to get light down into the stand.

ANNUALS

Production practices for annuals are very similar whether grown for seed increase or for commercial feed and grain production. Small shifts may be made in the population level to get the maximum benefit of scarce seed and an increase in seed size and uniformity.

Cereals. Eighty to 90% of the wheat, soybean, and oat seed production is done by the farmer himself. These self-pollinated crops require a minimum of effort to retain varietal purity in the field. However, quality of the seed may be low because poor harvest and storage practices contribute to low germination, heavy weed content, and mechanical mixtures from varietal mixtures in the bin.

Corn seed production is done almost exclusively to provide hybrid seed. After several (6–8) generations of inbreeding to fix desirable characteristics, inbreds are selected for particular attributes such as stalk strength and disease resistance as well as yield ability, and crosses are made to form single-cross, double-cross, and three-way hybrids. Hybridization of this monoecious plant is readily accomplished by detasseling, a mechanical emasculation, or by use of male sterility. The practicing farmer expresses a concern when one or two leaves are removed with the tassel. However, careful studies suggest (Chinwuba et al., 1961) that yield enhancements of 5–50% may result from mechanical emasculation or male sterility. Male-sterile single crosses outyielded their fertile counterparts, especially at high populations. Sarvella and Grogan (1965) noted that the lack of pollen production in male-sterile plants was correlated with shorter internodes, suggestive of less gibberellin formation by the tassel. Shortening was observable 10–12 days after meiosis. This combination of events likely allows the ear shoot to command a greater percentage of the plant's photosynthate and represents a reduced apical dominance by the male-sterile tassel.

In recent years cytoplasmic male sterility has become a key factor in seed production of corn, sorghum, and certain other crops. The male sterility factor most commonly used is carried in the cytoplasm. Crossing such a sterile inbred to a male-fertile inbred results in seed that becomes a male-sterile single cross. Sterility-inducing factors have

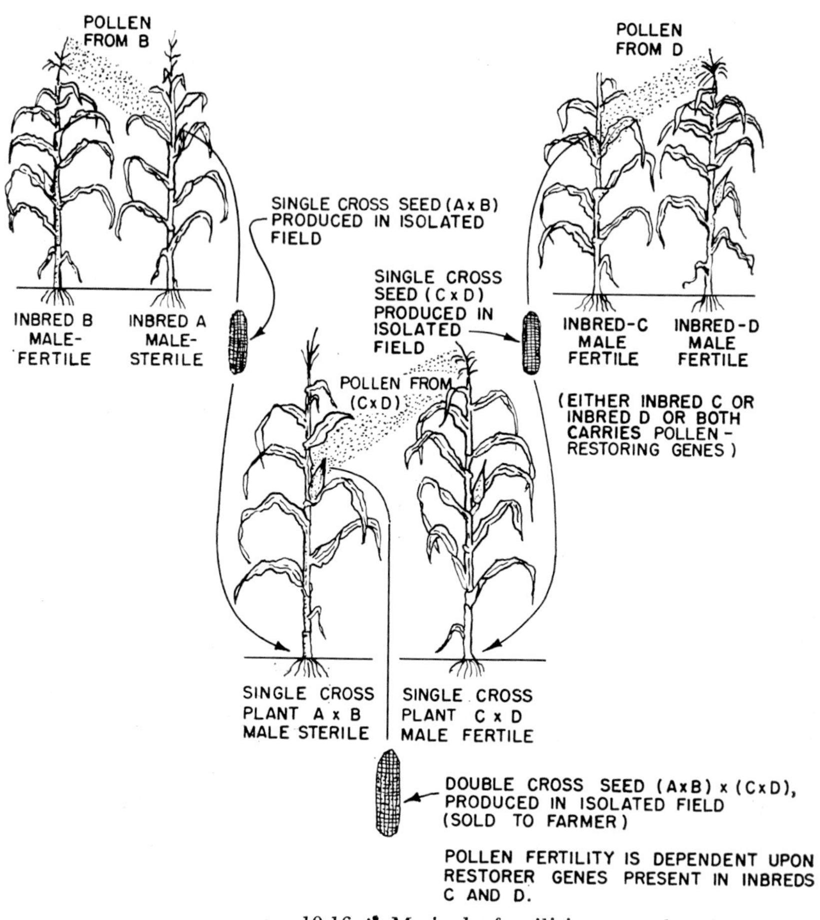

FIG. 10.16 ❧ Method of utilizing cytoplasmic male sterility in production of single-cross and double-cross hybrid corn seed (Poehlman, 1959).

been found in several lines of Texas corn. As long as normal corn is crossed to such a male-sterile line, the offspring are male sterile. However, genes which have fertility restoration ability have also been identified. These genes are incorporated into the line to be used as a male parent and the resultant offspring is male fertile. Sterility is used during seed production, then fertility is restored in seed sold to the farmer (Fig. 10.16). Because of the previous evidence that male-sterile plants produce higher yields, there has been interest in providing the farmer with seed that is, for example, 50% fertile and 50% sterile. This is now rather widely done and has provided tolerance to higher populations per acre.

Production of hybrid wheat and barley represents the first effort of wide-scale crossing of highly self-pollinated crops. Hybrid wheat production is potentially an enormous enterprise with an estimated 30 million acres of seed being required each year. The primary interest in any hybrid is for the greater yield potential it may provide; under experimental conditions, hybrid wheat has been reported to provide yield increases of 20–50% (Suneson, 1962; Holland, 1966). However, quality in the form of flour strength and mixing stability as well as quantity is important, because wheat is used for many baking and specialized food processes. Hybrids of the bread wheats generally have protein and bread-making qualities intermediate to the two parental lines.

To produce hybrid wheat, seed producers must grow three separate lines: a male-sterile line, a fertility-restorer line, and a maintainer male-fertile line. Each hybrid will be the product of two isolated seed fields. In one field the fertility-restorer line will pollinate both itself and the male-sterile line. These will be harvested separately. The seed from the male-sterile plants will be the hybrid seed for commercial sowing. In the other field the maintainer male-fertile line and male-sterile plants will be grown.

For crossing, alternate drill strips may be used, or up to two female drill strips per male drill strip since pollination is generally adequate several feet from the pollen source. One key problem in hybridizing wheat is the tendency for both fertility and sterility factors to act differently under different environmental conditions, thus making it difficult to be sure crossing will occur as planned in a particular season. Two major and several minor genes influence fertility restoration in wheat. One key problem related to this process is that with a flower that opens and allows pollen to blow to other plants, the wheat plant becomes more subject to infection by organisms such as loose smut and ergot than is the flower which is self-pollinated.

Barley hybridization may exploit two unique schemes in contrast to the cytoplasmic sterility-restorer combination described for corn and wheat. In barley some sterility factors are linked genetically to DDT resistance. Therefore, plants which are to be used as the female parent could have this DDT-resistance characteristic crossed into the population. Rather than select for a completely sterile population, a heterogeneous one could be sprayed with DDT; those surviving would be the male-sterile plants. This sterility does not require a restorer gene, and the plant bears seeds that are fertile in the next generation. This is an example of the possibilities of chemical roguing in crop management.

A second method which has been used successfully and of which the variety Hembar is the result is called the "balanced tertiary trisomic" method. It makes use of a genetic male-sterile gene and an extra chromosome. The extra chromosome regulates several critical genetic events but is eliminated automatically in one of the final steps.

Forages. Another recent innovation has been to cross grain sorghum with sudangrass to form a high-tonnage forage hybrid. Using the grain sorghum plant as the seed-bearing parent, a high seed yield is obtained which allows economical seed production. The sudangrass contributes desirable forage qualities. The sorghum-sudangrass hybrid will probably not replace sudangrass as a summer forage for grazing purposes, because it is higher in prussic acid and cattle tend to waste more of it because of its tall, stemmy growth. Nevertheless, it is very useful as a green chop material and has been widely used in southeastern United States where the alfalfa weevil has reduced the productivity of alfalfa.

Forage grass and legume breeders have considered F_1 hybrids of interspecific crosses, synthetic varieties, vegetative propagation of F_1 hybrids, and F_1 hybrids. If attainable, the F_1 hybrids will give the most heterosis. However, desirable parents and sources of male sterility are still sought. Alfalfa is most likely to be the first perennial forage in which a hybrid variety is available to farmers.

SEED STORAGE AND PROCESSING

Once quality seed has been grown and harvested, careful storage is necessary to retain the vigor of the material and to guard against mechanical contamination. Temperature and moisture content tend to control the rate of respiration and of deterioration and, therefore, must be carefully controlled (Hukill, 1963). Storage of seeds is complicated by the fact that seeds are hygroscopic. In bulk storage there is considerable moisture migration, but it can be controlled, if not eliminated, by aeration during periods when the atmosphere is cooler than the seeds. Texas studies suggest that seed quality and seedling vigor of commercial seed lots of corn may be retained for three or four years, and such seed can be used satisfactorily in years of short supply (Clark, 1965). In addition, wheat and sorghum seed treated with various insecticides and fungicides maintained their germination at original levels for one to three years after this treatment.

In preparation for planting, seed may receive a variety of treatments. One generally used for many grass seeds is a fungicide treatment, for which Captan is most widely used. These fungicides provide localized protection for the seed and for the young seedling as it emerges from the ground, but fungicide seed treatments do not give full-season protection nor do they protect the entire plant. In addition to the fungicides, seeds are also commonly treated with an insecticide to protect them against soil insects prior to and during early seedling development. Again, seed treatment is a short-term effect and does not protect the plant later in the season.

Scarification as a means of reducing the hard-seeded nature of many legume seeds (alfalfa, birdsfoot trefoil, red clover) has been previously discussed in the section on dormancy. Chemical treatment with potassium nitrate has been used successfully. Mechanical scarification is done with a shaft of air carrying the seed against an abrasive surface. Infrared light and radio-frequency electrical treatment have the least likelihood of damaging the embryo. These methods will increase germination from the 40–60% range to the 80–90% range and are about equal in their effects. Such seed suffers no damage and can be stored for several years.

Still another seed treatment includes pelleting seeds with coatings of fertilizer nutrients. This has been done extensively in Australia and New Zealand. Methyl cellulose has proved to be a useful adhesive for pelleting. The pelleting puts the fertilizer with the seed, improves the distribution of small-seeded species, and facilitates seeding by air (Brockwell, 1962).

INOCULATION

All legume species used as economic crops can be inoculated with bacterial cultures and thus support colonies of bacteria for nitrogen fixation in a symbiotic relationship. This fixation is still the dominant source of nitrogen for plant growth. Rain and snow may add 1–30 pounds of nitrogen per acre per year. In the annual harvests of the world, 3–5% of the nitrogen originates from nitrogenous fertilizers produced by industry; the remainder comes directly from nitrogen fixation or is released from organic matter formed originally from nitrogen fixed by bacteria.

The presence of bacteria in legume nodules and their action in nitrogen fixation was first suggested in 1888, but significant sales of inoculum have been made in the United States just since 1932. One of the early observations of workers in this area was the specificity of the bacteria infecting a given species. As a result of this work cross-inoculation groups were defined to describe legume genera infected by the same rhizobium species. The alfalfa group, for example, includes members of the *Medicago* and *Melilotus* genera; the true clover group consists of members of the *Trifolium* genus. Other groups are pea and vetch, bean, soybean, cowpea, and lupine. Some species, like birdsfoot trefoil and crownvetch, have individual rhizobial requirements. There is continual effort to include the most efficient strains for each group. Recent work suggests that for certain crops (e.g. soybeans) strains already in the soil, while not the most efficient, are nevertheless the ones which successfully infect the plants. The strain most common in 1905 in Iowa was again most prevalent in nodules in 1965, and recovery rates of strains applied in the inoculum were rather low (Johnson et al., 1965). There is considerable work yet to be done to make inoculation efficient in terms of getting the most useful strains to infect the plant.

Preparations of commercial inoculum may be placed in agar, sand, peat, or broth cultures. The peat carrier has been a standard one for many years. With interest in preinoculation (inoculation done more than 48 hours in advance of planting) and the nature of machinery available to apply the inoculant, liquid or broth cultures have been popular. However, Burton and Curley (1965) suggest that the broth carrier is not satisfactory for use when inoculation is done more than one day in advance.

The process of infection has received intensive study in recent years. It is known that most of the bacteria enter the plant through the root hairs and for this to occur the hair must be curled. It is suggested that root hair curling comes about from the following series of events. First, tryptophan is exuded by the plant. This is converted to indoleacetic acid by the bacteria and in turn IAA induces curling. Once the hair is curled, invagination by the bacteria proceeds. The bacteria do not initially penetrate the cell wall but merely push the wall inward. The infection thread travels to the cortical tissue and there infects tetraploid (not diploid) cells in the area of the endodermis. A rapid division of the bacteria follows with expansion of the area into a nodule which has vascular connections with the host plant. Food is transported to the nodule and fixed nitrogen is supplied to the plant directly. Nodules may also exude nitrogen into the soil, or as they slough off nitrogen is provided to the plant from nodule decay.

The intermediates in nitrogen fixation from the gaseous state to an amino form useful in amino acid synthesis are still in the category of postulates. Figure 10.17 outlines one well-accepted hypothesis. Ini-

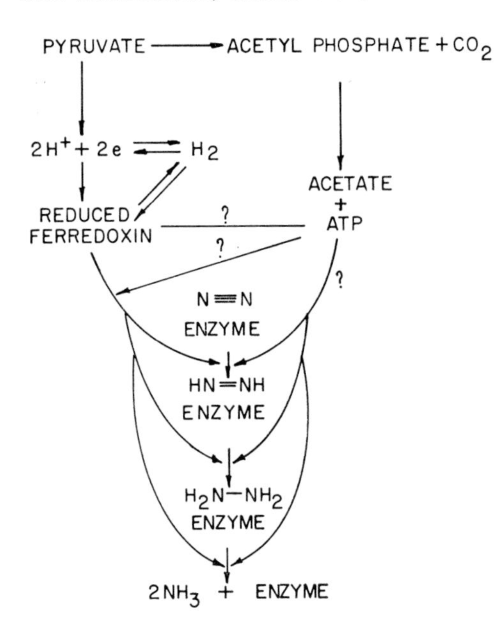

FIG. 10.17 ✤ A postulated reaction sequence in nitrogen fixation (Burris, 1965).

tially, N_2 is bound to an enzyme and then is reduced in a stepwise fashion until NH_3 is formed and released. Hydrogen is activated and reduces ferrodoxin. Both ATP and NAD appear involved to energize the reaction and provide hydrogen-carrying activity necessary to reduce N_2.

In an attempt to better understand the nodulation process, Tanner and Anderson (1963) studied nodulation with non-nodulating and nodulating soybean isolines (genetically the same except for this characteristic). They found no difference in the amount of tryptophan exuded. Through grafting experiments they concluded that the non-nodulating strain contained the cotyledon factor required for nodulation. Therefore, the control of the nodulation response appeared to be local in effect and not determined by the top of the plant.

The active functional nodule has a pink coloring due to the pigment leghemoglobin, and the most effective nodules appear on the main roots. This is most likely because such nodules obtain an adequate carbohydrate supply. As the plant matures or sets seed, the interior of the nodule turns green and presumably ceases to fix nitrogen. Light intensity, daylength, and temperature have been shown to influence nodulation. In the dark, nodules form on some species and not on others. On alfalfa and vetch seedlings only a few nodules are formed in the dark, and if nodulating plants are placed in the dark, nodule formation ceases and those already formed degenerate. Crops such as red clover and soybeans are quite resistant to infection in the dark unless provided with sugar, whereas in peas the number of nodules formed in the dark is reduced by sugar but their size is increased.

At very high light intensity, fixation is completely inhibited in the young seedlings of certain legumes. This apparently is related to the "nitrogen hunger" phase, which has been identified under a variety of conditions as resulting from too little nitrogen in the potential host to allow infection. The depressing effect of high light intensity can be overcome either by a short period of shading or by adding nitrate. Conversely, nitrate levels in the soil can play a critical role in infection. The best infection occurs with a moderate amount of nitrogen present. A high level of nitrate overstimulates top growth and tends to reduce the carbohydrate level in the root. High nitrate levels may also reduce infection by conversion to nitrite, which in turn is known to destroy IAA. If nitrite does destroy the IAA, less root hair curling occurs and poor invagination results.

The pH of the soil has a very distinct effect on bacteria. The rather narrow pH requirements of legumes are essentially the pH sensitivity of the bacteria in the nodules. In many cases the plant is tolerant of a more acid pH than are the bacteria. Sweet clover bacteria do best at pH 6.2; true clover bacteria require only pH 5.2. Finally, bacteria are quite sensitive to water stress and may die in sandy soils or under drouth conditions (Vincent et al., 1962; Alexander and Chamblee, 1965).

For many years standard inoculation techniques have required applying the inoculum a few hours before planting. This was to prevent desiccation of the bacteria. It was also done to reduce the likelihood of exposure of the bacteria to ultraviolet light. Alexander and Chamblee (1965) emphasize that 3–5 hours in sunlight or 2 or more weeks in dry soil significantly reduce nodulation in alfalfa and birdsfoot trefoil.

More recently there has been great interest in the practice of inoculating seed weeks or months in advance of planting. The first preinoculation treatment consisted of placing the seed and bacteria in a chamber and developing a vacuum in the chamber. The exact reason why bacteria survive better after this treatment is not known; they may be drawn under the seed coat in the region of the micropyle. Several commercial methods of preinoculation are now available. Some rely on a surface treatment designed to minimize bacterial death in the presence of sugarlike adhesives. These systems have potential merit but are not perfected and provide adequate nodulation in only about 50% of the cases.

To summarize, quality seed of known genetic composition is the very keystone of much of the world's agriculture. Agronomic practices have been described whereby this seed quality may be measured, genetic purity maintained, and seed production accomplished. ✤

LITERATURE CITED

Abu-Shakra, S. S., and T. M. Ching. 1967. Mitochondrial activity in germinating new and old soybean seeds. *Crop Sci.* 7:115–17.

Alexander, C. W., and D. S. Chamblee. 1965. Effect of sunlight and drying on the inoculation of legumes with rhizobium species. *Agron. J.* 57:550–53.

Amen, Ralph D. 1963. The concept of seed dormancy. *Am. Scientist* 51:408–24.

Anslow, R. C. 1962. Quantitative analysis of germination and early seedling growth in perennial ryegrass. *J. Brit. Grassland Soc.* 17:260–63.

Beevers, H. 1961. Metabolic production of sucrose from fat. *Nature* 191:433–36.

Bonner, J. F., and J. E. Varner. 1965. *Plant Biochemistry*. Academic Press, New York.

Brockwell, J. 1962. Studies on seed pelleting as an aid to legume seed inoculation. I. Coating materials, adhesives and methods of inoculation. *Australian J. Agr. Res.* 13(4):638–49.

Brown, R. 1965. Physiology of seed germination. In W. Ruhland (ed.). *Encyclopedia of Plant Physiology*. Springer Verlag, Berlin, 15(2):894–98.

Burton, J. C., and R. L. Curley. 1965. Comparative efficiency of liquid and peat-base inoculants on field-grown soybeans (*Glycine max* L.). *Agron. J.* 57:379–81.

Canode, C. L. 1965. Influence of cultural treatments on seed production of intermediate wheatgrass (*Agropyron intermedium* [Host] Beaur). *Agron. J.* 57:207–10.

Carlson, I. T. 1964. Effects of row spacing and fertilizer treatment on seed yield and related traits of "Sterling" orchardgrass. *Agron. J.* 56:565–69.

Ching, T. M., and W. H. Foote. 1961. Post-harvest dormancy in wheat varieties. *Agron. J.* 53:183–86.

Chinwuba, P. M., C. O. Grogan, and M. S. Zuber. 1961. Interaction of detasseling, sterility, and spacing on yields of maize hybrids. *Crop Sci.* 1:279–80.

Clark, Lewis E. 1965. Performance of carry-over hybrid seed corn. *Texas Agr. Exp. Sta., Progr. Rept.* 2353.

Crane, P. L. 1964. Effects of the gene *ae* on seed quality in maize. *Crop Sci.* 4:359–60.

Evanari, M. 1961. Light and seed dormancy. In W. Ruhland (ed.). *Encyclopedia of Plant Physiology*. Springer Verlag, Berlin, 15(2):804–47.

Eyre, J. V. 1931. Notes on oil development in the seed of a growing plant. *Biochem. J.* 25:1902–8.

Garrison, C. S. 1951. Producing seed for better forage crops in the United States. *Canada Seed Growers Assoc., Ann. Rept.,* pp. 21–28.

Grabe, D. F. 1967. Low seed vigor: Hidden threat to crop yields. *Crops and Soils* 19(6):11–13.

Grabe, D. F., and K. J. Frey. 1966. Seed quality and planting rates affect oat yields. *Iowa Farm Science* 20(8):7–9.

Holland, R. F. 1966. Hybrid wheat—When? *Crops and Soils* 18(5):7.

Hoveland, C. S. 1964. Some grass root extracts slow legume germination. *Crop and Soils* 16(7):18.

Howell, R. W. 1963. Physiology of the soybean. In A. G. Norman (ed.). *The Soybean.* Academic Press, New York.

Hukill, W. V. 1963. Storage of seeds. *Proc. Intern. Seed Testing Assoc.* 28(4):871–83.

Hyde, E. O. C. 1954. The function of the hilum in some papilionaceae in relation to the ripening of the seed and permeability of the testa. *Ann. Botany* (London) 18:241–56.

Ingle, J., L. Beevers, and R. H. Hageman. 1964. Metabolic changes associated with the germination of corn. I. Change in weight and metabolites and their redistribution in the embryo axis, scutellum and endosperm. *Plant Physiol.* 39:735–40.

Ingle, J., D. Beitz, and R. H. Hageman. 1965. Changes in composition during development and maturation of maize seeds. *Plant Physiol.* 40:835–39.

Ingle, J., and R. H. Hageman. 1965a. Metabolic changes associated with the germination of corn. II. Nucleic acid mechanism. *Plant Physiol.* 40:48–53.

———. 1965b. Metabolic changes associated with the germination of corn. III. Effects of gibberellic acid on endosperm metabolism. *Plant Physiol.* 40:672–75.

Jennings, A. C., and R. K. Morton. 1963. Changes in carbohydrate, protein, and non-protein nitrogenous compounds of developing wheat grain. *Australian J. Biol. Sci.* 16:318–31.

Johnson, H. W., U. M. Means, and C. R. Weber. 1965. Competition for nodule sites between strains of *Rhizobium japonicum* applied as inoculum and strains in the soil. *Agron. J.* 57:179–85.

Klostermeyer, E. C. 1964. Using alkali bees and leaf-cutting bees to pollinate alfalfa. *Wash. Agr. Exp. Sta., Circ.* 442.

Koller, D. 1964. The survival value of germination-regulating mechanisms in the field. *Herbage Abstr.* 34:1–7.

Leopold, A. C. 1964. *Plant Growth and Development.* McGraw-Hill, New York.

Lobb, W. R. 1958. Seed pelleting. *New Zealand J. Agr.* 96:556.

Nittler, L. W., G. W. McKee, and J. L. Newcomer. 1964. Principles and methods of testing alfalfa seed for varietal purity. *N.Y. State Agr. Exp. Sta., Bull.* 807.

Mertz, E. T., O. E. Nelson, and L. S. Bates. 1965. Growth of rats fed on *Opaque-2* maize. *Science* 148:1741–42.

Paleg, L. G. 1960. Physiological effects of gibberellic acid. I. On carbohydrate metabolism and amylase activity of barley endosperm. *Plant Physiol.* 35:293–99.

Percival, J. 1921. *The Wheat Plant.* Duckworth, London.

Rosenow, D. T., A. J. Casady, and E. G. Heyne. 1962. Effects of freezing on germination of sorghum seed. *Crop Sci.* 2:99.

Sarvella, P., and C. O. Grogan. 1965. Morphological variations at different stages of growth in normal, cytoplasmic male-sterile and restored versions of *Zea mays* L. *Crop Sci.* 5:235–38.

Simon, E. W., and A. Meany. 1965. Utilization of reserves in germinating *Phaseolus* seeds. *Plant Physiol.* 40:1136–39.

Smith, Dale. 1962. *Forage Management in the North,* 2nd ed. Wm. C. Brown Publishing Co., Dubuque, Iowa.

Sprague, G. F. 1936. The relation of moisture content and time of harvest to germination of immature corn. *Agron. J.* 28:472–78.

Suneson, C. A. 1962. Hybrid barley promises high yield. *Crop Sci.* 2:410–11.

Tanner, J. W., and I. C. Anderson. 1963. Investigations on non-nodulating and nodulating soybean strains. *Can. J. Plant Sci.* 43:542–46.

USDA. 1961. *Plant light—Growth response.* ARS Spec. Rept. 22.

Vincent, J. M., J. A. Thompson, and K. O. Donovan. 1962. Death of root-nodule bacteria on drying. *Australian J. Agr. Res.* 13:258–70.

Weibel, D. E. 1958. Vernalization of immature wheat embryos. *Agron. J.* 50:267–70.

Wiesner, L. E., and R. C. Kinch. 1964. Seed dormancy in green needlegrass. *Agron. J.* 56:371–73.

Works, D. W., and L. C. Erickson. 1963. Infrared irradiation, an effective treatment for hard seeds in small seeded legumes. *Idaho Agr. Exp. Sta., Res. Bull.* 57.

❧ CHAPTER ELEVEN ❧ SEEDING

❧ STAND ESTABLISHMENT by seeding is one of the primary challenges faced annually by most farmers. Prerequisites for successful stand establishment include: (1) quality seed of an adapted variety; (2) uniform seed distribution and placement; (3) firm seed-soil contact; and (4) for forages, appropriate competition from the companion crop.

Quality seed and the influence of processing on that seed quality were discussed in Chapter 10. In addition, the availability of new varieties means the farmer needs comparative information as is made possible by variety trials on most crops.

Once these considerations are met, the factors involved in seeding management may be divided into the broad areas involving mechanical and biological aspects of stand establishment. How depth of planting, seed size, seedbed texture, the companion crop, fertilization practices, and date of planting influence successful stand establishment will be discussed. In addition, seeding mixtures for cereal and forage crops and their management will be considered.

MECHANICAL FACTORS

DEPTH PLACEMENT

This factor is of prime importance although specific requirements are modified by seed size and moisture availability. Seed size is strongly a function of species; the values in Table 11.1 provide information on how several species respond to planting at different depths. Surprisingly, seeding practices for the small-seeded crops (represented by all those in Table 11.1 except corn) have sometimes been less exacting than for the large-seeded ones. The forage crops have often been sown on the

TABLE 11.1. Species variation in percent emergence from increasing depth placement (Smith, 1962)

Species	Percent Emergence			
	½ in.	1 in.	1½ in.	2 in.
Kentucky bluegrass	43	27	4	1
Timothy	89	81	39	12
Smooth bromegrass	78	69	51	24
Red clover	56	62	22	14
Alfalfa	64	53	45	19
Corn	25	70	90	90

soil surface and then covered to variable depths with a range of secondary tillage implements such as harrows, disks, and cultipackers. In contrast, the large-seeded species are given exact and accurate placement. It is obvious from this table that there is a most desirable depth to plant each crop, and seeding practices should be used that accommodate these needs.

Emergence Habit. A further separation must be made in depth placement relative to the emergence habit of the species. Species like soybeans and alfalfa, which exhibit an epigeal emergence (that is, the seed reserve is brought above ground), must be planted less deeply than the hypogeous grasses for successful emergence. The hypogeal emergence (seed reserve remains below ground) of the grass family generally allows successful emergence from a deeper depth on approximately the same food reserve. However, some species (like Kentucky bluegrass) require light for germination, and therefore shallow placement is beneficial in providing light as well as ease of emergence for this small seed. The large quantity of water which must be imbibed by a seed the size of corn means that such a seed is favored by a deeper placement into more moist soil.

Seed Size-Seedling Weight. Size of seed and seedling weight are closely correlated. Kalton et al. (1959) found that, in general, the heaviest seed within a lot produced the most vigorous seedlings. Large-seeded strains within a species emerge faster from deeper planting depths and produce better stands than small-seeded strains. An example of this is represented by vetch seed from which seedling vigor was greater as seed weight increased (Fig. 11.1). Clark and Peck (1968) make two key points in this regard. First, when large- and small-seeded lots of snap beans (a large-seeded dicotyledon) were planted with equal numbers of seeds per foot of row, the large-seeded lots outyielded the small-

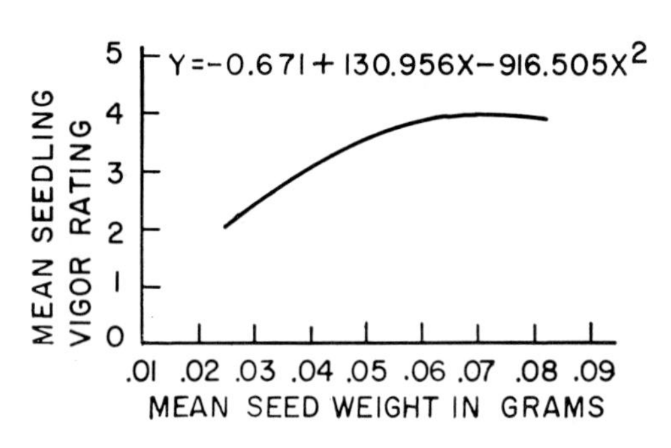

FIG. 11.1 ☘ Regression of seedling vigor on seed weight of 100 F_4 lines from *Vicia sativa* x *V. angustifolia* (field planting). The vigor rating increases with increasing seed weight (Allen and Donnelly, 1965).

seeded. If the same weight of seed were planted per row, the small-seeded rows generally outyielded the large-seeded. Secondly, if the seeds were large enough to develop numerous transverse cracks, germination was lowered; therefore, the large-seeded lots could be so large that they gave poor stands as a result of this cracking. In these few instances, large seed size gave some of the lowest yields of any seed lot size.

Classic experiments on seed size were carried out by Montgomery (1912) at Nebraska. He demonstrated that crops from small or even underdeveloped seeds gave only slightly lower yields than crops from large plump seeds (Table 11.2). Nevertheless, there was an advantage for the larger seed. Oat studies by Frey and Wiggans (1956) demonstrated that dry weight of seedlings was positively correlated with test weight. Plants from light seed were smaller even after ten weeks of growth, and while plant weights usually were equal by harvest, light oats produced light kernels. Black (1956) found that seed size gave rather linear effects on cotyledon weight and emergence. He observed that an increasing depth of planting had an influence on cotyledon weight but did not influence cotyledon area (Table 11.3). Dry weight at emergence appeared to be a function of planting depth and seed size; cotyledon area at emergence was primarily a seed size factor.

It is obvious that deep placement of most small-seeded forage crops is unsatisfactory. One result of this is that common agricultural

TABLE 11.2. Yield in relation to seed size in grams per 100 plants (Montgomery, 1912)

	Wheat	Oats
Grown alone		
Small	219	117
Large	236	129
Grown together		
Small	187	96
Large	234	125

TABLE 11.3. Effect of depth of sowing and seed size on seedling weight and cotyledon area of subterraneum clover (Black, 1956)

Depth of Sowing (in.) (large seed)	Weight (mg)	Area (mm²)
½	3.3	16.4
1¼	2.9	16.3
2	2.5	16.3
Seed size (planted at ½" depth)		
small	1.2	7.8
medium	2.1	11.8
large	3.3	16.8

practice with forages and small grains is to plant much more seed (e.g., 4–5 times more) than the farmer expects to have in the final population. In contrast very small losses are anticipated with seeding rates used for cereals like corn, as shown in the following comparison of corn and alfalfa.

	Suggested Seed/Acre	Expected Final Stand (plants)	Loss Expected
Corn	20,000	16,000–18,000	20%
Alfalfa	2,250,000	435,000	80%

To summarize the relationship of seed size and planting depth, it must be concluded that larger seed is an advantage only in establishing a stand under less than optimum conditions. In fact, vigor is defined by the seed worker as all attributes that favor stand establishment under unfavorable conditions. Embryo size likewise can be related to the ability to emerge vigorously and may be associated with the initial photosynthetic area for legumes (i.e. cotyledon size), but embryo size is of relatively little consequence after emergence. Only under optimum spacing can relative differences between plants of different seed size be maintained; it is therefore concluded that under conditions of the standard population for an economic species, seed size is significant only as it influences successful stand establishment under stress.

SEEDBED TEXTURE AND MOISTURE RELATIONSHIPS

Seedbed Surface. The soil particles in the seedbed should be left large enough to minimize any crust formation and to maximize aeration which in turn can promote favorable soil temperature development. This means the soil should not be worked too finely, or if it is naturally fine and prone to crust, some artificial treatment may be needed to provide the above attributes. For example, considerable work has been done with petroleum as a mulch. Petroleum significantly increases soil temperature but does not always increase cotton yields. Waxes may be even less expensive than petroleum and are easily liquified for ease of application. Polyethylene strips between corn rows have given yield increases and earlier maturity on certain crops. For example, Willis et al. (1963) observed these additional benefits of black polyethylene film on corn in North Dakota: reduced evaporation and conserved moisture, increased temperature, increased germination and seedling growth, and weed control. As a result, yields were also increased.

Seed-Soil Contact. Seed-soil contact is important for rapid water imbibition. A pulverized seedbed contributes to this contact, and as an auxiliary practice, seed firming or soil firming press wheels are useful. Stickler (1964) found that press wheels increased stands of grain sorghum 14% and yields 8% in Kansas. When Stickler and Fairbanks (1965) studied the minimum tillage practice of seeding grain sorghum in wheat stubble, surface firming press wheels gave the larger stand percentage increase when used alone, but the seed firming effects (Fig. 11.2) were additive to the soil firming action and further improved the stand. Many planters now incorporate one or both of these firming techniques.

In order to get the seed into moist soil or to place the seed in soil

FIG. 11.2 ☘ Three types of press wheels. The two on the left firm the soil; the one on the *right* firms both the seed (with the *center rib*) and the soil (Photo courtesy Allis-Chalmers Company).

beneath a sod surface, furrow openers may be used. A sweep-shaped furrow opener may move aside an inch or two of dry soil in front of the planter shoe to assure placement of the seed in dry soil. For grassland drills such furrow openers are much superior to single- or double-disk furrow openers when used on unplowed ground where the disk may ride up over the residue. These furrow openers are commonly used in conjunction with a seed firming wheel.

BIOLOGICAL FACTORS

COMPANION CROP

Stand establishment of forage crops has commonly included the use of a companion crop such as one of the small grains (oats, wheat, or barley) or flax. From time to time new crops are considered for use as a companion crop, such as sorghum for alfalfa (Scott and Patterson, 1962), corn for alfalfa, or berseem clover for alfalfa (Nelson et al., 1965). Corn planted in 60–80-inch rows has been used for the establishment of red clover and alfalfa. Leaving the crops in the field for a full season of competition reduced the forage stand levels and provided less than optimum yields of corn. Berseem clover was erratic as a companion crop; in some cases good stands of alfalfa were achieved, but in others the weed growth was extensive and a poor alfalfa stand resulted.

Competition for Light. The companion crop is expected to give weed control as well as erosion control and a dollar return in the seeding year. The expectation is to obtain weed control by "controlled competition"; that is, enough light is intercepted by the companion crop to reduce weed growth, but sufficient light filters through to allow the forage seedling to develop. As described in Figure 11.3, the various companion crops differ widely in the intensity and timing of their competition with weeds.

This competition is primarily for light but may also be for nutrients and water. That the competition is primarily for light is supported by the values in Table 11.4. The total dry matter production varies relatively little whereas percentage composition of the three components of the stand (legume, weeds, oats) varies widely. Also note that while legume yields are not statistically different, the trend favors a moderate to low seeding rate of oats for more legume dry matter production.

Light competition has been described by several workers. For

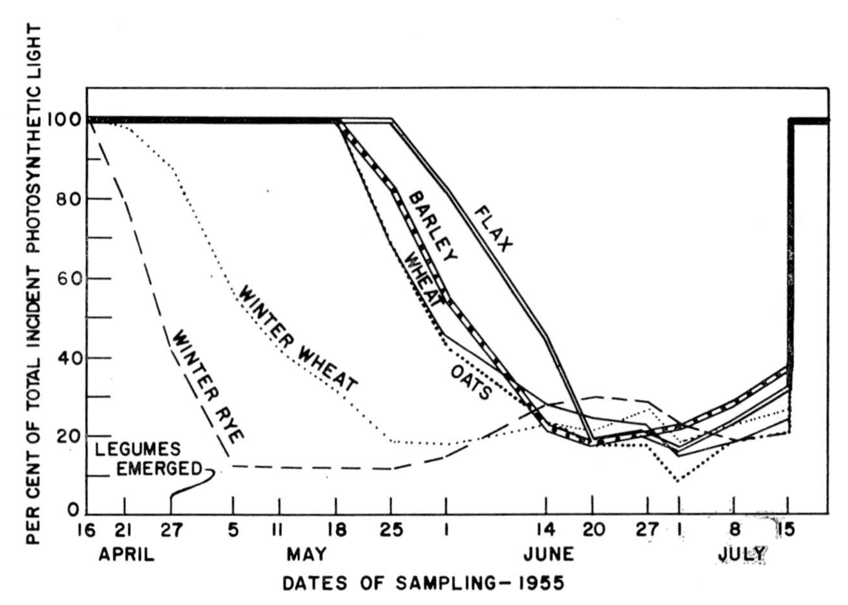

FIG. 11.3 ❧ Average percentages of total incident light effective in photosynthesis that penetrated through six different companion crops to 4 inches above the soil surface from mid-April until all companion crops were harvested at maturity on July 15 (Klebesadel and Smith, 1959).

example, the total dry weight of alfalfa was affected more by shading than was height, root length, or root to top ratio, while root dry weight was much more sensitive to shading than top dry weight (Matches et al., 1962). Vertical and lateral root growth of alfalfa, birdsfoot trefoil, and orchardgrass were all reduced when established with a companion crop (Cooper and Ferguson, 1964). These observations reemphasize the ideas presented in the discussion on shoot-root ratios (Chapter 9) that competition for light would have a relatively greater influence on root than on shoot dry matter production.

TABLE 11.4. Dry matter production of several components of a forage seeding mixture (Smith, 1962)

Oats Seeded	Pounds of Dry Matter Produced per Acre			
(bu/acre)	Legume	Weeds	Oats	Total
0.5	199	2102	2289	4391
1.0	239	1102	3177	4279
1.5	140	849	3525	4374
2.5	150	429	4066	4495
LSD: 5%	N.S.	338	829	N.S.

Managing the Companion Crop. In many cases it is desirable to reduce the competition provided by the companion crop. Several methods are available for doing this and yet obtaining the weed control benefit of the companion crop. For example, in a crop such as wheat, the farmer may have the choice between a fall-seeded and a spring-seeded variety. Since the forage seeding must be made in the spring (forage seedlings cannot get established sufficiently if planted as late in the fall as a winter cereal), the spring-seeded companion crop provides less competition. Secondly, the companion crop selected should be as lodging-resistant as possible and then nitrogen fertilizer used sparingly to minimize the chance of oversucculent stems lodging. If the companion crop does lodge, it may smother the underseeded forage and therefore should be removed as quickly as possible. Thirdly, where the underseeded forage is known to be very sensitive to light competition, early removal of the companion crop is desirable, because this allows the companion crop to supply competition and reduce weed growth early in the season and yet not compete too long with the seedling. An optimum time to remove the small-grain companion crop is just before the head emerges ("boot stage"). As an example, if oats are removed when they are 12–16 inches tall, the light four inches above the ground is never less than 45% of incident light. Therefore, removal of oats as hay or pasture provides good conditions for forage stand establishment and weed control.

The results from a western Canada experiment (Table 11.5) re-

TABLE 11.5. Dry matter yields (tons/acre) of grass-legume mixtures in year following seeding (Baenziger, 1966)

	Establishment Method*					
	A		B		C	
Mixture	1st cut	2nd cut	1st cut	2nd cut	1st cut	2nd cut
Brome-alfalfa	2.46	2.31	1.77	1.89	1.09	1.88
Brome-trefoil	2.34	2.07	1.80	1.63	1.09	1.90
Crested wheatgrass-alfalfa	3.07	1.71	1.77	1.41	1.15	1.25
Crested wheatgrass-trefoil	2.98	2.04	1.81	1.59	1.16	1.50
Int. wheatgrass-alfalfa	3.06	1.93	1.85	1.79	1.13	1.59
Int. wheatgrass-trefoil	2.88	2.04	1.97	1.78	1.27	1.53
Timothy-alfalfa	2.53	1.24	1.90	1.54	1.76	1.81
Timothy-trefoil	2.75	1.58	2.06	1.57	1.66	1.76
Average for cuts	2.76	1.87	1.87	1.65	1.29	1.65
Average for season	4.63		3.52		2.94	

* A = Seeded without companion crop.
B = Seeded with a companion crop and clipped three times through the season.
C = Seeded with a companion crop, harvested as silage.

emphasize that many forage species benefit by the absence of a companion crop or by removing it early (Baenziger, 1966). The highest yield obtained was for crested wheatgrass and alfalfa seeded without a companion crop. Yields for all the forages were higher after establishment without a companion crop. This work does not provide a comparison for the case in which the companion crop was grown for grain; presumably that would have given the lowest forage yield of all if the trends in the table continued. These results suggest there was little need for weed control, whereas in other management situations in which weeds were a greater problem, the three clippings or silage harvest might have been optimum.

ALTERNATIVE MEANS

Fall Seeding. Two possible alternatives to the use of a companion crop for forage establishment are available for some species. First, if the species or variety is sufficiently winter-hardy, it may be seeded in the fall without a companion crop. This practice can be used in the north central states for Kentucky bluegrass, alfalfa, and bromegrass. It is not satisfactory for species which have weak seedlings (e.g. birdsfoot trefoil) or limited winterhardiness (e.g. orchardgrass).

Herbicides. A second alternative of increasing usefulness is the application of selective herbicides. The development of reliable chemicals has been somewhat slower for forages than for row crops, but there is a growing group of possible herbicides for forage stand establishment. In many cases small differences in metabolism between a broad-leaved weed (i.e. pigweed, lambsquarters) and a broad-leaved forage (i.e. alfalfa, birdsfoot trefoil) make possible the use of selective herbicides such as 2,4-DB (see Chapter 13). However, the mechanism of selectivity is a very delicate one and requires great care in timing of application and dosage rate if the end result is to be satisfactory. Increasing numbers of stand establishments have been achieved using TCA (trichloroacetic acid), dalapon, and EPTC (ethyl N, N-dipropylthiol carbamate) for grassy weeds and 2,4-DB (4-2,4-dichlorophenoxy butyric acid) for broad-leaf control (Peters, 1964; Vengris, 1965; Wakefield and Skaland, 1965). However, one unexpected result of the Wakefield and Skaland study was that while first-year yields were improved by the herbicide, second-year yields were highest for the unweeded plots. This sort of variability has been observed elsewhere and suggests that the major benefits of herbicide use must be obtained during the year of establishment. Two- and three-ton yields of hay achieved the seeding year with herbicide establishment does provide promise for this practice.

BAND FERTILIZATION

Considerable discussion has been directed toward the desirability of placing a band of fertilizer near the seed drill row when planting forage crops (Fig. 11.4). Using a modified drill, fertilizer may be banded directly beneath the forage seed, a practice designed to favor early interception of the fertilizer by the downward growth of the root. This "band seeding" practice results in dry matter increases in the seeding year, but during the second year there are frequently no measurable differences between band-fertilized and broadcast-fertilized stands.

DATE OF PLANTING

Agronomists and practicing farmers have evaluated the effect of planting date on yield for many years. In temperate climates the

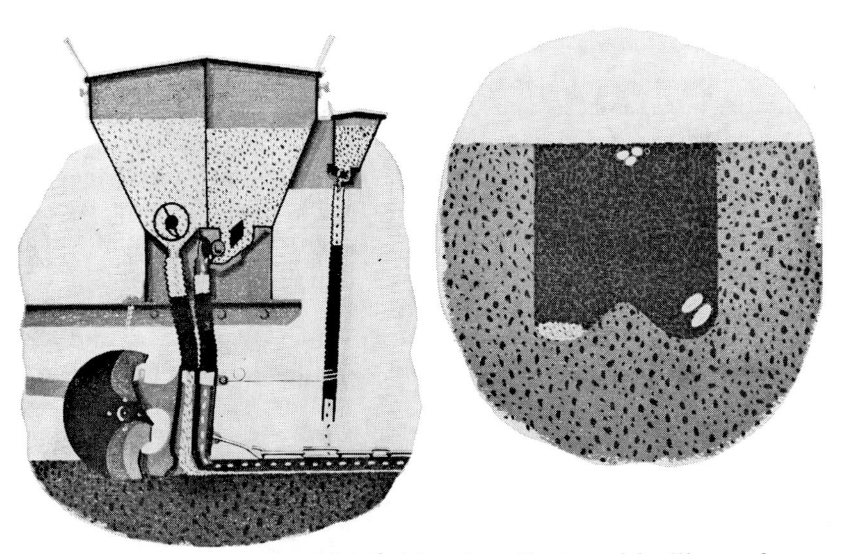

FIG. 11.4 ✤ A band application of fertilizer and seed. The fertilizer is level with the cereal seed (right) and below the forage seeding (Photo courtesy Allis-Chalmers Company).

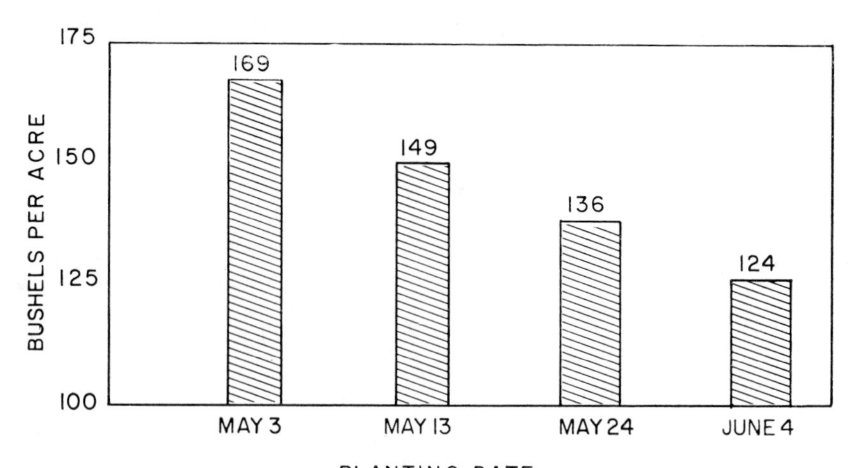

FIG. 11.5 ❧ Influence of early planting on corn yields under high fertility (Pendleton and Mulvaney, 1966).

farmer seeks to utilize as much of the warm season as possible. The date he plants a crop is modified by the danger of frost, soil temperature, and the danger of soil microorganisms damaging the seed, as well as the problem of weeds growing more rapidly than the economic crop early in the season. If these factors can be controlled, earlier dates of planting result in higher yields and a drier crop earlier, as indicated in Figure 11.5 and Table 11.6.

The values in Table 11.6 indicate that while emergence takes longer at an earlier date of planting, the plant silks earlier (thereby escaping a drouth stress frequently occurring in late July and early August) and results in a higher total yield. It is pertinent to note that the rate of kernel moisture decline, while lower on any given date for the earlier-planted material, is much more rapid for the late-planted material. Earlier-planted corn is shorter than late-planted corn and not only gives a greater yield of both grain and silage but also has a higher

TABLE 11.6. Effect of different times of planting upon yield of a full-season corn (1961–65 average yields). (Iowa OEF Report 65-29, 1965)

Planting Date	Date 75% Emerged	Date 75% Silked	% Kernel H_2O on			Yield (bu/acre)
			Sept. 13	Oct. 4	Oct. 25	
5–2	5–14	7–23	38	29	19	130
5–10	5–21	7–25	41	31	20	128
5–20	5–30	7–31	48	36	23	120
5–30	6–7	8–5	55	41	27	111

percentage of grain. Finally, there is a positive interaction between early date of planting and high fertility level (Fig. 11.5). This combination provides the maximum return for the two practices.

Recommending early dates of planting for a given region often raises the question of what will happen if the crop is frosted. As indicated in Table 11.7, frost-damaged corn still outyielded late replanted material; in the case of soybeans no yield difference occurred. Unless the crop is planted much too early, the advantages far outweigh the drawbacks.

FROST SEEDING

Frost seeding has been advocated off and on over the years and received renewed emphasis recently by Stoskopf et al. (1967) when a barley crop frost-seeded April 4 gave a good yield under Ontario conditions. This procedure uses the freezing and thawing of the soil to work the seed into the soil surface, and it allows the opportunity for seeding while the ground will support machinery. It works well for cool-season crops that outcompete weeds for available light and nutrients by starting before the weeds and thus shading out the weeds. For best results the farmer should plan to use high-quality seed treated with an insecticide and a fungicide. The soil should have a complete fertilizer applied in the fall and then disked to leave the surface rough to hold moisture. A topdressing of nitrogen in the spring completes the fertilization.

Interest in getting spring crops started earlier has caused some researchers and farmers to try fall seeding either with extra pesticides added or with a plastic seedcoat added that would survive the winter and allow imbibition only in the spring. No practical successes have

TABLE 11.7. Frost damage studies on corn and soybeans*

	Stand	Silking Date	% H_2O at harvest	Yield (bu/acre)
Corn				
Frozen	16,300	8–6	33.5	85
Replanted	18,700	8–11	39.5	72
Soybeans				
Frozen				25
Replanted				25

* A freezing temperature May 29–30 of 27° F. Corn planted May 11 was 4–5 inches tall, killed to ½ inch below ground surface with a 13% stand loss. Part of the field was replanted June 1. Soybeans planted May 14; 10% were killed by the frost. Part of the field was replanted June 1. (Iowa OEF Report 65-11) Similar data favor early planting for other spring-seeded crops (Leonard and Martin, 1963).

been achieved to date. These ideas may yield useful results sometime in the future, but an early spring planting on a properly prepared seedbed as discussed above will likely continue as a key management practice for many years.

SEEDING MIXTURES

CEREAL AND OIL SEED CROPS

Most species grown for their reproductive (seed) yield are commonly grown in a pure stand (e.g. corn, sorghum, soybeans, wheat, oats, barley, rice). Interest regularly reappears in small-grain production to use mixtures composed of oats and barley, oats and wheat, or a mixture of all three. The majority of quantitative studies report no increase in the yield of nutritive value per acre from such mixtures, and no advantage in yield is shown for mixing two varieties together (McKee and Pfeifer, 1964; Table 11.8). The mixture does not outyield the best variety.

One of the major problems in growing two species or two varieties together for seed is to get them to mature at the same time. Further, there is no complementarity arising from growing a combination of grains together, as is found in the symbiotic benefits from the mixture of a legume and a grass (see below). An additional reason for pure stands of seed crops is that they are commonly sold for industrial processing. The seed mixture would be undesirable to the manufacturer; as examples, soybeans containing corn or wheat mixed with rye are unsuitable for processing and lower in market value.

FORAGE CROPS

When these species are grown for seed or dehydration, pure stands are again most desirable. In certain unique situations where

TABLE 11.8. Yield of mixtures of oat varieties compared to a pure stand

Percent of Variety or Mixture	Yield (bu/acre)		
	1960	1961	Average
100% Clinton	40	74	57
95% Clinton: 5% Garry	44	75	60
50% Clinton: 50% Garry	48	75	62
5% Clinton: 95% Garry	52	82	67
100% Garry	54	82	68
LSD: 5%	9	N.S.	

wet soils make legume production difficult or where bloat is an extreme problem, pure stands of grasses may be utilized for forage production. However, more profitable results usually occur with a mixture of grass and legume. The legume can host bacteria to fix nitrogen for both species, and the grass provides a more complete ground cover, thus reducing the danger of frost heaving. The mixture provides a higher quality product than a pure stand because the presence of the legume increases protein content while the grass enhances palatability. In most cases grass-legume mixtures in which 30–40% or more are legume plants will give yields equivalent to pure grass stands heavily fertilized with nitrogen (Wedin et al., 1965), and the mixture has the advantage of requiring no expenditure for nitrogen. It should be pointed out that under well-drained, very well fertilized conditions in the north central states, some alfalfa varieties give yields in pure stands equal to mixtures, but the pure alfalfa does not have as much livestock acceptability as a mixture.

For use by grazing livestock (or where a herd is to be fed green chop forage) the grass-legume mixture offers a longer production season and more uniform production through the growing season. The grass starts early, commonly slows in production during midsummer, and then produces well again in the fall. The legume provides a much better midsummer production. A pure grass stand with four applications of nitrogen through the season will have an improved distribution of yield and greater dry matter production, but midsummer production is frequently not as good as with a legume present. Conversely, the legume does not provide as much early season production. For direct harvest by grazing, the mixture offers the advantage of reducing bloating problems over pure legume stands. If several species in separate fields are to be used to extend the grazing season, they may be used effectively as schematically described in Figure 11.6. Such a pasture calendar and carrying capacity prediction is a key part of quality pasture management.

PRINCIPLES OF PREPARING MIXTURES

The three primary bases on which to build a forage seeding mixture are: (1) simplicity, (2) similar maturity, (3) use to be made of the seeding.

Simplicity. For most conditions the first principle of preparing a seeding mixture is simplicity, not complexity (Volkart, 1929; Blaser et al.,

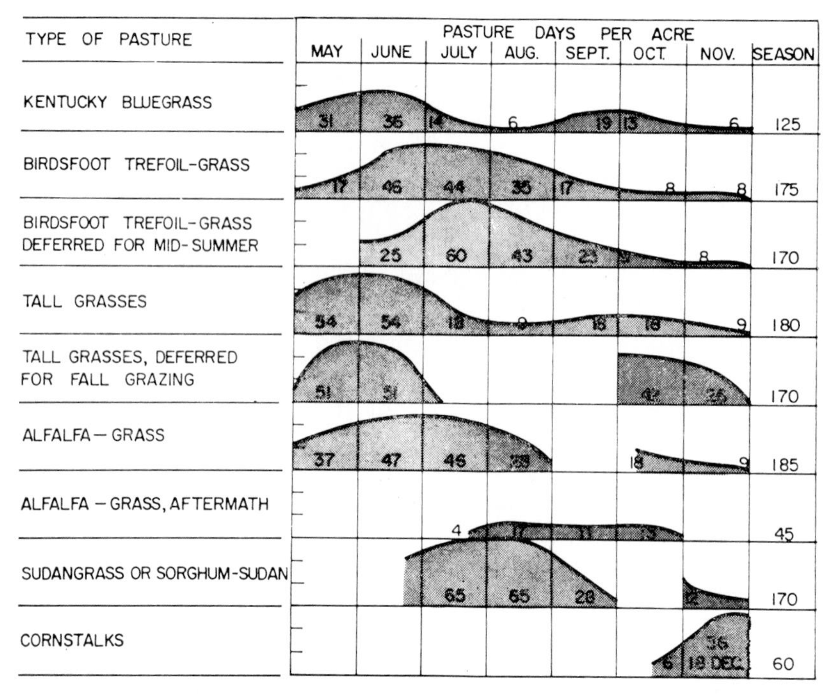

FIG. 11.6 ❧ A pasture calendar describing the carrying capacity (cow-days per month) for several species (courtesy F. W. Schaller and W. F. Wedin).

1952, 1956). The ratio of seeds in the most desirable mixture will vary according to latitude, altitude, date of planting, and the relative aggressiveness of the species included.

Similar Maturity. As nearly as possible, plants in the mixture should mature together. If soil conditions are uniform, one grass and one legume will suffice. If drainage or fertility problems are present, two legumes and/or two grasses may be needed.

Use. Selection of the species should be done according to the use to be made of it. For example, alfalfa is an excellent hay crop while ladino clover is an excellent component of a pasture mixture. Orchardgrass is good for pasture or silage use but matures early and most commonly before good haymaking conditions are present. Bromegrass, a later-maturing species, is more suitable for hay but is not as good as orchardgrass for pasture because the brome has a distinct midsummer production slump. As an example of what might be used

ABLE 11.9. Possible forage mixtures

rpose	Rotation	Species Chosen	Rejected
Hay	corn (C), oats (O) meadow (M)	1. Alfalfa 2. Alfalfa, Timothy 3. Red Clover, Timothy	1. Orchardgrass—matures too early 2. Ladino Clover—hard to mow, cures black 3. Brome—slow start
Hay	CCOMM	1. Alfalfa, Brome	1. All above 2. Timothy—short lived 3. Red Clover—short lived
Pasture	COM	1. Alfalfa, Orchardgrass (OG) 2. Red Clover, OG, Ladino	1. Brome—too slow 2. Bluegrass—too slow to start
Pasture	COMMM	1. Vernal Alf., OG, Ladino 2. Alf., Brome, Ladino	1. Red Clover—short lived 2. Timothy—short lived
Permanent	Hay	1. Viking Birdsfoot Trefoil (BFT), Timothy, & Brome 2. Vernal Alf., Brome	
Permanent	Pasture	1. Empire BFT, Bluegrass 2. Alfalfa, Brome	
Green manure		1. Sweet Clover 2. Southern Common Alfalfa	

for different purposes, a few forage mixtures are described in Table 11.9.

In general, the legume should be favored in the stand in eastern United States, the grass in the West. The cool-season grasses grow very vigorously and give the legume excessive vertical competition (shading) in the East. In addition the legumes are commonly more sensitive to grazing pressures. In the West somewhat drier and warmer conditions favor the legume relatively more.

MANAGING MIXTURES

While the several factors noted above and additional special ones play a role in mixture selection, it must be emphasized that the ensuing management is as important in determining the continuing nature of the stand as was the original mixture.

Grazing Management. Martin Jones, in work reported from the Rothamsted station in England in 1933, provided an example of how grazing management shifted the composition of a pasture mixture. His work has been supported many times by later researchers. He reported as follows:

> The following mixture was seeded per acre for pasture use: 24 pounds perennial ryegrass; 6 pounds roughstalked meadow grass; 2

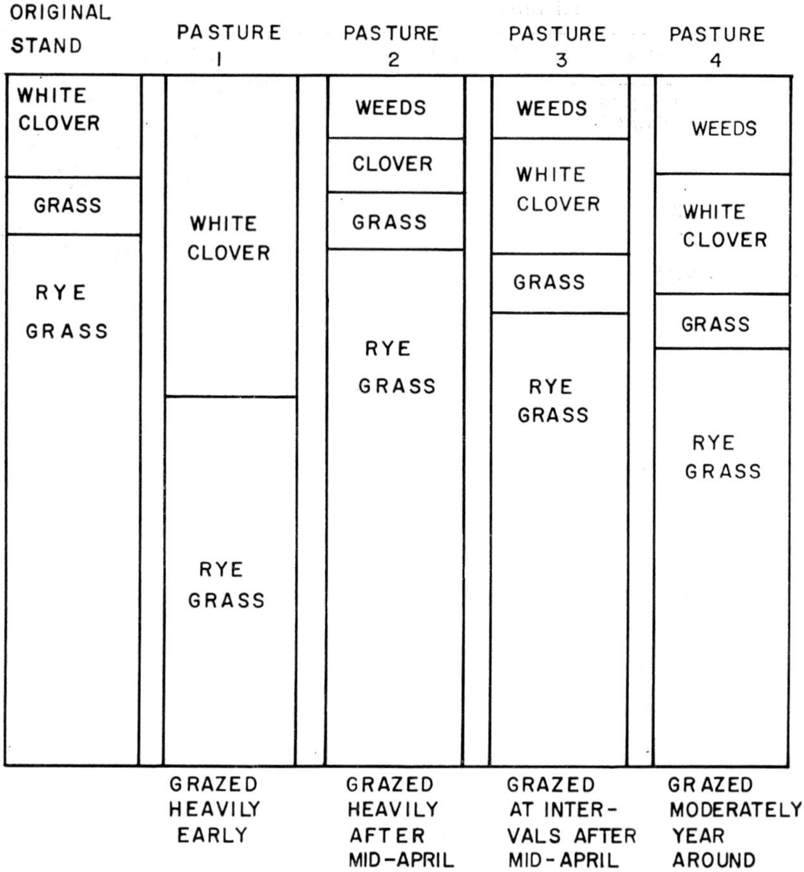

ORIGINAL STAND
WHITE CLOVER
GRASS
RYE GRASS

PASTURE 1
WHITE CLOVER
RYE GRASS
GRAZED HEAVILY EARLY

PASTURE 2
WEEDS
CLOVER
GRASS
RYE GRASS
GRAZED HEAVILY AFTER MID-APRIL

PASTURE 3
WEEDS
WHITE CLOVER
GRASS
RYE GRASS
GRAZED AT INTER-VALS AFTER MID-APRIL

PASTURE 4
WEEDS
WHITE CLOVER
GRASS
RYE GRASS
GRAZED MODERATELY YEAR AROUND

FIG. 11.7 ☙ Grassland management and its influence on the composition of the pasture mixture (Jones, 1933).

pounds white clover. After two years the pasture was divided into four sections and each grazed with sheep as follows:

Pasture 1—heavily stocked through the season but not in winter.
2—not grazed before mid-April, lightly thereafter.
3—not grazed before mid-April, but grazed bare each time and then left to rest one month.
4—grazed as customary in the area; i.e. the year around stocking rate varied by a factor of 2, production of forage by a factor of 20.

The stands after one year of the above treatments are shown as percentage of composition in Figure 11.7. The fact that the sward was grazed closely in March, April, and May (Pasture 1) seriously weakened the ryegrass in its critical period in the spring. White clover, on the other hand, suffered little under such grazing. Grazing the year around (Pasture 4) allowed more weeds to encroach upon the stand.

Fertilization. Fertilization is a second key management factor in maintaining satisfactory grass-legume mixtures once they are established (MacLeod, 1965). Nitrogen fertilization at rates in excess of 50 pounds per acre caused a decrease in the alfalfa content of mixtures and changed the mixture, but the percentage change differed depending on the grass species used. Potassium applied alone increased alfalfa content but did not prevent a reduction in the percent of contribution of alfalfa with heavy nitrogen fertilization. MacLeod reports that a competition index (yield in mixture/yield in pure stand) indicates orchardgrass was the most competitive associate (1.5) for alfalfa, brome next (1.2), and timothy third (0.8).

In summary, the nature of the species mixed together influences the original population of a forage seeding mixture, but this may be modified significantly by grazing management and fertilization. ❧

LITERATURE CITED

Adams, J. E. 1965. Effect of mulches on soil temperature on grain sorghum development. *Agron. J.* 57:471–74.

Allen, L. R., and E. D. Donnelly. 1965. Effect of seed weight on emergence and seedling vigor in F_4 lines from *Vicia sativa* L. x *V. augustifolis* L. *Crop Sci.* 5:167–69.

Baenziger, H. 1966. Forage stands hurt by companion crop. *Crops and Soils* 18(7):20.

Black, J. N. 1956. The influence of seed size and depth of sowing on pre-emergence and early vegetative growth of subterraneum clover *(Trifolium subterraneum* L.). *Australian J. Agr. Res.* 7:98–109.

Blaser, R. E., W. L. Griffith, and T. H. Taylor. 1956. Seedling competition in compounding forage seed mixtures. *Agron. J.* 48:118–23.

Blaser, R. E., W. H. Skrdla, and T. H. Taylor. 1952. Ecological and physiological factors in compounding forage seed mixtures. *Advan. Agron.* 4:179–219.

Burris, R. H. 1965. Nitrogen fixation. In J. Bonner and J. E. Varner (eds.). *Plant Biochemistry.* Academic Press, New York.

Clark, B. E., and N. H. Peck. 1968. Relationship between the size and performance of snap bean seeds. *N.Y. State Agr. Exp. Sta., Bull.* 819.

Cooper, C. S., and H. Ferguson. 1964. Influence of barley companion crop upon root distribution of alfalfa, birdsfoot trefoil and orchardgrass. *Agron. J.* 56:63–66.

Frey, K. J., and S. C. Wiggans. 1956. Growth rates of oats from different test weight seed lots. *Agron. J.* 48:521–23.

Iowa OEF Report 65–29. 1965. Northwest Iowa Experimental Farm.
Jones, M. G. 1933. Grassland management and its influence on the sward. *J. Roy. Agr. Soc. Eng.* 94:21–41.
Kalton, R. R., R. A. DeLong, and D. S. McLeod. 1959. Cultural factors in seedling vigor of smooth bromegrass and other forage species. *Iowa State J. Sci.* 34:47–80.
Kneebone, W. R., and C. L. Cremer. 1955. The relationship of seed size to seedling vigor in some native grass species. *Agron. J.* 47:472–77.
Klebesadel, L. J., and Dale Smith. 1959. Light and soil moisture beneath several companion crops as related to the establishment of alfalfa and red clover. *Botan. Gaz.* 121:39–46.
Leonard, W. H., and J. H. Martin. 1963. *Cereal Crops.* Macmillan Company, New York.
MacLeod, L. B. 1965. Effect of nitrogen and potassium on the yield, botanical composition, and competition for nutrients in three alfalfa-grass associations. *Agron. J.* 57:129–34.
Matches, A. G., G. O. Mott, and R. J. Bula. 1962. Vegetative development of alfalfa seedlings under varying levels of shading and potassium fertilization. *Agron. J.* 54:541–43.
McKee, G. W., and R. P. Pfeifer. 1964. Oat mixtures give no yield advantage. *Crops and Soils* 16(9):22.
Montgomery, E. G. 1912. Competition in cereals. *Nebr. Agr. Exp. Sta., Bull.* 127.
Nelson, C. J., A. R. Schmid, and C. H. Cuykendall. 1965. Performance of berseem clover (*Trifolium alexandrium* L.) as a companion crop. *Agron. J.* 57:537–39.
Pendleton, J. W., and D. L. Mulvaney. 1966. The early word in planting corn early. *Crops and Soils* 18 (5):11–12.
Peters, E. J. 1964. Preemergence, preplanting and postemergence herbicides for alfalfa and birdsfoot trefoil. *Agron. J.* 56:415–19.
Phillips, R. E., and R. E. Frans. 1965. Effect of petroleum mulch on growth and yield of cotton in eastern Arkansas. *Ark. Agr. Exp. Sta., Bull.* 699.
Poehlman, J. M. 1959. *Breeding Field Crops.* Holt, Rinehart and Winston, Inc., New York.
Scott, W. O., and F. L. Patterson. 1962. Grain sorghum as a companion crop for alfalfa. *Agron. J.* 54:253–56.
Smith, D. 1962. *Forage Management in the North.* 2nd ed. Wm. C. Brown Co., Dubuque, Iowa.
Stickler, F. C. 1964. Stand establishment and yield of grain sorghum as affected by method of planting and use of press wheels. *Agron. J.* 56:53–56.
Stickler, F. C., and G. E. Fairbanks. 1965. Grain sorghum stands and yields as affected by tillage methods and use of press wheels. *Agron. J.* 57:497–500.
Stoskopf, N. C., E. Reinbergs, and G. E. Jones. 1967. Frost seeding. *Crops and Soils* 19:12–13.
Vengris, J. 1965. Use of grass-killing herbicides in establishing grass-legume mixtures. *Agron. J.* 57:59–61.
Volkart, A. 1929. The principles of compounding mixtures of grass and clover seeds. *Sci. Agr.* 9:510–21.

Waggoner, P. E., P. M. Miller, and H. C. DeRoo. 1960. Plastic mulching: Principles and benefits. *Conn. Agr. Exp. Sta., Bull.* 634.

Wakefield, R. C., and N. Skaland. 1965. Effects of seeding rate and chemical weed control on establishment and subsequent growth of alfalfa (*Medicago sativa* L.) and birdsfoot trefoil (*Lotus corniculatus* L.). *Agron. J.* 57:547–50.

Wedin, W. F., J. D. Donker, and G. C. Marten. 1965. An evaluation of nitrogen fertilization in legume-grass and all-grass pastures. *Agron. J.* 57:185–88.

Willis, W. O., H. J. Haas, and J. S. Robins. 1963. Moisture conservation by surface or subsurface barrier and soil configuration under semiarid conditions. *Soil Sci. Soc. Am. Proc.* 27:577–80.

❧ CHAPTER TWELVE ❧ WINTER AND DROUTH SURVIVAL OF CROP PLANTS

❧ THE WIDE RANGE of crop-growing conditions which exist in agricultural areas of the world require that the plant survive short or long periods of cold or moisture stress. The forage crop producer and the winter cereal producer especially are faced with keeping a crop alive through winter and drouthy periods.

Under Midwest conditions an average of 5% of the perennial forage stands are lost each winter. Losses in a particular state may range from 10 to 35%, with as much as 75% losses occurring in some localized areas. Management practices may be adjusted to aid survival, and such practices will be considered later in the discussion. However, as a first step, the basic plant responses to these two stress conditions will be described.

WINTERHARDINESS

Smith (1962) states that "plant losses during the winter usually are the result of temperature in association with free water in or on the soil or in the plant itself." This observation provides a very useful basis for evaluating the wide range of conditions which cause winter injury to a plant. These conditions will be classified under two major headings: climatic conditions and soil conditions.

CLIMATIC CONDITIONS

Plants exhibit a broad range of responses to low temperature, and the responses may or may not be confounded with other environmental factors. Chilling injury is used to describe temperatures above the freezing point of water which affect plant metabolism adversely. For example, Bermudagrass exhibits a loss of chlorophyll under 40° F. Sorghum, corn, sugarcane, and many of the tropical forages may exhibit a yellowing of leaves, a necrosis of leaf parts, and finally leaf loss. In some tropical and subtropical plants death may occur at these temperatures. Sellschop and Salmon (1928) provide a detailed classification for many crop plants and their response to chilling.

Freezing Injury. Freezing injury is related to plant temperatures below 32° F. It is well established that the freezing damage is caused by the formation of ice crystals within the plant. Numerous theories have been advanced to explain why this injury may occur in some cases and not in others. An early hypothesis suggested that ice forms internally in the cell, thus causing the cell to rupture, but currently this is considered to be a minor factor. Levitt (1956) states that "tissues actually contract instead of expanding on freezing" and he observed that ice more commonly forms outside the cell wall. The protoplasm is well adapted to undercooling because it is small and protected by a lipid layer which prevents "seeding" of the aqueous phase of the protoplast. When water moves out to the intercellular layer, the concentration of cell fluid increases and the cell can withstand lower and lower temperatures. The most commonly accepted hypothesis for freezing injury relates to the following two theories: (1) protein precipitation and a coagulation of the protoplasm that is irreversible; or (2) formation of ice crystals that subject the protoplasm to stresses and strains that lead to mechanical injury. The protein precipitation theory is supported by microscopic observations of plasmolysis and by physiological measurements of impaired metabolism.

Levitt (1962) proposes that freezing injury is finally due to an unfolding and denaturation of the protoplasmic proteins. This results from the formation of intermolecular disulfide bonds (S-S) induced by the close approach of the protein molecules during severe dehydration associated with intercellular ice formation. Conversely frost resistance is a resistance to sulfhydryl (SH) oxidation and SH ⟷ S-S interchange. Thus frost resistance in his terms is the resistance to the formation of S-S bonds and subsequent protein denaturation. More recent work by Levitt's group continues to support this hypothesis, although work by Schmutz (1962) raised certain questions about this concept.

As an end result of the several factors discussed above, winter

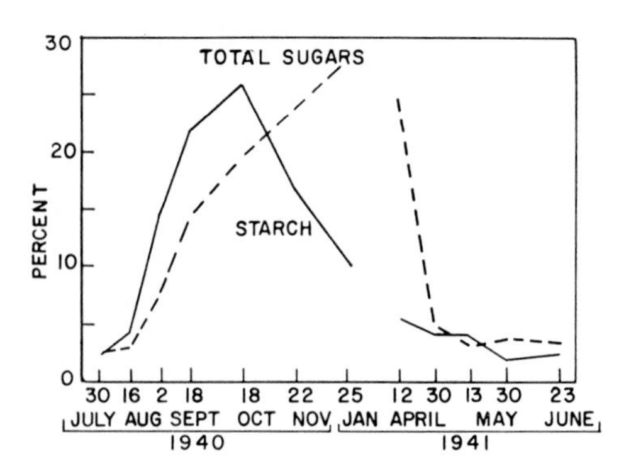

FIG. 12.1 ☙ Changes in the sugars (– –) and starch(—) contents of the roots of biennial sweetclover during the seedling and second years of growth in Wisconsin (after Smith, 1964).

injury causes the cell to exhibit a loss of differential permeability and ultimately to lose its turgidity.

Plant Response to Cold Temperature. Plant responses which reduce freezing injury include the following: an increase in sugar concentration, an increase in the osmotic potential of the cell sap, increased permeability of the plasma membrane, an increase in soluble proteins, and an increase in bound water.

As a first step, a high level of carbohydrates accumulated by the plant during the fall provides the necessary constituents for hardiness. With an adequate level of carbohydrate, these major steps occur in the fall hardening process.

1. Starch and other long-chain molecules go to sugar, thus forming a larger number of molecules (Fig. 12.1). This increases the osmotic potential in the cell, especially in the vacuole. Thus water is held more strongly in the vacuole.
2. Simultaneously, there is an increased synthesis of protein in the protoplasm which results in an increased binding of water and the reduction of free water. The protein acts as a hydrophilic colloid and water thus bound on its surface resists crystallization by freezing.
3. There is an increased permeability of the cell membranes which supposedly aids in a decrease of total water content.
4. Toman and Pauli (1964) studied the trends in nitrate reductase activity during cold hardening of winter wheat. There was no significant correlation between specific conductance and nitrate reductase or between nitrate reductase activity and water-soluble protein. However, total amino acid content in the crowns of winter wheat closely followed the trends of cold hardiness as measured by specific conductance (Pauli and Zech, 1964).

The development of cold resistance is stimulated by three major factors interacting: (1) shortening daylength, (2) lowering temperature, and (3) inherent genetic capacity of the plant. Adequate light for a high rate of photosynthesis and a wide spread between day and night temperatures enhances the hardening process. As shown in Figure 12.2, cold resistance increases as temperature declines, with a near maximum developed shortly after permanent soil freezing. Declining temperatures increased the enzymes responsible for conversion of glucose and fructosans to fructose in a hardy variety, but these enzymes were lacking in a nonhardy variety (Young and Feltner, 1966).

Plants do lose cold resistance upon exposure to warmer temperatures as such temperatures occur in the late winter and early spring. However, hardiness may be regained upon alternate exposure to a low temperature if growth has not occurred to any great extent and provided that sufficient carbohydrates are still available. However, alternating temperatures are basically harmful during the late winter

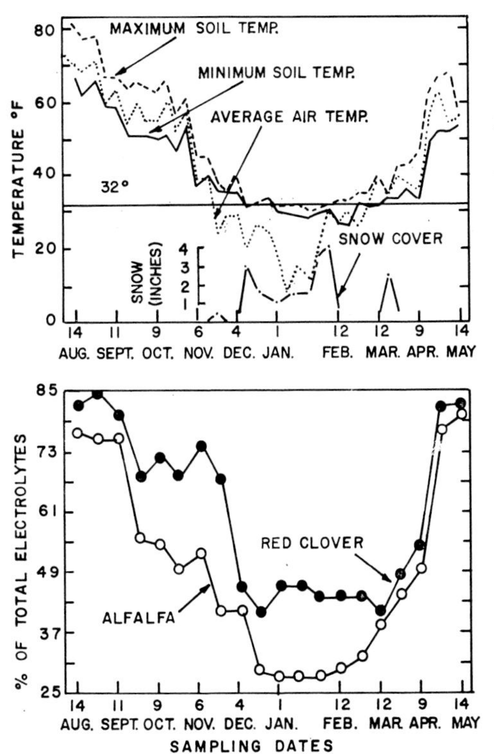

FIG. 12.2 ☙ Average biweekly air and soil (1 inch below surface) temperatures related to inches of snow cover in the field. Plant response as measured by cold resistance in the roots and crowns of alfalfa and medium red clover. Cold resistance was based on percent of total electrolytes (measured by electrical conductance) in the plant tissue that diffused into distilled water after 4 hours of freezing at −8° C (after Smith, 1964).

and early spring, and management practices discussed later are designed to minimize temperature fluctuation.

Development of hardening may be hindered for plants weakened by disease or limited in their mineral nutrition as discussed below. In addition fall defoliation can be harmful. Finally, the stage of plant growth influences hardiness, and the emerging seedling is especially sensitive.

SOIL FACTORS

Numerous factors external to the plant influence its ultimate winterhardiness. Climatically low temperature plays a dominant role. For most overwintering plants, the "thermal death point" is considered to be between 5 and 15° F. The internal factors stated above are responsible for lowering the temperature tolerance this much below the freezing point of water. Once the air temperature drops lower than the 5–15° range, survival depends upon insulation by the surrounding environment. Weather Bureau measurements for prediction and reporting of weather are taken at the five-foot height. Figure 12.3 indicates the wide variation in temperature at the five-foot height compared to that three inches below ground level. Thus the soil does provide insulation for the plant root and some protection for the crown. It is important to note, however, that the point of widest temperature fluctuation is in the few inches above the soil surface; thus aboveground plant parts experience a very rigorous climate.

Further insulation effect is achieved, in addition to that afforded by soil, from the presence of vegetation and/or snow cover. The ideal condition maximizes dead air spaces for insulation; as shown in Table 12.1 (e.g. dates 15 and 29), the combination of uncompacted snow and vegetation accomplishes this goal.

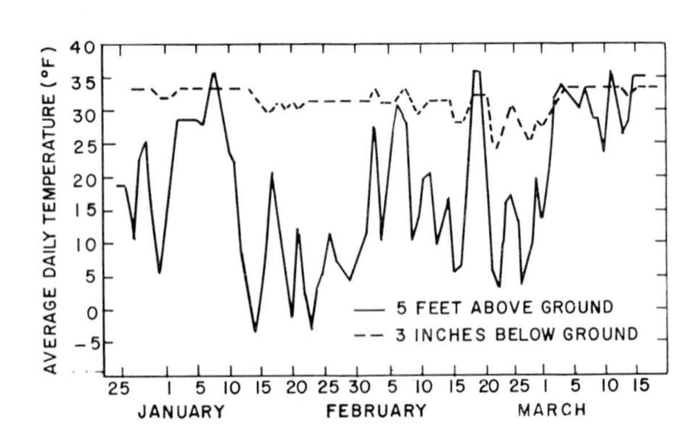

FIG. 12.3 The average daily soil temperature at 3 inches below ground and the air temperature at 5 feet above ground from Dec. 25, 1962, to March 16, 1963 (Beard, 1965).

TABLE 12.1. Effect of snow cover on soil temperatures three inches below soil surface in Michigan (Table 4.1, Smith, 1962)

Date	Average Daily Temperatures, °F				
	Air	Bare Soil	Compacted Snow	Uncompacted Snow	Uncompacted Snow and Vegetation
Jan. 15	30	30.8	31.6	32.1	35.4
20	18	28.5	30.4	31.8	35.1
21	15	26.7	29.7	31.4	34.9
22	11	22.2	28.6	31.1	34.6
23	8	25.6	28.8	31.1	34.1
25	20	24.5	27.6	30.9	33.5
26	14	22.9	26.3	30.5	33.5
27	16	22.0	24.8	30.3	33.2
28	2	20.8	22.7	29.6	33.0
29	0	17.1	21.4	29.2	32.7

Ice Sheets. The formation of ice sheets directly over bare soil can have a very damaging effect on overwintering plants (Fig. 12.4). Two factors are involved. First, the ice sheet has a smothering effect when the carbon dioxide released by the plant's respiration is trapped under the sheet and the carbon dioxide may rise to toxic levels. In addition, there may be internal accumulations of toxic products of aerobic and anaerobic respiration. Secondly, ice has a thermal conductivity four times greater than water and 90 times greater than air. Therefore, temperature fluctuations of plant material in or under the ice are tremendous.

FIG. 12.4 ❧ The effect of an ice sheet on stand recovery. Three-week recovery growth of ladino (left) and common white clover (right) following 12 days storage at −3° C. All stolons stored in air grew vigorously. With ice encasement, 15% of the ladino and 100% of the common white clover stolons grew and produced 5% and 107% respectively of the leaf weight of the samples stored in air (Smith, 1962).

Sprague and Graber (1943), working with hardened alfalfa frozen and maintained at —3° C, found injury after 12 days and complete kill after 26 days for plants under an ice crust. Beard (1965) found little injury to Kentucky bluegrass or bentgrass under similar conditions. He found no injury to these grasses after 51 days under two inches of ice or from an ice layer over snow. A slush that was compacted and then frozen did increase winterkill in his experiments. These experiments point out that plant response varies with species, but ice does increase the danger of winterkill for all plant material, and one goal is to prevent its formation over any part of the field.

Drainage Conditions. Soil drainage conditions also play an important role in overwintering. Taprooted plants are particularly sensitive to the plant heaving (Fig. 12.5) or soil heaving which may occur if excess water is present in the soil. Heaving occurs chiefly in the early spring on organic soils or those soils of high water-holding capacity. The fundamental principle involved in heaving is the growth of ice crystals at or near the soil surface due to the withdrawal of soil

FIG. 12.5 ☙ An alfalfa plant partially heaved from the soil by freezing and thawing of the soil surface (courtesy H L. Portz).

moisture from nearby areas by the force of crystallization. These crystals may be localized under the crown of taprooted plants and force the plant out of the ground. This is the most common type of heaving on heavy soils when moderate freezing temperatures prevail. Sheet heaving of the soil surface usually results from a sudden hard freeze followed by moderate freezing. Heaving results in plant damage by root breakage, direct temperature damage, and exposure of the crown to desiccation. Good surface drainage, surface insulation, and use of fibrous-rooted grasses will help reduce heaving.

Fertility. Soil fertility conditions also influence winterhardiness. Increased supplies of phosphorus and potassium usually increase hardiness. An increased supply of nitrogen may have either a positive or negative effect; if it is present in excess (that is, not in balance with phosphorus and potassium) decreased hardiness results. The important benefit of potassium in improving winterhardiness is not a new observation. A Frenchman, Couturier, writing in 1903 stated, "Potassium fertilizers have enabled lucerne (alfalfa) and numerous other crops to more successfully withstand the rigors of winter." Adequate fertility, especially potassium fertilization, allows plants to develop larger, more evenly distributed xylem vessels. This presumably results in a better flow of raw materials through the plant and especially to the storage organ.

Physiologic Drouth. One additional environmental factor, winter drouth or "physiologic drouth," may occur when the aerial portions of the plant are transpiring and yet the soil is frozen or at a low enough temperature that water absorption is slow and places the plant under water stress. This may occur for winter wheat quite frequently since it maintains leaves and a transpiring surface much of the winter. This problem is confounded by the increased viscosity of the water and thus an increased resistance to water movement across the living cells of the roots.

SUMMARY

There is often a close tie between maximal productivity and minimal winterhardiness. Therefore, the production manager of perennial crops seeks to strike a balance between high yields and sustained yields over a period of years. Certain management practices will favor winterhardiness: (1) variety selection, (2) well-drained and fertile soil,

(3) fall management to favor carbohydrate storage, (4) a solid stand rather than row culture of a species or variety that is marginal in winterhardiness, (5) age of stand (an old stand is more susceptible to killing than a young stand).

DROUTH RESISTANCE

The term drouth usually refers to a deficiency of available soil moisture which produces water deficits in the plant sufficient to cause a reduction in growth. However, many definitions of drouth are of limited value because they provide a qualitative indication of the amount of available water in the root zone only and are less informative about the water status of the plant.

Under some conditions low soil moisture is accompanied by low humidity, high temperature, and wind, all of which increase the stress on the plant. However, these conditions may not occur where winter rainfall is inadequate; even so, the plant experiences a drouth stress. It must be emphasized that the term drouth has a relative meaning, depending on geographic location, the normal rainfall cycle, and the pattern of the rainfall. Thus a farmer in the Pacific Northwest knows he will have water and growing conditions in the winter months but a dry summer, so he plans for this and grows winter crops. Summer is a dry season, not a time of drouth.

The nature of drouth resistance has been controversial for many years. In early work, water loss was emphasized as the primary factor, while in later years the capacity to endure desiccation has been emphasized. Nevertheless, for field crops water loss is still of major importance. Levitt (1963) suggests that the term drouth resistance can be best understood by its relation to xerophytism (Fig. 12.6). There are three aspects of drouth survival.

1. Drouth evasion is the ability of the plant to complete its life cycle before being subjected to serious water stress. This commonly means the seed lies dormant during the dry season and grows again when rainfall occurs. A winter crop in a Mediterranean climate, for example, has an early maturity before spring and summer drouth.
2. Drouth endurance with a high internal water content is an avoidance of the drouth. This is accomplished by a deep root system or a reduced transpiration, such as alfalfa and cactus exhibit.
3. Drouth endurance with a low internal water content is tolerance. Such plants then exhibit the ability to recover and grow when

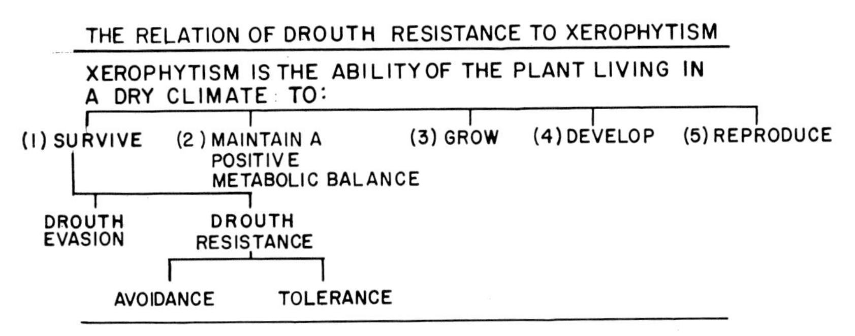

FIG. 12.6 ☙ A diagrammatic analysis of the relation of drouth resistance to xerophytism (Levitt, 1963).

soil water is replenished. It seems unlikely that the ability to survive low levels of turgidity for an appreciable period is important in determining differences in drouth resistance for economic crops.

It is important to observe that the highest drouth resistance may not invariably lead to the highest yield. The most successful water savers must sacrifice yield since closure of their stomata leads not only to water saving but also to a reduced rate of photosynthesis.

New supplies of water become available to the plant either from water supplied by precipitation or by the extension of the plant root system. While there may be some upward movement of water in the soil between periods of rewetting, this effect is relatively small.

RESPONSES OF PLANT TO WATER STRESS

All available evidence suggests that growth is reduced by a decrease of relative turgidity below 90%, is cut in half when the relative turgidity reaches 85 to 83%, and is moved to a very low value (— 10% of maximum) at 75%. However, Slayter (1957) does report that some elongation and dry matter accumulation occurred until the relative turgidity had fallen to 62%. This corresponded closely with the first appearance of permanent wilting. Visible wilting does not appear until considerable productivity has been lost.

Turgidity Changes. Experiments in which plants have been subjected to varying water stress show that the rate of growth decreases as the soil water decreases throughout the whole range below field capacity (water readily available for plant use). The reduction is greater in *fresh* than *dry* weight. This observation suggests that small reductions in relative turgidity do not greatly influence the rate of the main metabolic processes. The relation between production of fresh weight and dry weight of lettuce with different mean soil-water potential is described in Figure 12.7. A soil at field capacity represents desirable conditions for root growth and varies between 0.04 and 0.25 atm for most soils (May and Milthorpe, 1962).

The relative water loss rate is probably due to closing of the stomates or a rolling or drooping of the leaves. The relative rate of water loss is much more rapid above the level of 80% relative turgidity. There are so many different structural features which contribute to drouth tolerance that May and Milthorpe suggest it is rather futile to seek a single cause in any species.

Anatomical Changes. Although water stress leads to a reduction in the total growth of the plant, it influences the growth of different organs variably as follows: (1) There is normally a decrease in the ratio of shoot to root growth. (2) A decrease occurs in the proportion of lateral roots to total root length. (3) The ratio of leaf to stem is decreased. That is, during a period of soil water stress, the growth of organs is influenced in this order of decreasing severity: leaves > stems > roots.

Periodic water deficits in the plant lead to many anatomical changes: (1) smaller cells, (2) greater vein length, (3) more stomata per unit leaf surface, (4) thicker cell walls, (5) more wax deposition on walls bordering intercellular spaces, (6) greater development of mechanical tissues.

Metabolic Changes. Photosynthesis and respiration respond to water deficits in the following fashion. As turgidity decreases, the respiration rate first increases but finally declines below the level for turgid

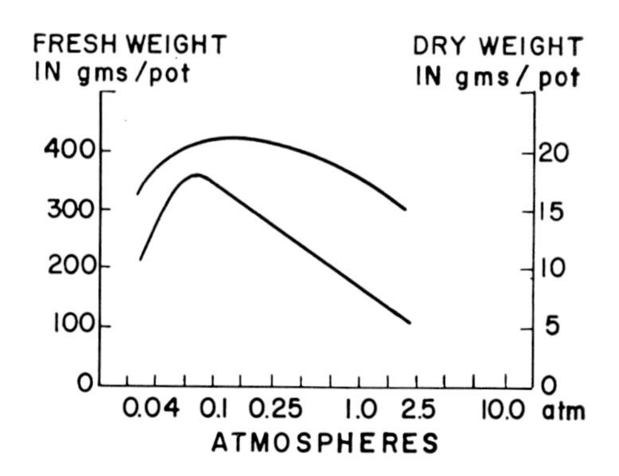

FIG. 12.7 ☙ The relation of fresh weight (top line) production and dry weight (bottom line) production of lettuce with different mean soil-water potential. The field capacity of most soils varies between 0.04 and 0.25 (after May and Milthorpe, 1962).

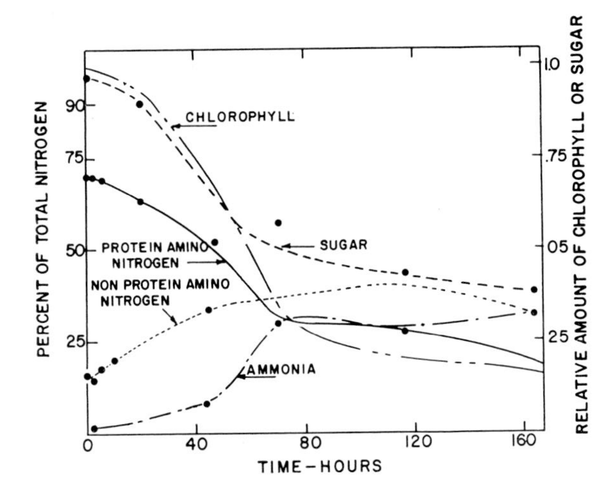

FIG. 12.8 The effect of wilting on the various constituents in turnip leaves (Thompson et al., 1966).

tissue. Photosynthesis commonly declines at first but then increases until the deficiency becomes very severe.

Almost all metabolic reactions are influenced by a deficit of water. In addition, as shown in Figure 12.8, both chlorophyll and sugar content decline markedly in wilted leaves. Hydrolytic reactions predominate under water stress. Protein amino nitrogen declines while ammonia increases sharply (Fig. 12.8). The hydrolysis of starch to sugar and protein to amino acids suggests the failure of energy-producing mechanisms. Amino acids are apparently exported to the nonwilted portions of the plant. In addition, amino acids are continually synthesized during water stress, but protein synthesis is inhibited and protein levels decrease. Water stress induced a 10- to 100-fold accumulation of free proline and a 2- to 6-fold accumulation of free asparagine in shoots; therefore, during water stress free proline probably serves as a storage compound (Barnett and Naylor, 1966). A supply of sugar appears essential for proline accumulation because proline accumulation is greater and most prolonged in wilted leaves with a higher sugar and starch content (Howell and Jung, 1965). These authors concluded that oxidation of the sugars furnished α-ketoglutarate and NADPH for proline synthesis. Various workers have found that RNA does not increase in water-stressed leaves in the linear fashion normally observed in well-watered leaves. In addition, one of the first effects of reduced turgidity is a disruption of the synthesis of ATP and an uncoupling of the ATP system.

Stress Effects on Different Growth Stages. In flowering plants some tissues are more resistant to drying than others. Meristematic tissue appears able to survive complete drying. However, most extended cells and tissues die when 50–70% of their water has been lost. In addition, certain stages in the growth cycle are more susceptible to drouth injury than others. The two most susceptible are (1) during stem elongation and spikelet differentiation and (2) at anthesis. Water stress during the period of fruit growth leads to a smaller fruit size, either by a direct effect on photosynthesis or more likely by effects on the ATP system. Although in cereals this leads to small grain of poor quality, the total yield of grain does not appear to be reduced as much as by drouth at the earlier stages.

DROUTH HARDENING

After recovering from a period of drouth plants are usually much more resistant to the influences of further water stress. The hardened plant exhibits increased viscosity of the protoplasm, higher rates of photosynthesis and lower rates of respiration, greater xeromorphism, a higher proportion of root, and finally less reduction in yield on further subjection to water stress. Mattas and Pauli (1965) report that corn plants 6–9 inches tall exhibited the following metabolic changes under varying degrees of heat and moisture stress. (1) They decreased both in relative turgidity and moisture content. (2) Nitrate reductase activity decreased sharply. (3) An accumulation of nitrate resulted. (4) The total nitrogen per plant increased rapidly during the initial stress and remained fairly constant thereafter. (5) The general levels of molybdenum were low and decreased with increased moisture stress.

PRESOWING HARDENING

One interesting development in studies on drouth resistance has been the work on presowing hardening (Jarvis and Jarvis, 1964). Treated seeds resulted in a crop yield that was 10–20% more than for crops from untreated seeds. Seeds of a cereal like wheat are treated in the following manner. Water is added in a quantity equal to 30% of the dry weight and the seed is allowed to remain at this moisture level 14 hours at temperatures between 10 and 25° F. Following this the seed is air-dried and the cycle repeated two or three times. The physiological and morphological properties responsible for the increased yields achieved are not clearly understood, nor is the commercial practicality known, but presowing hardening presents an interesting possibility for plants to be sown on low-moisture soil and has been tried under field conditions in Finland and Russia.

SUMMARY

Winter and drouth hardiness appear to be related in that both result from a dehydration of the cell. Small cell size is correlated with both. Low temperature gives drouth hardiness, and under limited water, plants that become drouth hardy are also frost hardy. Thus the two plant responses exhibit many characteristics in common. A better understanding of these two phenomena is essential in extending the world's agricultural area.

LITERATURE CITED

Barnett, N. M., and A. W. Naylor. 1966. Amino acid and protein metabolism in bermuda grass during water stress. *Plant Physiol.* 41:1222–30.

Beard, J. B. 1965. Effects of ice covers in the field on two perennial grasses. *Crop Sci.* 5:139–40.

Bouyoucos, G. J. 1916. Soil temperature. *Mich. Agr. Exp. Sta., Tech. Bull.* 26.

Bugaevskii, M. F., and K. S. Zitnikova. 1936. Choking winter wheat plants under an ice crust. *Zbirn. Prac. (Rob) Agrofiziol.* 2:103–15 (Reviewed in *Herbage Abstr.* 1938. 8:1785).

Colby, W. G., M. Drake, D. L. Field, and G. Kreowski. 1965. Seasonal pattern of fructosan in orchardgrass stubble as influenced by nitrogen and harvest management. *Agron. J.* 57:169–73.

Howell, J. H., and G. A. Jung. 1965. Cold resistance of Potomac orchardgrass as related to cutting management, nitrogen fertilization and mineral levels in the plant sap. *Agron. J.* 57:525–29.

Iljin, W. S. 1957. Drought resistance in plants and physiological processes. *Ann. Rev. Plant Physiol.* 8:257–74.

Jarvis, R. G., and M. S. Jarvis. 1964. Presowing hardening of plants to drouth. *J. Exp. Botany* 21:113–17.

Jung, G. A., and D. Smith. 1959. Influence of soil potassium and phosphorus content on the cold resistance of alfalfa. *Agron. J.* 51:585–87.

Kramer, P. J. 1963. Water stress and plant growth. *Agron. J.* 55:31–35.

Lamb, C. A., and R. I. Grady. 1963. A review of soil heaving studies related to agriculture. *Ohio Agron. Dept. Ser.* 170.

Levitt, J. 1956. *Hardiness of Plants.* Vol. 6. Monographs of Am. Soc. Agron. Academic Press Inc., New York.

———. 1962. A sulfhydryl-disulfide hypothesis of frost injury and resistance in plants. *J. Thoret. Biol.* 3:355–91.

Levitt, J. 1963. Hardiness and the survival of extremes: A uniform system for measuring resistance and its two components. In L. T. Evans (ed.). *Environmental Control of Plant Growth.* Academic Press, Inc.. New York.

Mattas, R. W., and A. W. Pauli. 1965. Trends in nitrate reduction and nitrogen fractions in young corn plants during heat and moisture stress. *Crop Sci.* 5:181–84.

May, L. H., and F. L. Milthorpe. 1962. Drought resistance of crop plants. *Field Crop Abstr.* 15(3):171–79.

Pauli, A. W., and A. C. Zech. 1964. Cold hardiness and amino acid content of water soluble proteins in crowns of winter wheat. *Crop Sci.* 4:204–6.

Sellschop, J. P. F., and S. C. Salmon. 1928. The influence of chilling, above the freezing point, on certain crop plants. *J. Agr. Res.* 37:315–38.

Schmutz, W. 1962. Further investigations on the relationship between sulfhydryls and winter hardiness of 15 wheat varieties. 2. Acker-und Pflanzenbau 115:1–11.

Shaw, R. H., and D. R. Laing. 1966. Moisture stress and plant response. In W. H. Pierre, D. Kirkham, J. Pesek, and R. Shaw (eds.). *Plant Environment and Efficient Water Use.* Am. Soc. Agron. and Soil Sci. Soc. Am.

Slayter, R. O. 1957. The influence of progressive increases in total soil moisture stress on transpiration, growth and internal water relationships of plants. *Australian J. Biol. Sci.* 10:320–36.

———. 1967. *Plant-Water Relationships.* Academic Press, Inc., New York.

Smith, Dale. 1949. Differential survival of Ladino and common white clover encased in ice. *Agron. J.* 47:469–73.

———. 1962. *Forage Management in the North,* 2nd ed. Wm. C. Brown Co., Dubuque, Iowa.

———. 1964. Freezing injury of forage plants. In *Forage Plant Physiology and Soil-Range Relationships.* Am. Soc. Agron. Spec. Publ. No. 5.

Sprague, M. A., and L. F. Graber. 1943. Ice sheet injury to alfalfa. *Agron. J.* 35:881–94.

Thompson, J. F., C. R. Stewart, and C. J. Morris. 1966. Changes in the amino acid content of excised leaves during incubation. I. The effect of water content of leaves and atmospheric oxygen level. *Plant Physiol.* 41:1578–84.

Toman, F. R., and A. W. Pauli. 1964. Changes in the nitrate reductase activity and contents of nitrate and nitrite during cold hardening and dehardening of crowns of winter wheat (*Triticum aestivum* L.) *Crop Sci.* 4:356–59.

Vaadia, Y., F. C. Raney, and R. M. Hagan. 1961. Plant water deficits and physiological processes. *Ann. Rev. Plant Physiol.* 12:265–92.

Young, A. L., and K. C. Feltner. 1966. Managerial and physiological factors influencing the winter hardiness of barley in Wyoming. *Crop. Sci.* 6:547–51.

CHAPTER THIRTEEN WEED CONTROL

CONTROL OF WEEDS historically has been one of the major concerns of anyone engaged in crop production. A weed may be broadly defined as any plant growing where it is not desired (i.e. barnyard grass in rice; corn in soybeans). Further, it could be said that the plants man wants to grow are his crops; the ones he wishes to be rid of are his weeds. State and federal laws define certain problem plants as weeds, but in practical agriculture the list of weedy plants is much longer.

Weeds have a detrimental effect on crop yield by competing for many environmental factors, and they reduce crop quality by introducing unpalatable or toxic components to the harvested crop. In earlier years when a farmer relied more heavily on the natural fertility of the soil to nourish his crops, one of the major factors for which weeds competed was the plant nutrients. At present high levels of management this competition is relatively less severe, but the values in Table 13.1 emphasize that weeds can be very wasteful users of nutrients. Nevertheless, now that fertility levels are commonly higher, the competition of weeds for light and water has become relatively more important than that for minerals. In many of the major crop production regions of the world, competition for light (e.g. vertical competition) is of prime importance because the months of satisfactory temperatures, moisture, and radiation for plant growth are limited.

Weed growth may also have deleterious effects on crop quality. The weeds may have direct effects by remaining mixed with the crop at harvest and adding less nutritious, unpalatable, or in some cases toxic content to the seed or cause poorly filled, chaffy seed. Therefore, the agriculturist seeks at least to minimize weed growth, even though

TABLE 13.1. Comparison of relative yields and plant nutrient uptake by corn alone, corn with weeds, and different weeds grown alone (Vengris, 1955)

Plant	Relative Yields, Total Dry Matter					
	(Corn 100)	N	P	K	Ca	Mg
Corn alone	100	100	100	100	100	100
Corn with weeds	63	58	63	47	67	77
Pigweed	60	102	80	124	275	234
Lambsquarters	69	120	74	121	281	216
Crabgrass	67	100	64	157	131	228
Barnyard grass	91	105	60	138	430	337

it may be uneconomical to eliminate weeds completely with present technology.

One of the long-term goals of the agronomist is to seek a weed-free environment. Of the various pests requiring control, complete control of weeds should be most feasible because:

1. They represent a contained population. The weeds are usually in the field before the crop is planted; therefore, the farmer knows what his problems are beforehand.
2. Weeds have a longer generation time and are less likely to shift to resistant strains.

GROWTH HABITS AND LIFE CYCLES

Weed species are represented by the same range of growth habits and life cycles as those which occur in crop species. While weed laws have much to say about perennial, deep-rooted weed species (commonly classed as the primary noxious weeds), these are not the plants which cause the greatest losses and for which the greatest control effort is expended on our crop acres. The greatest effort is expended in controlling annuals, those plants exclusively propagated from seed.

In the cases where perennial weeds are a problem, they are commonly characterized by a high level of persistence. In order to eliminate such plants, both the accumulated carbohydrates and the potential axillary buds must be destroyed. It is therefore desirable to know the pattern of carbohydrate storage in a perennial weed and to seek the removal of its photosynthetic area when the stored reserves are at their lowest (Fig. 13.1), thus weakening the plant generally (as discussed in Chapter 4) and eliminating its ability to support axillary buds.

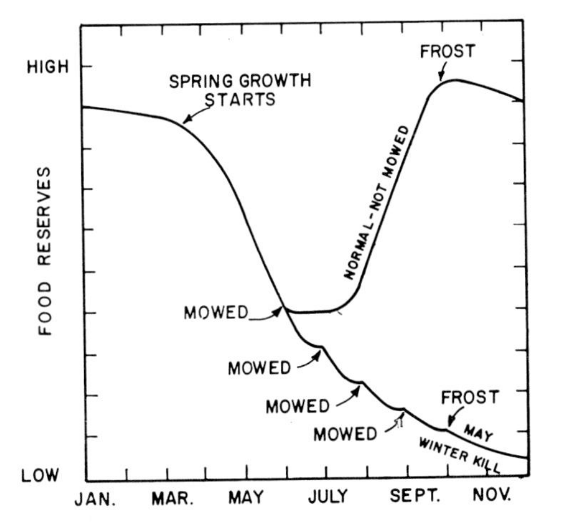

FIG. 13.1 ⁂ Food reserves of of a perennial weed of a normal unmowed plant compared with a repeatedly mowed plant (Klingman, 1961).

A reduction of weed competition is possible through prevention, control, and eradication of the weeds. Of these three, only control is economically feasible in most cases. Seed laws and quarantine do allow a degree of prevention in the introduction of new weed species; this stands as the most advocated and least practiced of all methods. Once a species is established, eradication is often too expensive to be justified, thus leaving control as the most practical solution.

METHODS OF WEED CONTROL

The basic principle underlying the several methods of weed control is to direct control at the plant survival mechanism. For annuals this means to prevent seeding, to deplete seeds in the soil, and to kill the germinating seed or young seedling. For perennials the goal is to destroy the underground organs and to prevent the plant from going to seed.

MECHANICAL

Man has sought for hundreds of years to reduce weed growth by the use of the hoe or the plow. For the past 200 years he has hitched a horse or a tractor to that hoe to assist him, and this is still the most widely used means of control. Seeds and seedlings can be buried by this method; however, previously buried seeds may be turned to the surface and activated. These in turn can be killed by a well-timed disking.

At present there is considerable emphasis on minimum tillage systems designed to reduce the number of trips across the field. Some of these minimum tillage systems leave most of the surface loose and less suitable for weed seedling growth (Fig. 13.2) and prepare a seedbed only in the row zone. Such systems are potentially of merit for intertilled crops. However, for some intertilled crops in which weed control is still quite difficult (i.e. soybeans), extra seedbed preparation in the form of additional tillage with a disk or spring-tooth harrow may continue to be a key means of good weed control. It is important to emphasize that minimum tillage systems favor perennial weed growth because these weeds do relatively better in undisturbed soil.

Cultivation disturbs the rooting system and seedlings dry out quickly. When the weed seedlings have not yet emerged but are germinated, a very slight soil disturbance by a rotary hoe, a finger weeder, or cultivator sweeps will destroy many seedlings.

Mowing is useful for the control of tall-growing annuals, especially to reduce their seed production. Very often annuals can be successfully controlled by mowing high the first time, thus removing

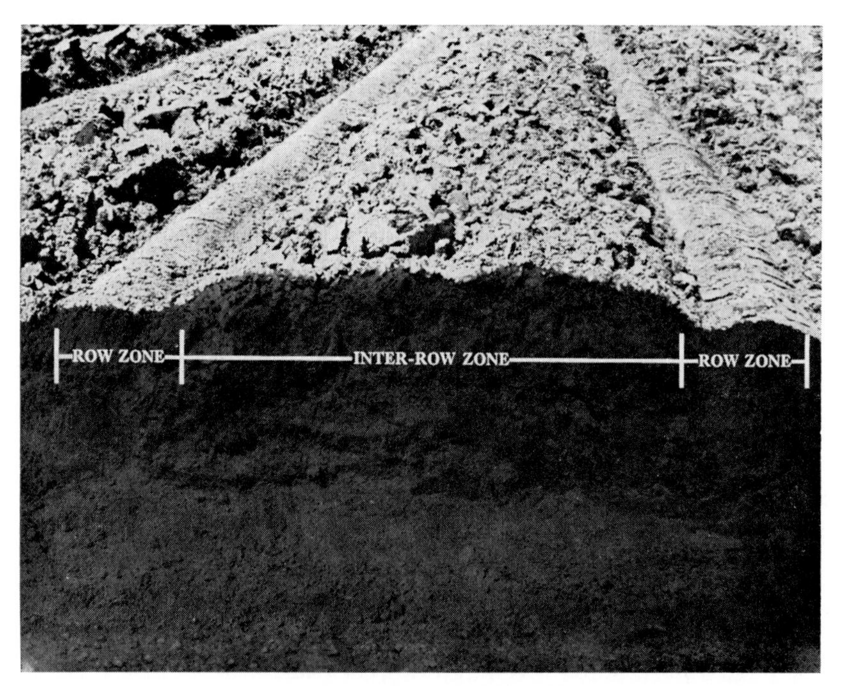

FIG. 13.2 ☙ The two-zone tillage concept. The row zone is prepared as a seedbed; the inter-row zone is loose and less suitable for weed growth (Photo courtesy Deere and Co. and W. E. Larson).

apical dominance and stimulating activation of the lower axillary buds. A low second mowing will remove these new shoots and greatly reduce the plant's vigor and seed production capacity. Well-timed mowing of the perennial plant when carbohydrate reserves are low may serve to starve it, eliminate its carbohydrate reserve, and result in death (Fig. 13.1).

CROP COMPETITION

High-yielding crops usually provide maximum competition for weeds. A dense, well-fertilized stand is often one of the best weed control systems for the major portion of the crop season. Staniforth (1957) points out that increasing the corn plant population from 8,000 to 20,000 plants with 70 pounds of nitrogen added reduced foxtail growth from 3,000 to 1,600 pounds and increased corn yield 1,500 pounds (Fig. 13.3). These corn yield benefits from increased fertility and plant population were more pronounced in a somewhat weedy field than in a weed-free field. A combination of higher plant populations and an herbicide also maximized weed control (Eplee and Klingman, 1968).

For most crop plants, early season weed competition is the most detrimental to yield. At this stage crop competition is not readily available. Therefore, many management systems are designed to give maximum mechanical or chemical control early and then rely on crop competition for late season weed control.

CROP ROTATIONS

Crop rotations are a useful means of weed control. Intertilled and broadcast stands alternating on a given field successfully reduce the growth of a varied group of weeds. For example, annuals are adversely affected by frequent cultivations of corn, whereas a perennial like Canada thistle is reduced in vigor by shading of alfalfa and the three or four defoliations which occur when the alfalfa is harvested.

FIRE

In grasslands vegetation fire is a common means of controlling woody weed species. This might be classed as natural control because

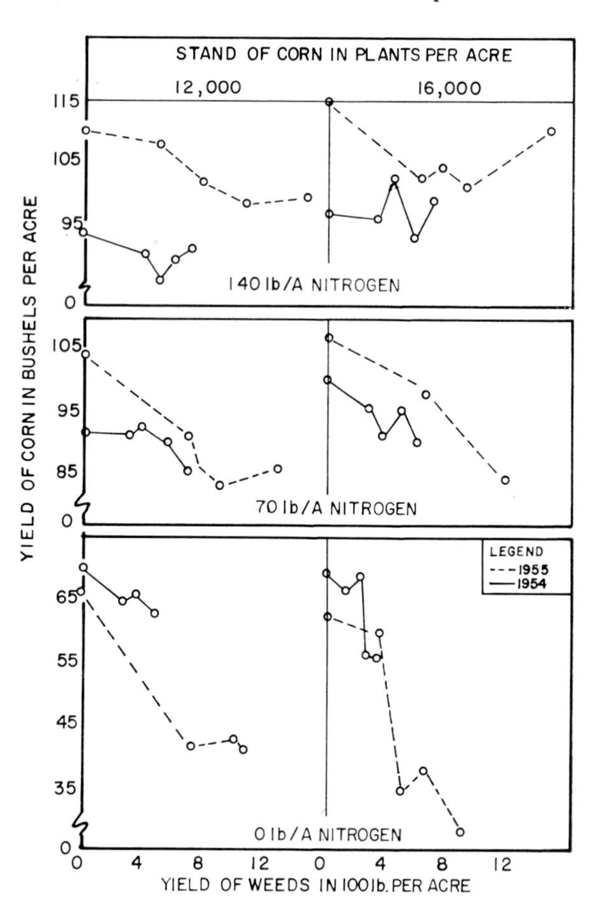

FIG. 13.3 ☙ Relationship of yield of mature foxtail to the resulting corn yield reductions with two rates of nitrogen fertilizer application and two corn plant populations. In 1955 the total rainfall for the May 15 to September 15 period amounted to 12 inches—approximately half of that for the same period in 1954 (Staniforth, 1957).

it was present before man's appearance. In many places, fire continues to be used by man for this same purpose. Flame cultivation, a modification of weed control by fire, has been studied on several crops and is frequently used in cotton. This procedure requires that a source of heat be directed at the weed and that the flame or heat band minimally contact the plant. Flaming has been tried although less successfully on succulent-stemmed crops like corn and soybeans.

BIOLOGICAL

Concern for chemical residues has served as a recent stimulus for the age-old interest in using biological means of weed control. The successful use of biological control for weeds often depends upon at least two acts of man: (1) The weed species must have been introduced to a geographical region and in the process have been freed of its natural predators. (2) Natural predators must be introduced into this geographical region, but in turn must have been freed of parasites.

It is known that root borers, stem borers, and internal seed or fruit feeders are more highly specialized than leaf or foliage feeders. Thus the former type was originally preferred for biological weed control. However, experience has proved that leaf and stem feeders can be highly selective and very useful in weed control. Both the weed and the insect predator populations may fluctuate for some time after the introduction of an insect, and several years may be required before an equilibrium is obtained. Complete eradication usually is not achieved, but rather a reduced and controllable weed population is created. The first notable success of biological control was achieved in Hawaii on the thorny shrub *Lantana camara*. Numerous flower-eating insects from Mexico and the South Sea Islands controlled this species when introduced to Hawaii. The classic example was control of the prickly pear in Australia by a moth borer introduced from Argentina. In a third instance, the larva of the leaf beetle *(Chrysolina quadrigeminal)* has successfully controlled Klamath weed, a serious weed of rangeland in California. Crafts and Robbins (1962) describe nine other weeds on which efforts at biological control have been made, but they report limited success was achieved. This will be a mode of control for some weeds in the future, perhaps most often in extensive, rangeland conditions.

CHEMICAL

Ashes and salt have been used since biblical times to eradicate plant growth nonselectively. However, not until the early 1900s were certain inorganic chemicals shown to have selective herbicidal action. After a flurry of interest in the early 1900s, research work in this area lagged in the United States until the 1940s, although there was continued activity in the interim in Europe. Since the advent of 2,4-D in the early 1940s, herbicide research has been sustained and intense. The modern era of chemical weed control was rapidly introduced with this evidence that herbicides could be: (1) effective in small quantities (therefore of relatively low application cost); (2) highly selective and thus used in the presence of crops; (3) systemic in action, translocated, and thus capable of killing underground organs.

Herbicides in effect have only one characteristic in common—

they kill plants. Their action may be described as: (1) acute or chronic, (2) contact or systemic, (3) selective or nonselective. On the point of selectivity it is important to note that this is rarely absolute but depends on dosage, the means of application, the herbicide formulation, and the stage and condition of the plant's growth. However, the importance of crop tolerance to herbicides is being recognized as new and improved herbicides replace old ones, and selectivity is becoming more and more discrete.

Factors of Selectivity. Contact and translocated herbicides may act in a selective or a nonselective manner. Selectivity is based upon many factors which may be categorized as follows: (1) morphology, (2) absorption, (3) translocation, (4) physiology.

MORPHOLOGY. Tall plants with chemically tolerant stems permit easy application of herbicides to weeds near ground level. Conversely, because alfalfa crowns are at or below ground level, a spray can be applied during dormancy to kill aboveground parts of annual weeds.

ABSORPTION. Initial leaf penetration may take place through the leaf surface or through the stomates. Probably the most important penetration is through the chemically nonpolar cuticle. Chemical salts (i.e. 2,4-D salt) are polar (± charges) and go through the cuticle slowly. Most organic substances are nonpolar (i.e. 2,4-D esters) and go through the nonpolar cuticle readily.

Wetting agents (surfactants) are used with polar compounds to get better wetting and to solubilize the wax and oil-like substances of the cuticle and cell wall. However, wetting agents may change selectivity patterns and require lower dosages to obtain the same toxicity. For example, sodium and amine salts of 2,4-D are five times more toxic with wetting agent added.

TRANSLOCATION. This may occur through the phloem, the xylem, or on an intercellular basis. The following points are of importance when considering translocation:

1. The phloem is living tissue. Chemical flow is commonly from the leaf to areas of most active food use, which may be a stem or root apex. This movement may be quite rapid; for example, 2,4-D has been shown to move 40 inches per hour in the phloem. This translocation parallels the movement of food materials and occurs most effectively when full leaf development has occurred.

2. Xylem transport most often occurs for materials absorbed by the roots and represents movement along with the transpiration stream and concentration in the leaves. However, under stress conditions downward transport may occur in the xylem.

3. Intercellular transport may occur for nonpolar substances with low interfacial surface tensions.

PHYSIOLOGY. Represented here are differences in enzyme systems, variable responses to pH, cell metabolism, cell permeability, variations in chemical constituents or polarity. For example, the heterocyclic nitrogen derivatives (simazine, atrazine) and the substituted ureas have been shown to inhibit the Hill reaction in susceptible weeds. Further, the economic crop may detoxify the herbicide (atrazine) or the weed may toxify it (2,4-DB).

Herbicide Categories.

CONTACT HERBICIDES. These herbicides have an acute or rapid effect and function like a chemical "mowing machine." Those which are selective foliage contact herbicides work for the following reasons: (1) There is differential wetting of the crop and the weed (e.g. sulfuric acid differentially wets chickweed and grass, thereby killing the chickweed). (2) The crop and the weed have a different location of the growing point (e.g. the crown of alfalfa is low in the ground, whereas annual weeds have their growing point fully exposed). (3) There are differences in plant tolerance to toxic chemicals (e.g. a very succulent weed absorbs more herbicide than a more mature crop plant). (4) A selective spray placement may be made (e.g. a spray directed on weeds below a taller cotton plant).

Nonselective foliage contact herbicides are used to kill all plants to which they are applied. They may be used (1) where no growth is wanted (in a parking area, along a railroad right-of-way); (2) in a growing crop with proper shielding or placement of the spray; or (3) in dormant perennial crops.

TRANSLOCATED HERBICIDES. These chemicals offered a new dimension in weed control when they became available to agriculture. Systemic herbicides may be divided into: (1) those absorbed by the roots and moved in the xylem, (2) those absorbed by the leaves and moved in the assimilate stream in the phloem.

The leaf surface is not an efficient structure for entry because it has a waxy cuticle and interfacial water tension to prevent entry via the stomate. Nevertheless, foliar applications are a very important method of herbicide treatments. A surfactant is desirable to enhance entry through the leaf. A major representative of the leaf-absorbed group is 2,4-D.

A unique, practical characteristic of these leaf-applied herbicides is that they can be applied in very low volume. Instead of concentrating in the leaves, as do many of the xylem-transported herbicides, this phloem-moving group travels to the regions where food is being actively utilized, particularly the meristematic zones.

The roots seem to have little ability to discriminate among organic molecules of low molecular weight and, in addition, serve to accumulate concentrations higher than those found in the soil. Nevertheless, root applications may also function in both selective and nonselective fashions. Soil application is now the most important way of applying herbicides. Root applications have advantages: (1) in convenience of applying during seeding or shortly after; (2) because effectiveness is less dependent upon the stage of growth or physiological condition of the weeds; (3) because roots are usually more sensitive to toxic chemicals (as they are to auxins—Chapter 7) than are the tops. However, soil application may: (1) be ineffective because of the tremendous dilution effect of the soil; (2) allow fixation of the herbicide; (3) allow residues to accumulate in the soil; and (4) need additional water from rain or irrigation to activate the chemical.

Herbicides as a group have a very low mammalian toxicity. It is important to handle such chemicals with care and minimize exposure to them, but in comparison to insecticides, the herbicides are not very poisonous. However, there is enough concern for any possible problems to stimulate studies on how residuals might be reduced. For example, in a study on forage crops, 2,4-DB residue was found to be reduced 53–63% by 30 days under silage fermentation, and most of this occurred in the first 15 days (Linscott et al., 1965).

There is also the question of soil residue effects on the following crop in a rotation. This greatly affects the management of species in a cropping system. As a means of checking for potentially harmful residuals, studies have been made to describe the effect of long-residue phytotoxic compounds on an indicator plant (Lynd et al., 1967). Such a procedure may frequently be useful if there are unknown factors of application rate or rate of breakdown from the previously treated crop.

Herbicide Persistence. Klingman (1961) suggests six factors which affect the persistence of an herbicide in the soil: (1) microorganism decomposition, (2) chemical decomposition, (3) absorption on the soil colloids, (4) leaching, (5) volatility, and (6) photodecomposition.

A population that can use the specific herbicide as a food may increase temporarily, but no long-time effect on the microorganism population of the soil is expected nor has any been observed to date (Kearney and Kaufman, 1967).

Chemical decomposition by the soil destroys some herbicides and activates others.

Soil colloids have unusual absorptive capacities. High colloid content means high herbicide absorption, slow release, and reduced effectiveness as an herbicide. However, even though various soils show large differences in their absorptive capacities, the adjustment in range of application rate required is not as great as might be expected.

Leaching may determine the effectiveness of an herbicide, explain its selectivity, or account for its removal from the soil. The strength of "absorption bonds" between the herbicide and the soil is considered to be just as important as water solubility in determining the leachability of the herbicides.

Volatility of an herbicide may result in its effectiveness being lost, or the vapor produced may prove toxic to nearby plants.

Photodecomposition by sunlight of chemicals applied to the soil surfaces may result in their loss, especially if they remain there for an extended period without rain. Both volatility and photodecomposition are involved in such losses.

Herbicide Families. Herbicide specialists and plant physiologists have spent extensive time and effort in an attempt to characterize the properties of chemicals useful as herbicides (Audus, 1964). This approach to finding new herbicides is called prediction; it has been done in the hope of stating what properties should be built into an herbicide for a particular purpose. While these studies have been very instructive on the nature of present herbicides, they have not yet contributed significantly to the discovery of new ones. Therefore, commercial firms follow the alternate, laborious, but effective procedure of screening massive numbers of materials and selecting promising materials by a combination of chance, shrewd guesses, and previous knowledge or experience. Such screening is done by about 50 industrial concerns in the United States and Europe. Nearly all new chemicals synthesized by such companies are screened for their herbicidal activity, and this results in a very diversified group of herbicidal materials. Thus a unified classification of herbicides is not possible. There are now over 30 groups of herbicides, thousands of known

2-METHYL-4CHLOROPHENOXYACETIC ACID

O-CH_2COOH

CH_3

Cl

FIG. 13.4 ☙ MCP, the first phenoxy herbicide, developed in 1940 by the British.

active compounds, more than 100 in active use, and 50 or more in an advanced state of development.

This extensive group is catalogued in various reference sources such as the Weed Society's *Herbicide Handbook* (1967) and NAS's *Weed Control* (1968). A few examples are included here to describe how structural modifications make an herbicide more useful for a particular purpose.

PHENOXYS. The first family of translocated, selective herbicides was the 2,4-D group which was developed in the early 1940s. The British developed methylchlorophenoxyacetic acid. This chemical had an acetic acid end group to substitute for the H in the phenol OH, a methyl at position 2, and a Cl at position 4 (Fig. 13.4).

FIG. 13.5 ☙ Reproduction of the first published photograph of 2,4-D activity (Zimmerman and Hitchcock, 1942). Lanolin preparations of different concentrations of 2,4-D acid were applied to stems of tomato plants. (1) Control; (2) 0.03 mg 2,4-D/g of lanolin; (3) 0.06 mg/g; (4) 0.125 mg/g; (5) 0.25 mg/g; (6) 10.0 mg/g. Photographed after 24 hours. Strong epinastic responses are evident at these low dosages (courtesy Boyce Thompson Institute).

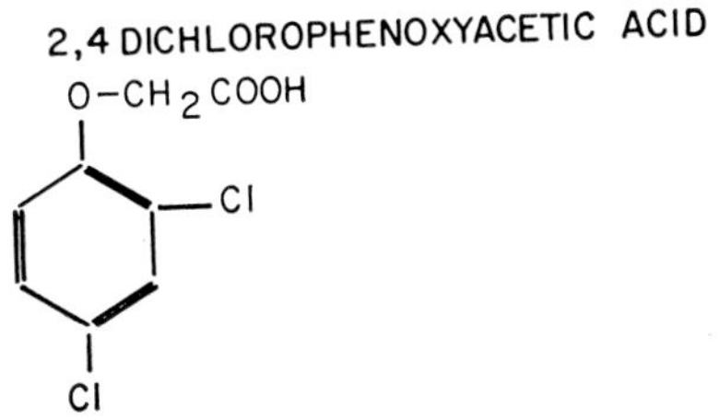

2,4 DICHLOROPHENOXYACETIC ACID

$O-CH_2COOH$

Cl

Cl

FIG. 13.6 2,4-D, the first phenoxy herbicide developed in the United States. An excellent foliar-applied, translocated, selective herbicide.

About the same time, USDA researchers described the selectivity of 2,4-D in the now classic paper of Zimmerman and Hitchcock (1942; see Fig. 13.5). In this case, Cl was substituted at both positions 2 and 4 (Fig. 13.6).

These two materials have very similar properties. MCP is somewhat less active and is useful where the crop plant sensitivity is high. Therefore, it is recommended for broadleaf weed control in small grains and in cases where the small grain has an underseeding of legume. 2,4-D is most useful where the economic crop is a more tolerant member of the grass family (i.e. corn, sugarcane, sorghum).

The key features of these herbicides chemically are: (1) length of the acid end group and (2) placement and number of chlorines. Placement of a third Cl at position 5 produces 2,4,5-trichlorophenoxyacetic acid (2,4,5-T) which is useful as an herbicide on woody species. In contrast, 2,4,6-T has no growth-regulating value.

A further modification of herbicidal activity may be made by changing the acid chain length. Substitution of a butyric acid for the acetic acid to give 4-(2,4-dichlorophenoxy) butyric acid–2,4-DB–produces a chemical with different herbicidal characteristics. It is more selective than 2,4-D; that is, while 2,4-D will kill most broad-leaved species, 2,4-DB differentiates between broad-leaved plants. Certain weed species (lambsquarters, pigweed) have a high level of activity in enzymes causing β-oxidation (2-carbon stepwise breakdown) of fatty acids. Hence the two end carbons of the butyric acid are removed by the weed plant enzymes, causing the formation of 2,4-D and thereby producing a chemical with herbicidal properties. Alfalfa and other economic legumes have a much lower activity of this enzyme and at the recommended rates are not harmed by applications of 2,4-DB.

Commercially it is not feasible to sell the pure acid of these herbicides, nor is the acid the form most readily absorbed by the plant. Therefore, commercial formulations are prepared in the following groups of salts, amines, or esters.

The salts (Fig. 13.7) have low volatility and are safe to use in the vicinity of sensitive species. However, they have low solubility, cannot be used in low volume sprayers, and tend to "salt out" rather

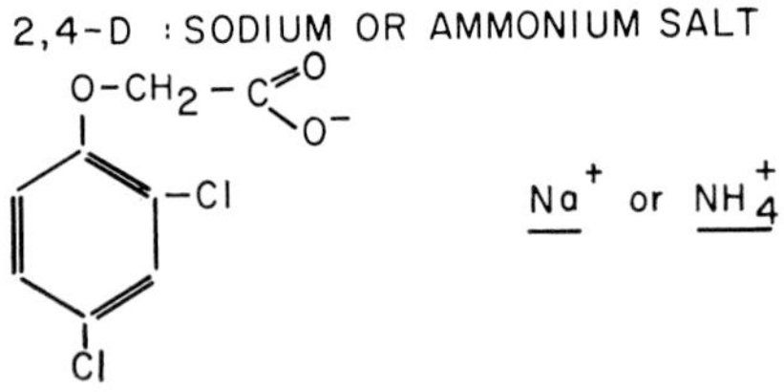

2,4-D : SODIUM OR AMMONIUM SALT

Na^+ or NH_4^+

FIG 13.7 2,4-D salt, the first commercial formulation of 2,4-D.

quickly on the leaf surface. Because they are polar, their ability to penetrate the cuticle is low.

For the amine, the hydrogen ions of the ammonium are replaced by longer sidechains, as for example with the addition of ethyl alcohol to form an ethyl amine (Fig. 13.8). The amine form is intermediate in cost, volatility, and drift.

The combination of an acid and an alcohol yields an ester linkage (Fig. 13.9). The ester form is the most active and often the least expensive.

Ethyl esters were some of the first produced and have been used in large quantities for over a decade. However, the ethyl ester is volatile and develops a spray drift that may travel long distances from the site of application. This may be a serious problem when the target crop (for example corn or wheat) is grown in an area with a susceptible crop like grapes or tomatoes. As a result of such problems longer-chain alcohols have been substituted for the ethyl alcohol; this increased chain length (Fig. 13.9), butoxyethyl ester, reduced the volatility of the material and made it possible to continue the use of the effective ester form. Esters are prepared in oil emulsions and give excellent penetration through leaf cuticle.

As a group the phenoxy compounds speed the rate of respiration and slightly reduce the rate of photosynthesis. Treatment with these herbicides gives a steady loss in total dry weight, a simultaneous increase in turgor pressure and cell brittleness, a resultant epinasty or twisted plant growth, reduced root development, and finally death of the susceptible plant.

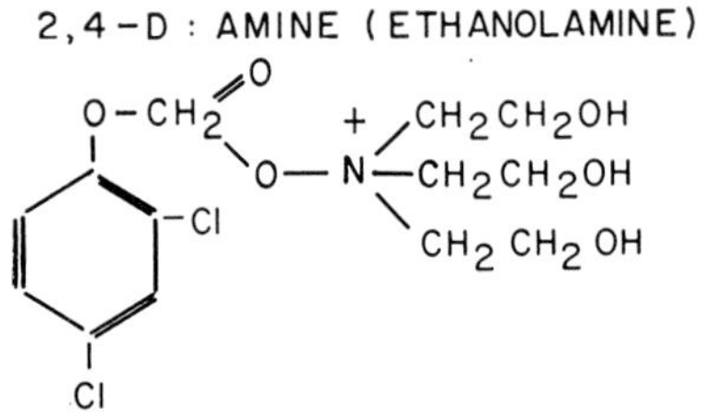

2,4-D : AMINE (ETHANOLAMINE)

FIG. 13.8 2,4-D amine.

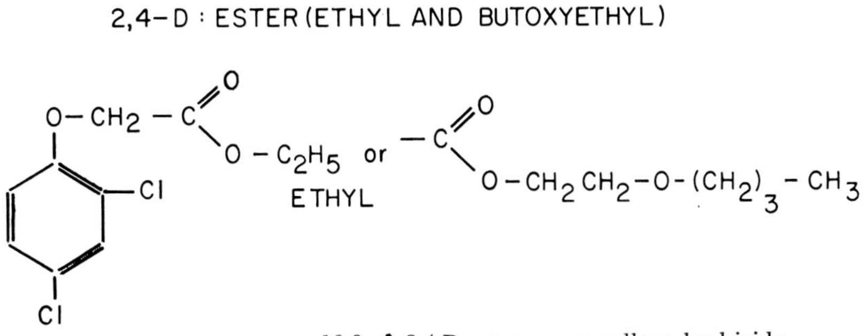

FIG. 13.9 2,4-D ester, an excellent herbicide. The ethyl ester is high in volatility. The longer side chain of the butoxyethyl ester reduces the volatility.

HETEROCYCLIC NITROGEN DERIVATIVES. A second family of very important herbicides are the heterocyclic nitrogen derivatives. These herbicides are either azines (6-membered rings) or azoles (5-membered rings) (Fig. 13.10).

Triazines have a six-membered ring with two or more nitrogens present. The symmetrical triazines are described in Figure 13.10. Table 13.2, adapted from Bartley (1959), characterizes certain members of this group.

Most of the chloro-substituted triazines are limited by lack of phloem mobility to soil application and translocated in the transpiration stream. However, atrazine has been used successfully as a foliage spray on very small weeds because its increased solubility kills them by contact action. The structural differences also affect residual tendencies. Simazine is more likely to carry over to the next year than atrazine although both must be managed carefully and at the recommended dosage rates to minimize damage to a susceptible crop the next season.

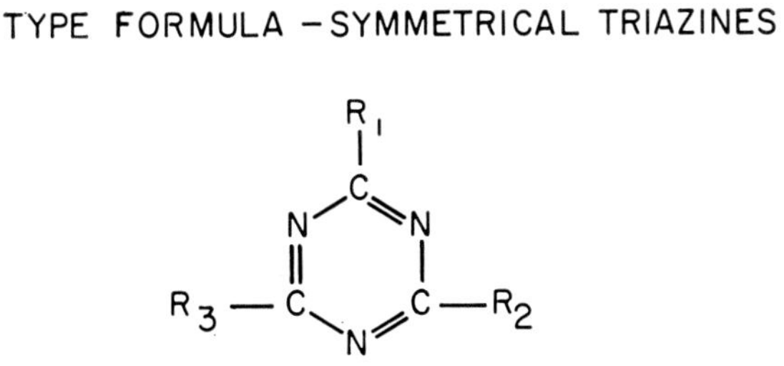

FIG. 13.10 The type formula of the triazines (see Table 13.2).

TABLE 13.2. Properties of triazine herbicides (C. E. Bartley, Triazine Compounds, *Farm Chemicals* 122, May 1959)

Compound	R_1	R_2	R_3	Water Solubility at 20° C.
Simazine	chloro	ethylamino	ethylamino	5
Propazine	chloro	isopropylamino	isopropylamino	8.6
Atrazine	chloro	ethylamino	isopropylamino	70

The triazines function to inhibit photolysis and the transfer of hydrogen to NADP in susceptible plants. Tolerant plants metabolize the chemical. It is of interest to note that simazine has the side effect of increasing the nitrate reductase activity of susceptible species (i.e. beans) and thereby increasing nitrogen content and growth (Tweedy and Ries, 1967).

In summary, the above representatives from the phenoxy and triazine families serve as a few examples from the many herbicides to represent ways in which weed-killing activity, residual tendency, and selectivity are affected by small changes in the structural characteristics. Chlorine or a related halogen frequently serves as an activating side group. Organic acids are important side chains, and both amino and nitrate-nitrite nitrogenous groups are frequently useful. But it is appropriate to repeat that chemists are still not able to predict what specific structure will be useful, and so each new chemical synthesized is screened for possible herbicide activity, causing the categories to expand in many directions.

Formulation of Herbicides. Herbicides are commonly combined with a liquid or solid carrier to improve their uniformity of application. Formulations for field application are available as sprays, granular preparations, and fumigants.

The NAS publication *Weed Control* (1968) outlines the following types of formulations.

SPRAYS. These materials are commonly prepared as emulsions, suspensions, or solutions. More specifically, the herbicide spray concentrates may be classified as one of the following five types: soluble herbicides, solution concentrates, emulsifiable concentrates, wettable powders, and flowable formulations.

The soluble herbicide requires very little formulation and may be put directly into the carrier, most commonly water. Solution concentrates must first be dissolved in a solvent which in turn is soluble in the carrier. Water is most commonly the solvent and a water concentrate can be prepared. Emulsifiable concentrates are most commonly based on organic solvents. Then an emulsifier is added to allow an oil-in-water emulsion. Wettable powders are used when the three previous formulations are not possible. In this formulation the finely divided solid is suspended in water and then agitated to provide a homogeneous mixture during application. Finally, flowable formulations represent a two-phase system of a solid or a liquid suspended in any liquid (but most commonly water) that provides a stable concentrate.

GRANULAR. Granular formulations have the herbicide impregnated into a solid carrier and are applied in this way rather than in solution. These carriers may include clays, minerals, shales, corncobs, perlite, and many other compounds that have good liquid retention capacity.

FUMIGANTS. Fumigants are very volatile materials and are most often used for perennial weed control by injecting them into the soil surface. They have several drawbacks and are primarily for very specialized uses.

ADJUVANTS AND SURFACE-ACTIVE AGENTS. Several types of adjuvants may be added to herbicides and improve their action. Surface-active agents may reduce surface tension and improve the spread of the herbicide over the leaf surface. Emulsifiers maintain the stability of a mixture such as an oil-water combination. Thickening and sticking agents improve adherence of the herbicide to the leaf surface. It is crucial to recognize that a single surfactant may increase, decrease, or not affect the action of an herbicide. For example, one such reagent increased the activity of dalapon sevenfold, trebled amitrole activity, and left the activity of 2,4-D and DNBP unchanged on corn. On soybeans, the same surfactant doubled the action of dalapon and amitrole and trebled the action of 2,4-D and dinoseb. Therefore, it is exceedingly important to know what interaction to expect with a particular combination before using it.

In some cases, herbicide activity may be increased by improved translocation once it has entered the plant. For example, amitrole activity on quackgrass is enhanced by ammonium thiocyanate (NH_4SCN). The thiocyanate does not increase herbicide absorption but improves translocation. In addition, the light regime prior to treatment with an herbicide can influence the relative toxicity to the plant. For example, ultraviolet radiation is effective in bean plants in preventing epinasty normally caused by 2,4-D. Apparently ultraviolet light prevents the auxin action (see auxin-ethylene interactions, Chapter 7).

MAJOR WEED CONTROL QUESTIONS

Questions related to weed control often focus on the following three areas.

1. When is competition from weeds most detrimental to the crop? Commonly, early in the crop plant's growth. In the north central states the grassy annual weeds and the small grains germinate rapidly between 58 and 62° F. Many of the broadleaf annuals also germinate well at cool temperatures. In contrast, corn is favored by temperatures of 60–65° and soybeans and sorghum by 65–70°. Therefore, while some of the small grains and forage crops compete well with weeds at cool temperatures, many of the intertilled crops do not.

2. Where are these major problem weeds, the annuals, found in the soil profile? The seeds may be distributed throughout the upper several feet. However, the requirements of all for adequate moisture and by some for light and oxygen to stimulate germination are met only when they are near the surface. Therefore, the major weed problems arise in the upper few (often one or two) inches of the soil. If cultivation equipment is available to work this area when the weed seedlings are very small, excellent control can be achieved.

Therefore, for intertilled crops methods are sought that (a) kill the weeds before planting the crop, or (b) leave a seedbed rough and porous, poor for weed germination (Fig. 13.2), or (c) allow rapid and shallow working of the field by a rotary hoe or spike-tooth harrow after the crop is up, or (d) provide an herbicide residual that continues to kill weeds as they germinate.

3. The third major question is relative cost and benefit of various weed control systems. A 1966 report on costs from Illinois suggests that a band application of herbicide plus one cultivation costs $4.64 to $7.25 per acre, depending on the acres treated and the herbicide used. In contrast, conventional cultivation costs (one rotary hoeing and two cultivations) were from $2.97 to $5.83 for the same acreage range. This study concluded that if weeds can be controlled by cultivation, this is the least costly method. However, as farmers push for higher yields and narrower rows, band applications (i.e. a 14-inch band on a 20-inch

row) are not strikingly different in cost from a blanket application, and such blanket treatments will probably become more common.

METHODS OF WEED CONTROL FOR SPECIFIC CROPS

The availability of chemical weed control has had a major impact on farm management and crop production. However, the use of such chemicals does not preclude the necessity of careful cultural practices and good general crop management. Seedbed preparation, seed selection, cropping systems, and other means of weed control all have a role to play in preventing weed growth from reducing potential yields.

CORN

Seedbed preparation, cultivation, and herbicides should be combined to develop an effective weed control program for corn. Seedbed preparation should be done to favor rapid, even germination of the corn. When possible the field surface should be left smooth and level to favor the effectiveness of the shallow cultivation implements like the rotary hoe and the finger weeder. Recent interest has been strong on minimum tillage of corn seedbeds for the purpose of cost reduction. A maximum reduction of tillage is represented in studies like those of Triplett (1966) who observed that an herbicide may be used to kill the sod and provide a no-tillage system of corn production. By use of combinations such as atrazine and 2,4-D, Triplett obtained results similar to conventional tillage. Such systems will not likely increase yields but may provide a method for reducing cost and erosion. One important side benefit of several minimum tillage systems is the loose, open inter-row zone that is a poor medium for weed growth (Fig. 13.2). In these systems only the row zone is worked finely for the crop. This row zone can be treated with a band of herbicide to combat the weeds which are favored to grow there.

Early shallow postemergence weed control is done very effectively with the spike-tooth harrow when the corn is 2–4 inches tall or with the rotary hoe for corn from the time it has just emerged until it is 12 inches in height. These implements allow speed and timeliness in weed control and are low-cost methods. The rotary hoe functions best when operated at speeds of 6–12 miles per hour in a slightly crusted soil and with weed seedlings not yet emerged from the soil. Rotary hoeing gives some weed control benefit even though a crust is not present. It should be noted here that minimum tillage systems do preclude the use of the rotary hoe. Sweep cultivation should always be shallow; roots extend horizontally as far as the extended leaves are tall, and the roots are so close to the surface that cultivation of any depth may cause undesirable root pruning.

As indicated previously, early-season weed competition is very critical to the crop plant. In the Corn Belt this is when weeds germinate and can be killed by preemergence herbicides. Therefore, much of the recent emphasis in herbicide work on corn is for this period, and preemergence herbicides may substitute for shallow and early sweep cultivation. In some seasons very satisfactory mechanical weed control is achieved. In others the field may be too wet for mechanical activity, and it is then that chemical control is particularly attractive.

As a general rule, 0.5 inch of rain in the two weeks following herbicide application is necessary to leach an herbicide into the preemergence zone of germinating weed seeds. From the averages of a 4-year study in Iowa, Staniforth and Lovely (1964) found that various herbicides gave 4–6 more bushels of corn than cultural weed control. This study compared early-season weed control and used two sweep cultivations for later-season control. They do point out that the margin of extra return is not as great as corn yields increase. Therefore, the farmer who is using high rates of fertilizer and high plant populations may achieve completely satisfactory early weed control by mechanical means. However, a band application to supplant early cultivation and then using sweep cultivation is likely to be the optimum combination economically.

In those situations where broadcast applications of herbicides appear desirable, the triazines (simazine and atrazine) give the best full-season control under Midwest conditions. Simazine is used in the higher rainfall areas of the eastern Corn Belt; atrazine is desirable in lower rainfall areas. Care in application rate adapted to soil type has minimized residue problems and makes atrazine one of the best preemergence or early postemergence chemicals. Problems of residual or carryover still do occur, however, and require very careful planning as to how much to apply and what crop is expected to follow in the cropping system the next season.

Postemergence applications of 2,4-D have given excellent control if applied when corn is in the 2–3-leaf stage; 2,4-D may be used later in the corn plant's growth, but great care must be taken not to get it on the plant. Propachlor, CDAA, CDAA-T, linuron, and dalapon may all be used under certain conditions to handle weed problems in corn. As a late rescue operation against grassy weeds, dalapon and 2,4-D amine may be used. Leaf lifters are used to minimize the spray getting on the corn, but this practice is potentially very detrimental to the corn plant and must be used with great care.

Postemergence applications of atrazine when the corn is 2–3 inches tall may be made with a standard atrazine-water solution or may include atrazine in water plus oil. The oil must be nonphytotoxic and contain 1–3% emulsifier.

The use of flaming for corn weed control recently has received renewed attention. Best results appear to be achieved with 2 or 3 flamings when the corn is 12–18 inches tall. However, flaming will most likely be used as a rescue operation when weed growth has gotten out of hand and not under standard weed control conditions.

SOYBEANS

At the present time soybeans are less tolerant to the available herbicides than many other crops. Therefore, emphasis remains heavy on weed control by mechanical means. An initial disking of the seedbed two weeks before planting, delay, then reworking just before planting is useful in minimizing weed growth. The rotary hoe is even more important to early-season weed control for soybeans than for corn. Preemergence chemicals are now available; amiben, naptalam + chloropropham, lasso, and trifluralin are suitable for grassy weeds and for some broadleaf weeds. CDAA is excellent for grass control. These chemicals are not as fully predictable as desired. However, Staniforth et al. (1963) found band applications of the best residual preemergence herbicides were equal to timely shallow cultivations and superior when wet weather delayed the early shallow cultivations. Over a 10-year average they measured a net loss of 3.8 bushels despite normal control measures; this would suggest that the herbicide could reduce weed losses and pay for itself. A combination of band-applied herbicide or rotary hoeing and sweep cultivation will be important in soybean culture for the foreseeable future.

SORGHUM

Weed control practices are similar to those useful for corn. However, the seedling of sorghum is weaker than that of corn and very susceptible to 2,4-D. Spraying with this chemical is best done when the plant is 4–12 inches tall. The triazine propazine is considered to be the most desirable one for this species, although careful use of atrazine in reduced rates is also successful.

COTTON

With the advent of mechanical harvesting, weed control is doubly important for cotton to assure ease of harvest. In addition, in earlier years the cotton farmer would plant the crop rather thick, then "chop cotton" to the right population and at the same time achieve intrarow weed control. Now the practice is to plant to a stand and therefore weed control is a separate concern.

For preemergence treatments trifluralin, diuron, and CIPC are used. Trifluralin is banded in the top inch above the seed. Postemergence directed sprays include diuron, varosol, and Stoddard's solvent, which rely in part on the woody stem of cotton to minimize potential damaging effects.

Flaming is a useful weed control system on cotton. The practice has been used successfully with this woody-stemmed plant after it is ¼ inch in diameter. The flame is developed from the combustion of natural gas and is directed at the ground between the rows. It has been much more difficult to show regular benefits in using flame for weed control in the more herbaceous species (see corn).

SMALL GRAINS

Most of the small grains are seeded broadcast or in drill rows that are narrow enough to result in solid stands very quickly. Therefore, good seedbed preparation and fields as weed-free as possible are important because there is little opportunity for later work in these fields. If weeds do not get started early, those which germinate later will be shaded and stopped in growth. In the Great Plains area, wild oats and members of the mustard family are the major problems for the small-grain producer. Crop rotation reduces these problems, but if the weed growth is heavy herbicides may be used. In fact, more acres of small grains are sprayed with 2,4-D than any other group of crops. The best stage to spray is when the crop is 5–8 inches tall, has five or more leaves, and is fully tillered. Spraying should be completed before jointing (stem elongation) begins. Sometimes MCP is favored over 2,4-D for safety or where a legume is present as an underseeding. For wild oats control in wheat and barley the very selective carbamates, such as diallate and barban, may be used. Dicamba and bromoxynil are used for control of polygonums, especially wild buckwheat.

RICE

For centuries flooding has been a major means of weed control. This works well on most grassy weeds but actually favors some broadleaf weeds. Therefore, 2,4-D found early favor as a broadleaf control measure. For control of one grassy weed now favored by shallow flooding—barnyard grass (*Echinochloa* spp.)—certain propionanalide herbicides show promise. Herbicides are used to control weeds on more than half the rice acreage in the United States. As with all other crops, appropriate fertilization, weed-free seed, a rotation (a rice-pasture or a rice-soybean rotation in the South), and good seedbed preparation all provide an important measure of weed control.

FORAGE CROPS

After the crop is well established, competition by the dense stand provides excellent weed control. To reduce weed competition in the fall of the seeding year, clipping or mowing is desirable. In lieu of a companion crop, early season weed control by EPTC, 2,4-DB, and dalapon looks promising. ⁂

LITERATURE CITED

Audus, L. J. 1964. *The Physiology and Biochemistry of Herbicides.* Academic Press, Inc., New York.

Bardsley, C. E., K. E. Savage, and V. O. Childers. 1967. Trifluralin behavior in soil. I. Toxicity and persistence as related to organic matter. *Agron. J.* 59:159–60.

Bartley, C. E. 1959. Triazine compounds. *Farm Chemicals* 122 (May).

Burnside, O. C., G. A. Wicks, and C. R. Fenster. 1964. Influence of tillage row spacing and atrazine on sorghum and weed yields from nonirrigated sorghum across Nebraska. *Agron. J.* 12:211–15.

Crafts, A. S., and W. W. Robbins. 1962. *Weed Control,* 3rd ed. McGraw-Hill Book Company, Inc., New York.

Eplee, R. E., and G. C. Klingman. 1968. Effect of corn population and simazine on weed growth. *Agron. J.* 60:87–89.

Kearney, P. C., and D. D. Kaufman. 1967. Biochemistry of herbicide decomposition in soils. In A. D. McLaren and G. H. Peterson (eds.). *Soil Biochemistry.* Marcel Dekker, Inc., New York.

King, L. J. 1966. *Weeds of the World.* Interscience Publishers, New York.

Klingman, G. C. 1961. *Weed Control as a Science.* John Wiley and Sons, Inc., New York.

Leasure, J. K. 1963. The mode of action of dalapon. *Down to Earth* 19(1):19–22.

Lee, G. A., A. K. Dobranz, and H. P. Alley. 1967. Preliminary investigations of the effect of Tordon and 2,4-D on leaf and root tissue of Canada thistle. *Down to Earth* 23(2):21–23.

Linscott, D. L., R. D. Hagin, and M. J. Wright. 1965. Effect of ensiling on the decomposition of several herbicides. *Crop Sci.* 5:455–56.

Lynd, J. Q., C. Rieck, D. Barnes, D. Murray, and P. W. Santelmann. 1967. Indicator plant aberrations at threshold soil herbicide levels. *Agron. J.* 59:194–96.

Mann, J. D., L. S. Jordan, and B. E. Day. 1965. A survey of herbicides for their effect upon protein synthesis. *Plant Physiol.* 40:840–43.

National Academy of Sciences. 1968. *Weed Control.* Vol. II of *Principles of Plant and Animal Pest Control.* Publication 1597. Washington, D.C.

Nieto, J. H., and D. W. Staniforth. 1961. Corn-foxtail competition under various production conditions. *Agron. J.* 53:1–5.

Prihar, S. S., and D. M. VanDoren, Jr. 1967. Mode of responses of weed-free corn to post-planting cultivation. *Agron. J.* 59:513–16.

Santelmann, P. W., C. J. Scrifes, and J. Murray. 1966. Influence of post-emergence herbicides on the fiber quality of selected cotton varieties. *Crop Sci.* 6:561–62.

Smith, R. J., Jr., and W. C. Shaw. 1966. *Weeds and Their Control in Rice Production.* USDA Agriculture Handbook No. 292.

Staniforth, D. W. 1957. Effects of annual grass weeds on the yield of corn. *Agron. J.* 49:551–55.

Staniforth, D. W., and W. G. Lovely. 1964. Preemergence herbicides in corn production. *Weeds* 12:131–33.

Staniforth, D. W., W. G. Lovely, and C. R. Weber. 1963. Role of herbicides in soybean production. *Weeds* 11:96–98.

Triplett, G. B., Jr. 1966. Herbicide systems for no-tillage corn (*Zea mays* L.) following sod. *Agron. J.* 58:157–59.

Tweedy, J. A., and S. K. Ries. 1967. Effect of simazine on nitrate reductase activity in corn. *Plant Physiol.* 42:280–82.

Vengris, J. 1955. Plant nutrient competition between weeds and corn. *Agron. J.* 47:213–16.

Weed Science Society of America. 1967. *Herbicide Handbook.* W. F. Humphrey Press, Inc., Geneva, N.Y.

Zimmerman, P. W., and A. E. Hitchcock. 1942. Substituted phenoxy and benzoic acid growth substances and the relation of structure to physiological activity. *Contrib. Boyce Thompson Inst.* 12:321–43.

❧ CHAPTER FOURTEEN ❧ INSECT AND DISEASE PROBLEMS

❧ THE HISTORY OF AGRICULTURE is a history of man's battle with the environment as he sought to produce an adequate food supply. A considerable part of that battle has been waged against some of the insect and disease pests which are the subject of this chapter. A crop production system, whether intensive or not, faces the problem of pest control. In agriculture throughout the world any system of crop production must provide sufficient pest control either to eliminate pests as a limiting factor or in some cases to prevent complete loss of a crop by a pest buildup that destroys the crop. The farmer's long standing battle with the losses from these pests is well summarized by the following ditty:

> one for the bug,
> one for the crow,
> one to rot and
> two to grow.
> —PFADT, 1962

The two pest groupings of insects and diseases cause similar types of injury to economic crops. These effects may be described as: (1) a loss to photosynthetic or root surface, (2) a plugging of the translocation system, (3) an injection of toxic poisons or diseases, and (4) damage to the reproductive plant structures. A unique insect-disease interrelationship occurs when (5) the insect serves as a vector to carry a disease from one plant to another.

Control for both types of pests may be broadly organized under the headings of natural, biological, legal, cultural, genetic, and chemical. With these generalizations, it seems most fruitful to discuss each pest category separately.

INSECTS

The majority of insect problems are caused by the order Insecta which represents that group of Arthropods with three body sections and six legs. The anatomy of this order is useful knowledge as a basis to discuss control. The head serves a sensory function, the thorax is locomotory, and the adbomen is the site of metabolic activity. Each of these three sections is further subdivided into segments which are rings of chitin that give the insect an exoskeleton of great strength. This exoskeleton contains a lipoid layer almost impervious to water; therefore the contact insecticides are chosen on their ability to penetrate it.

Most insects obtain air conducted to the respiring cells through tubes called tracheae. These tracheae, which are most concentrated on the abdomen, open to the exterior through spiracles which usually have valves or a closure apparatus. During respiration oxygen is carried directly to the organs through the tracheae. Oils used as control sprays block the spiracles and fumigants enter at this point as well. Insect blood, hemolymph, possesses no respiratory pigments and plays no part in oxygen transport, but the hemolymph does serve to transport food, excretory products, and hormones.

Insect mouthparts may be adapted for chewing (grasshopper, beetles), chewing-lapping (bees), rasping-sucking (thrips), piercing-sucking (aphids, bugs), siphoning (adult moths), or sponging (fly). The nature of the insecticide formulation and mode of action may be selected specifically to adapt to these mouthpart characteristics, but more frequently the formulation and mode of action are determined by the environment in which the insect must be attacked (i.e. closed grain bin, soil, aerial plant portion, inside leaf whorl, etc.).

The immature insect does not grow continuously but has stages called instars. At the end of these stages the epicuticle and exocuticle split off, a process referred to as moulting. Each moult is followed by a rapid increase in size before the hardening of the new epicuticle. One key feature of the insect life cycle is that the attack of a particular species on field crops may end abruptly because of the onset of moulting or of pupation.

The life cycle of an insect may be classified either as complete or incomplete metamorphosis. Under complete metamorphosis the larva does not resemble the adult, may have a different habitat, usually feeds in an entirely different way, and never possesses wings. The larva has one key function—growth—and is seldom good at locomotion. The sequence moves from egg to larva to pupa to adult. In contrast, incomplete metamorphosis consists of a sequence from egg to nymph to adult, and the nymph is very similar to the adult. This means that the larval stage of complete metamorphosis and the nymph (and often the adult) stage of incomplete metamorphosis are the most destructive but at the same time less mobile and therefore more susceptible to control. Insects of both the incomplete and complete metamorphosis patterns are usually characterized by two additional features: (1) wings on the adult, suggesting a high degree of mobility, and (2) an immobile, nonfeeding resting period. This resting period may be the pupal stage or it may occur during the adult stage (e.g. beetles).

The order Insecta contains an extensive range of species and a great diversity of genotypes, a complexity beyond the scope of this reference. The reader may wish to study any current applied entomology text to note the complexity of the order (Pfadt, 1962; Rolston and McCoy, 1966).

METHODS AND PRINCIPLES OF CONTROL

Natural. Natural control is viewed as a system uncontrolled by man. Some call this the "balance of nature." The ecologist is particularly interested in this system. Since agronomic practices frequently call for maximum control of the crop environment, this method is not very appealing for use with intensive cropping practices (Richter, 1966). Natural forces do not necessarily work to man's advantage unless given direction. Yet it is important to emphasize that man must carefully select a variety of control measures and not shift the natural environment more than absolutely necessary for food production. A balance of environment is still very much desired.

Biological. Biological control is related to natural control and is that phase of natural control in which man plays an important part by selective application of the biological agent. Biological control has a strong appeal because it represents a minimal hazard to human health. Natural enemies may be used on the basis of introducing them to the environment, augmenting those already present, or taking care to conserve those present by care in application of chemicals toxic to a favorable natural enemy.

Many insects are kept in check by parasites, predators, and diseases. However, if an insect is introduced to an area accidentally, without its natural enemies, it may cause great problems. Of the 183 most important insect pests in North America, 81 are of foreign origin. Parasitic insects have been most useful in biological control, but certain diseases are also being developed. Obviously the control organism must be well understood lest it also become a pest.

The classic example of biological control was the introduction of a predatory beetle (*Rodolia cardinalis* Muls) from Australia into California to combat the cottony-cushion scale (Doutt, 1958). The scale was introduced from Australia and threatened to destroy the entire citrus industry of California. The predator beetle was established from 140 individuals and has now nearly eliminated the pest. Such predators are most successful where the pest is isolated topographically or geographically.

Diseases can also be used to control undesirable insects. Spores of the bacteria *Bacillus thuringiensis* have been isolated, impregnated into an inert carrier, and found to be effective against some Lepidoptera and Coleoptera larvae, including the European corn borer, the small cabbage white butterfly, the diamondback moth, and the confused flour beetle. For European corn borer control, this method is as effective as chemical insecticides but more expensive and therefore not used extensively. Studies by USDA also point to the possible use of viruses to kill the cabbage looper and the corn earworm (cotton bollworm larva). However, viruses appear to be slower in action than the insecticides, a disadvantage if the pest population has reached a high level before the control measure is applied.

The release of irradiated males shows promise as a control measure. The males are sterilized by irradiation, and when they mate with a female an infertile egg results. This technique has been used successfully for screwworm control on livestock and shows promise for control of some of the insect pests of field crops. This appears to be most feasible in isolated areas.

Still another example of biological control is represented by studies in which the photoperiodic environment is changed (Fig. 14.1). Intense, mobile (even airborne) sources of light of narrow frequency ranges may control insect pests such as the cabbageworm selectively by inhibiting their diapause. Photoflashes from a xenon lamp applied to larvae of cabbageworms in the dark period of a 24-hour cycle inhibited diapause. The wavelengths involved seem to be different from those involving plant reproduction and should not inhibit a photoperiodic response.

This rather extended statement on biological control systems has been presented to emphasize that they are difficult to perfect, and they may not act as quickly as the chemical insecticides, but they do have the advantage of leaving no residue and are expected to be increasing-

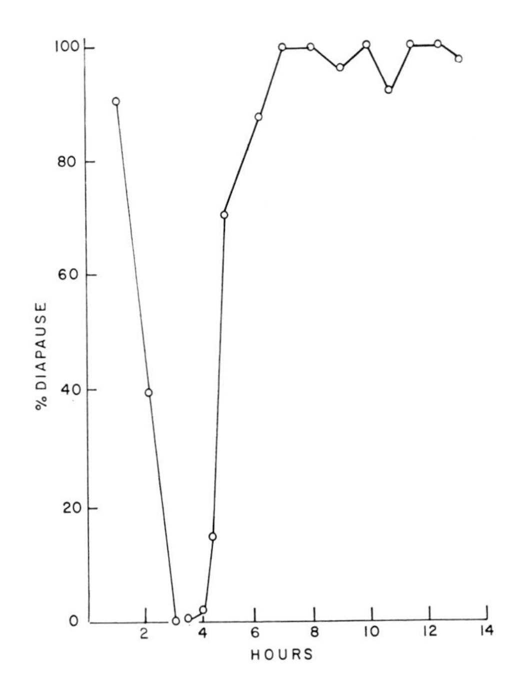

FIG. 14.1 The percent of pupae in diapause (slowed growth) when photoflashes were applied daily to *Pieris rapae* larvae. "Hours" are measured from the end of a 10-hour light period (Barker et al., 1964).

ly important in the future as man becomes more concerned about retaining or improving the quality of the environment.

Legal. Legal control is the control of insects by controlling human activity. Legal control may include (a) inspection and quarantine laws, (b) laws to enforce application of control measures, (c) insecticide and poison residue laws.

Cultural. Cultural practices have been used by man for centuries to control insects. The use of crop rotations was stimulated by the need for pest control. The benefit of this practice is limited to the distance the egg-laying adult can travel and the time of the year when the female lays her eggs. For example, a corn borer moth may fly a mile before depositing her eggs on an actively growing crop. Rotation of crops on an individual farm thus has little benefit in corn borer control. In contrast, the corn rootworm adult lays her eggs primarily in corn stubble or in a field containing large quantities of pollen late in the season. Eggs are not deposited in soybean or oat fields, and corn grown following these crops would not have rootworm infestations. Planting corn or sorghum some distance from fields of small grains to avoid chinch bug damage is useful because the chinch bug nymph cannot travel far from where it was hatched in the small grain stubble. Spatial separation of this pest and the host therefore may be helpful.

A second cultural practice useful in insect control is the timing of the planting or sowing date. The objectives are either to avoid the egg-laying period or to have the crop mature before the pest appears. Generally early planting is beneficial but in a few cases late sowing provides the primary benefit. Early planting of corn favors strong root development which can tolerate the corn rootworm better when it hatches in June. For many years winter wheat has been planted after the "Hessian fly-free" date in the fall. The adult females lay their eggs during a given photoperiod; therefore, after the daylength is short enough, the flies cease laying eggs, and wheat which emerges after that date will not be infested (Fig. 14.2).

Plowing under residue may give some insect control and is often referred to as sanitation. Special cultivations are useful against soil insects where, for example, turning the soil or rolling to consolidate the soil may bring insects to the surface where they are eaten by birds.

Genetic. Genetic control of insect infestation is essentially a type of biological control. Relatively few varieties have been developed that are resistant to a particular insect pest (e.g. corn borer), but the problems for the plant breeder are many.

The enormous complexity and several basic aspects of breeding for insect resistance are illustrated in Figure 14.3 as an outline of the

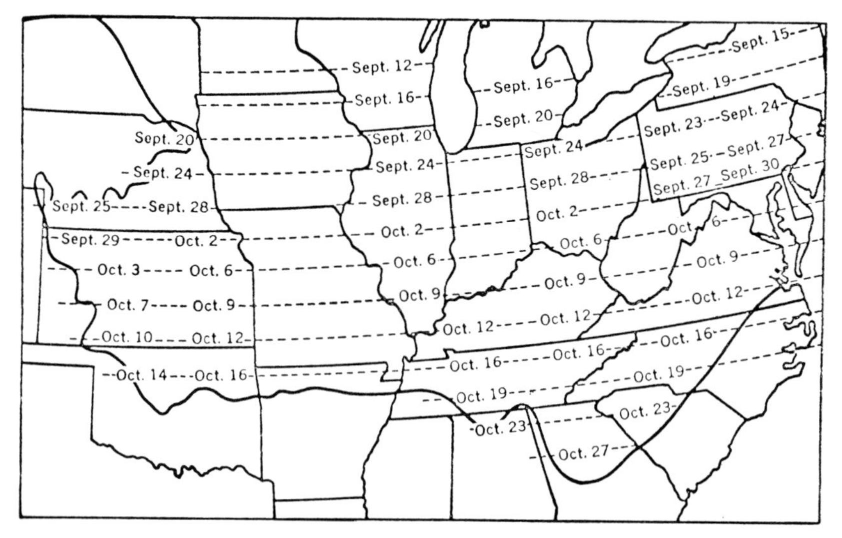

FIG. 14.2 ☙ Hessian fly-free dates for sowing winter wheat (Martin and Leonard, 1967).

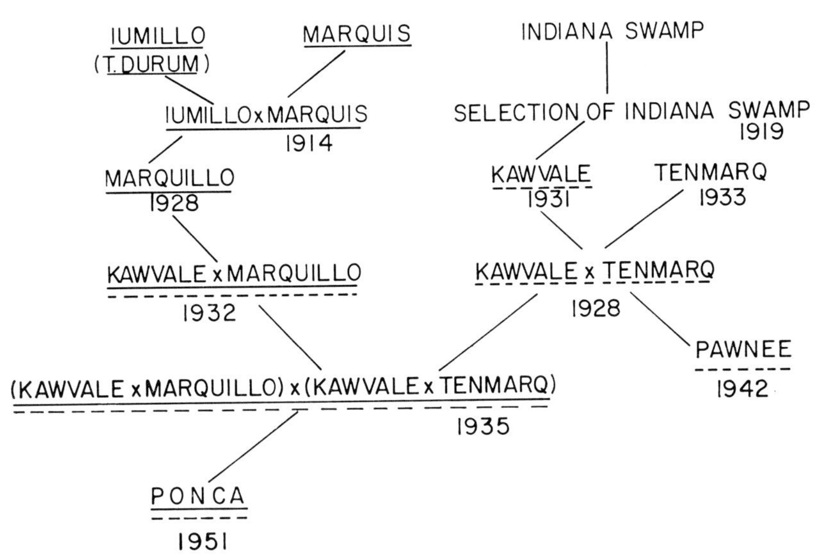

FIG. 14.3 ✤ Pedigree of Ponca wheat. Two sources of resistance to Hessian fly indicated by the solid and dotted lines (after Rolston and McCoy, 1966).

breeding done to obtain the Hessian fly-resistant Ponca wheat. The varieties Iumillo and Marquis gave two unlike sources of resistance while Marquis and Tenmarq gave other desirable wheat qualities. Kawvale also carried a source of resistance, and crossing it with Tenmarq gave an excellent Hessian fly-resistant hard red winter wheat—Pawnee. Pawnee had both a considerable degree of tolerance to Hessian fly damage and a fair degree of antibiosis resistance.

Crossing Kawvale and Marquis combined the two sources of resistance, but the progeny was deficient in some respects. A further cross with a sibling line of Pawnee overcame the difficulties and produced a hard red winter wheat with outstanding Hessian fly resistance. Ponca, therefore, represents nearly four decades of work; and a decade after its release, Ponca was in serious disease difficulty. The process is arduous and constant. Host plant resistance is no quick or easy answer to insect control problems.

A truly resistant plant has physiological or physical features which enable it to withstand an insect attack. The mechanism of resistance to an insect pest can be classified in three groups: tolerance, antibiosis, and nonpreference.

1. Tolerance. The plant survives and grows even when heavily infested, either by great vigor or rapid repair of damaged organs. For example, under English conditions, rapidly tillering oat varieties

are able to withstand heavy frit fly attacks in effect by outgrowing the fly attack. Corn lines which produce adventitious roots readily more adequately withstand attacks by root feeding pests. Sheer plant strength provides tolerance as found in some wheat lines which do not break easily even when infested with Hessian fly.

2. Antibiosis. In this case the variety may be an inferior food source and reduce the rate of aphid reproduction, or a thicker cuticle may prevent sucking insects from feeding. The chemical dimboa has been shown to repel feeding of the European corn borer in certain lines of corn.

3. Preference or nonpreference. The texture, aroma, color, or taste of the plant may reduce insect attack. For example, the sweetclover weevil is attracted to sweetclover high in coumarin; conversely, the weevil finds low-coumarin lines unpalatable. Tightly sheathed leaves resist larva burrowing in them and hairy leaves slow larva movement.

Chemical. This method of control has had a great upsurge since World War II and was primarily stimulated by the availability of DDT, a chemical which has been used widely to control insects causing man direct and indirect problems. This insecticide is looked upon less favorably in recent years because of its long residual, and DDT points up the need to use chemical control very carefully and discreetly.

A good insecticide is toxic to the pest insect; as specific as possible and not toxic to parasites and predators of the pest or bees; harmless to man, birds, and mammals; not harmful to the crop; does not taint or leave residues; is soluble or suitable for formulation as a dust or wettable powder; and is cheap. So far, no chemical discovered possesses all these qualities—each one represents a compromise.

Insecticide tolerances, the maximum amount permissible in the end product, are set by the United States Food and Drug Administration based on pharmacological tests of the insecticide and upwards to two years of animal feeding trials. The Secretary of Agriculture has defined as "highly toxic" any compound with an acute oral LD_{50} value (50% of the animals killed by that rate) of 50 mg/kg or less for laboratory animals.

Before considering the insecticides available and their mode of action, it is well to review some of the characteristics of insects. For insects with complete metamorphosis, the larval stage is the most immobile, the most susceptible to control, and frequently the most damaging. In contrast the adult often flies, is very mobile, and may not have destructive mouthparts. Insects with chewing mouthparts may be effectively treated with a plant surface treatment that acts as a stomach poison. Insecticides applied to the foliage must usually be insoluble so they will not be absorbed and kill the plant. The insect with sucking mouthparts must be treated either with a systemic, internal poison or a contact poison that enters through the integuments into the blood or through the spiracles into the tracheae. Under closed conditions fumigants may be used.

INSECTICIDE FORMULATIONS. A wide range of possibilities exists for insecticidal formulations. They may be purchased as aerosols in high-pressure containers and used primarily for spot treatment procedure. A fumigant is a gas or liquid that volatilizes readily in the closed area in which it is applied. An oil solution has its most common application as a residual spray and often on a dormant crop. Where water is readily accessible and a high-volume spray is convenient, the wettable powder provides a formulation technique that will accommodate a wide range of synthesis problems. The emulsifiable concentrate provides a means for placing the insecticide and a surfactant (a chemical included to improve wettability and reduce surface tension) together during factory processing. The low-volume spray has been a recent innovation stimulated by aerial applications in which a minimum quantity of water is desired. Because this material is extruded under extremely high pressure and develops a very small particle size, it has sometimes caused undesirable problems of drift. A granular formulation is popular for its handling convenience and for getting the insecticide down through a dense growth. Addition of the insecticide to a fertilizer mixture is appealing from the convenience aspect but seldom does an effective job of insect control because the fertilizer is not placed where the insecticide is most needed.

INSECTICIDE ACTION. Three types of insecticidal action may be described: (1) those which have a massive effect of destroying the digestive system, acting as an abrasive and precipitating protein, (2) those which damage the respiratory mechanism and inhibit the electron transport needed for respiratory synthesis of ATP, and (3) those which inhibit acetylcholinesterase and disrupt nerve function. The third group's action will be expanded because it represents a high degree of selectivity.

The organophosphates and the carbamates are frequently referred to as the "anticholinesterase insecticides" since they inhibit the enzyme acetylcholinesterase and lead to disruption of nerve function as described below. The transmission of nerve stimuli is described as transfer of electrical charges between nerve endings—the synapses. The rather complex fatty material acetylcholine apparently mediates the nerve impulse transfer and then must be removed from the synaptic region. Removal is accomplished by action of acetylcholinesterase, which hydrolyzes the acetylcholine. The buildup of excess acetylcholine causes excitation and the presence of neuroactive substances in quantities that disrupt normal nerve function and cause tremors, convulsions, paralysis, and death. Thus, the action of these anticholinesterase substances is to allow the buildup of the spent choline and disruption of nerve function.

(acetylcholine)

$$CH_3CO - OCH_2CH_2N^+(CH_3)_3 + H_2O \xrightarrow{\text{cholin} \downarrow \text{esterase}} \underset{\text{(acetate)}}{CH_3COO^-} + \underset{\text{(choline)}}{HOCH_2CH_2N^+(CH_3)_3}$$

CLASSES OF INSECTICIDES. The insecticides may be classified as follows. A few examples and the general mode of action are noted for each class.

1. Inorganics. Acid lead arsenate, sodium fluoride, sulfur, and copper sulfate-lime (Bordeaux mixture). The heavy metals inhibit many biological reactions by competing for sites with essential enzyme cofactors (Chapter 3).
2. Dormant spray oils. These oils act to plug the respiratory system, especially the spiracles.
3. Botanicals. Nicotine from tobacco, pyrethrum from members of the Chrysanthemum genus, and rotenone, which may be extracted from the roots of 68 species of legumes, are three of the most commonly used materials. The botanicals are thought to damage nerve functions. Their use is often replaced by more specific materials listed below.
4. Synthetic organics. Members of this group may act as contact or stomach poisons or function as cholinesterase inhibitors.
   (a) Chlorinated hydrocarbons. This subgroup, of which DDT continues as the dominant member, has been widely and successfully used for insect control around the world. They are generally persistent and lipophilic, with a resultant storage in fats and oils. The continued use of this group is very likely to diminish because of their residual tendencies.
   (1) DDT

$$Cl-C_6H_4-\underset{C-Cl_3}{\overset{H}{C}}-C_6H_4-Cl$$

(2) Methoxychlor
(3) Toxaphene
(4) Benzene hexachloride (BHC)
(5) Aldrin, heptachlor, dieldrin, chlordane

(b) Organic phosphates. These materials are characterized by a short residual and a relatively short period of effective kill. They are generally cholinesterase inhibitors. The earliest and most widely used representative is parathion. New formulations are designed primarily to extend the very short residual of parathion.

(1) Parathion

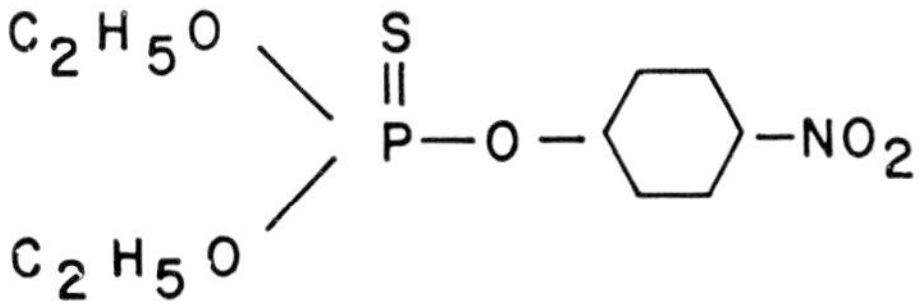

$$\begin{array}{l} C_2H_5O \diagdown \quad \overset{S}{\overset{\|}{P}}-O-\bigcirc-NO_2 \\ C_2H_5O \diagup \end{array}$$

(2) Malathion
(3) Diazinon, phorate (Thimet), naled (Dibrom), disulfoton (DiSyston)

(c) Carbamates: carbaryl (Sevin), Bux, Landrin

5. Fumigants. Hydrogen cyanide, carbon bisulfide (CS_2), methyl bromide. The fumigants commonly block respiratory processes.

6. Chemosterilants. Apholate, tepa, metepa. These substances sterilize the insect. This sterilized insect slows the reproductive rate of the population and thus prevents buildup of large insect numbers that could destroy a crop.

7. Attractants. Medlure, gyplure, eugenol. These attractants can be used in localized problems to attract the insects to a point where they may be killed by another means.

Insect control ideally represents a careful integration of the several methods described above. Chemicals are very useful but often have an accumulative effect and require constant vigilance by the user. When chemical control becomes necessary, the disturbance of natural control should be kept to a minimum. "Blanket" treatments should be minimized and the greatest selectivity possible sought out.

SPECIFIC INSECT PROBLEMS

Small Grains.

HESSIAN FLY. It is the most destructive pest of wheat in the United States. This pest has two main broods; one attacks winter wheat in fall and the other attacks both winter and spring wheat in the spring. During severe outbreaks there may be 20 or 30 larvae per plant. They cause injury by extracting juice and altering the stem tissue. Eggs are laid on young plants and the larvae migrate to a position under the leaf sheath next to the stem. When they pupate, the puparium is called the "flaxseed" stage, and the Hessian fly overwinters or oversummers in this condition. Release of resistant varieties since 1945 has greatly reduced losses, and several varieties carry resistance to at least some of the races of this pest.

GREENBUG. These aphids directly injure plants by injecting saliva and sucking up juices. They cause leaf injury, thin stands, poor tillering and serve to spread plant diseases in oats, barley, and wheat. The greenbug overwinters in the egg stage and exhibits an incomplete metamorphosis. Vigorous plants, resistant varieties, destruction of volunteer plants, and chemical treatment with parathion are possible means of control.

CHINCH BUGS. These are characterized by the enormous populations that build up in some years, especially in a dry season. The kind of crop attacked and the extent of injury may vary depending on the generation of chinch bug doing the damage. Normally two generations develop in the Midwest and three in the Southwest. The first generation is injurious to small grains and the second or third to corn and sorghum. Chinch bugs destroy plants principally by withdrawing enormous quantities of plant juices. They overwinter as adults and then nymphs hatched on new growth do the damage. As they grow to adult stage, chinch bugs fly and move to new fields. Barriers of poisons, insecticide applications, and planting crops not favored by chinch bugs all have been used as methods of control.

GRASSHOPPER. These insects may cause injury over all the Northern Hemisphere but do their most severe damage in areas with average precipitation of 10–30 inches. Grasshopper injury consists primarily of defoliation or destruction of the plant. Most species of economic grasshoppers have a single generation annually and overwinter as eggs in the soil. If adequate food is available, the nymph prefers small grains. Outbreaks seem to be related to favorable dry warm weather at hatching and abundant food supplies for the nymph and the adult. Both chemical (the chlorinated hydrocarbons) and cultural control methods have been used effectively for reducing grasshopper infestation.

ARMYWORM. An attack by this insect is notable for its suddenness and severity and results in plant consumption by chewing mouthparts. Armyworms prefer grasses and thus are particularly destructive to small grains, corn, and forage grasses. In the northern states two generations occur each year with the larvae overwintering. An essential requirement for effective and profitable control is the early discovery of infestations. Chlorinated hydrocarbons are useful insecticides.

Corn.

EUROPEAN CORN BORER. This is the most destructive pest of corn. During its spread in the 1910s the corn borer had only one generation, but in the 1930s two generations appeared. First-generation borers begin feeding in the whorl on the leaf surface but later bore into the stalk. This weakens the plant and starves the ears. In addition, saprophytic and parasitic fungi enter these tunnels and cause rots. Adult females lay their eggs on the underside of corn leaves, usually in clusters of 14 to 20. The corn borer passes the winter as a full-grown larva inside its tunnel in stubble.

The part of the plant upon which the borers feed is influenced by their age, generation, and the stage of development of the corn plant. Hatching when corn is in the whorl stage, the first instar of the first brood feeds on leaf surface. The third and fourth instars feed heavily on the sheath, midrib, and around the collar. The fifth and sixth instars bore into the stalk. The second brood bores at an earlier instar and feeds primarily on the stalk, the ear shank, and the base of the tassel.

Early control measures sought to reduce the overwintering larval population by stalk chopping and clean plowing. While this could be a beneficial practice, it had to be practiced by all corn growers to be of much benefit because of the wide ranging habits of the female. Therefore, chemical control has been looked to as a solution whenever an individual farmer wishes to control the corn borer. DDT was used extensively in earlier years; now Sevin, toxaphene, and diazinon would be favored for a best combination of maximum control and minimum residue. Biological control with the disease organism *Bacillus thuringiensis* is effective but also higher in cost and not used widely. Cultural control by correct timing of planting corn reduces infestation, but this means planting when most other growers plant. There is a yield advantage to early planting in the Corn Belt, so it is better to plant early and control the insect chemically. Genetic tolerance and antibiosis are possibilities, as mentioned earlier, with strong stalks to tolerate the attack and antibiosis from the chemical dimboa. These methods are not fully perfected and cannot be relied upon as yet as the sole means of control.

ROOTWORM. These pests represent the larval stage of three related beetles and present a serious problem where corn is grown continuously. They damage corn by feeding on and tunneling inside the roots. Losses result from lodging and from reduced water and nutrient uptake. The northern and western rootworms have one generation annually and overwinter in the egg stage. Eggs hatch in May and June, become full grown in July, pupate, and finally the adult beetles emerge in late July and August. The adults feed primarily on pollen but may consume the silk if they are present in high populations.

Crop rotations are useful for control because the adults do not lay their eggs in weed-free soybeans or meadow. The application of nitrogen fertilizers to enhance vigorous top growth and more roots gives added tolerance to rootworm attack. Early planting favors tolerance by allowing more root development before the larvae attack; when they do attack they feed on the young succulent roots. The chlorinated hydrocarbons were used earlier in chemical control, but as resistant strains developed, the organophosphates have become the important chemical mode of control.

Legumes. Insect pest problems of forage legumes may affect seriously both the production of the forage and the production of seed. Leafhoppers, aphids, and weevils may all be the source of reduced yields.

ALFALFA WEEVIL. The larvae of the alfalfa weevil have placed continued production of alfalfa in southeastern United States in doubt although genetic resistance may be achieved. The larvae feed in the stems, chew cavities in the buds, and skeletonize the leaves so that an infested field may appear to be frosted, grayish or whitish in color. Early cutting of both the first and second hay crops has helped to control the pest. Flaming the stubble and chemical control with DDT, methoxychlor, or malathion and its relatives have been useful, but control is difficult and surrounded by many problems ranging from chemical residual on the forages to inadequate control of the pest.

LYGUS BUGS. These insects have been called the most devastating insect pest of the alfalfa seed crop. The piercing-sucking mouthparts of the lygus nymph enter and leave plants without much immediate obvious damage, but later the plants develop short internodes, become stemmy, produce many branches and an unusual number of short racemes. When lygus bugs feed on alfalfa buds, the buds turn white and die in two to five days. If the bugs are sufficiently numerous, blooming may be completely prevented. Early cutting or chemical treatment with DDT, dieldrin, or malathion may reduce the population.

Stored Grain. The major pests of stored grain the world around are rice weevil, granary weevil, lesser grain borer, saw-toothed grain beetle, cadelle, confused flour beetle, red flour beetle, flat grain beetle, rusty grain beetle, khapra beetle, Angoumois grain moth, Indian-meal moth, and Mediterranean flour moth. The behavior and habits of stored grain insects are closely attuned to the moisture and temperature of their food media, with a narrow moisture band between 11.5 and 14.5% favoring their development. Grain-infesting insects are solely responsible for the heating which may develop in grain that has been adequately dried. This in turn frequently results in molding, caking, and spoilage. Grain-infesting insects do not survive in grain with a moisture content above 15%. If they do occur, fumigation with carbon bisulfide is effective for control.

DISEASES

Disease control has received intensive study by agronomists for many years. Nearly every crop-breeding staff has a plant pathologist working directly with the breeder. This teamwork is necessary because the occurrence of a given disease may increase production costs, reduce yields, or perhaps even render impossible the production of a crop in an area of previous importance.

METHODS AND PRINCIPLES OF CONTROL

Legal. Control by means of quarantine is much more effective than it is for insects. There is some state-to-state control and more from nation to nation. This is more effective for diseases than insects because the diseases are not "self-propelled." There are exceptions, however, where spores are transported many miles by wind. This is true on both the North American and Eastern Asiatic continents for cereal rust dispersal where the spores are moved northward through the season. Another group, the seed-borne diseases, are difficult to detect, and if legal control is to be accomplished the plants must be grown to determine if any diseases are present.

Chemical. Chemical treatment for disease control is often possible but not sound economically in most cases. Rusts, for example, may be controlled with fungicides, but the expense is too great under field conditions. For high-value crops or for localized treatment, chemicals may be used. For example, chemical pathogenicides are used for the fumigation of tobacco beds and for horticultural plots. Fungicides are commonly used to treat and protect the seeds of flax, corn, cotton, and sorghum to give temporary control in a localized area of the plant's environment. Systemic fungicides are of great promise in the future (Pederson, 1967) because they can be added to the soil, added only once per season (in contrast to the weekly spray of foliar applications), and they appear to be very effective.

Cultural. Crop rotation is useful because many diseases are host-specific, requiring not only a particular plant but a specific type of protein which they may parasitize. Crop rotations are a gross means of changing the plant host and reducing the parasitic home. Diseases that do not live well on dead material are controlled conveniently by this method, and it is thus most useful with obligate parasites.

Planting Date. Planting date adjustment often helps the crop escape or better tolerate disease attack. Early oat varieties are favored by Iowa farmers partly because they mature before the disease organism builds up to a high level. A reduction in damage by winter wheat streak mosaic, a disease for which no genetic resistance is known, can be attained in South Dakota by planting between September 10 and 14 rather than earlier. This date of planting reduces the activity of the wheat curl mite that carries the mosaic virus and provides an 80% control of the disease.

Fertilization. Fertilization practices have been shown to influence the types of disease problems. For example, Huber (1966) concluded from studies in Idaho on nitrogen effects on soil-borne diseases that it was not the carbon:nitrogen ratio of the soil or the crop residue that increased or decreased the disease level but rather the form of nitrogen used. Foot rot of wheat was more severe from spring applications of ammonia and could be reduced by fall applications of ammonia or spring applications of nitrogen. Nitrate nitrogen decreased the diseases caused by species of *Fusarium, Rhizoctonia,* and *Aphanomyces* but increased those caused by *Verticillium* and *Streptomyces.* Huber suggests that both host resistance and activity of the pathogen may be changed by the form of nitrogen it receives. And, as noted above, use of one form to reduce a particular disease may increase another disease. A knowledge of the expected disease problems and the effect of timing on the nitrogen fertilization used can be a useful disease control step.

Genetic. Genetic resistance stands as the most important, the most commonly used, and the most attractive means of disease control. Essentially this means breeding a new line with a protein type enough different that the disease organism cannot parasitize it. As a basis for better understanding genetic resistance, the types of organisms causing disease will be discussed.

DISEASE-PRODUCING ORGANISMS

Bacteria, fungi, viruses, and nematodes are all common causal organisms of plant diseases. The viruses and the fungi are the most important agents of plant diseases in general, while the fungi and nematodes are most important in causing soil-borne diseases.

Bacteria. Bacteria are unicellular organisms which reproduce by division. Approximately 1,800 species of primitive, single-celled bacteria are known. Many of these are essential to life on earth and most of them are beneficial. However, a considerable number induce disease. They may be carried from season to season on the host plant or in some cases in the seed. The bacteria have a capacity to form resistant coverings for protection under adverse conditions.

Some bacteria are parasites; that is, they live on living organisms. Not all of these parasites are pathogenic (the symbiotic bacteria in legume nodules are not) but most are harmful in the sense that they use a portion of the food supply from the host plant.

Fungi. Fungi are multicellar organisms. Their vegetative body is a mycellium with hyphae that function much like roots in higher plants. Some fungi have a sexual spore stage which allows the possibility of hybridization and the rapid development of new strains. There are three classes of parasitic and saprophytic plants collectively called fungi: *Phycomycetes, Ascomycetes,* and *Basidiomycetes.* The *Phycomycetes* are water-loving plants and thus may cause great destruction in wet years. The *Ascomycetes* are the most numerous of the fungi and can tolerate relatively dry conditions. Common leafspot of alfalfa, scab of barley and corn, many storage diseases of wheat, and the powdery mildews are in this group. The *Basidiomycetes* are also relatively tolerant of dry conditions. They cause the rusts and smuts of cereal crops.

Viruses. Viruses are complex proteins capable of multiplying in living cells and are so small they cannot be seen with an ordinary microscope. In many respects the viruses behave like bacterial and fungal pathogens, but there is a continuing debate as to whether they are living organisms or not. They do increase greatly in association with protoplasm of appropriate plants.

Nematodes. Nematodes are microscopic eelworms of the animal phylum *Aschelminthes.* Although minute in size (1/8 inch or less) they are very complex organisms with a well-developed and rather complicated digestive system, a nervous system, an excretory system, musculature, and a reproductive system. They feed by puncturing plant root cells with a stylet (spearlike mouthpart) and may simultaneously provide entry for disease-causing fungi into the plant. Some spend their lifetimes inside the host plants while others live in the soil and simply feed on the rootlets of the plants. The nematodes may cause grotesquely twisted roots (root knot nematode) or form large cysts on the root (as does the female of the cyst nematode).

PARASITISM

Disease organisms are classified commonly by the degree of parasitism which they exhibit. The obligate parasite lives and reproduces on living tissue only. Although the exact basis for this living tissue requirement is not well understood, it is a very important characteristic and made use of in developing resistant varieties. Most virus diseases, many fungi (mildews, rusts, and smuts), a number of nematodes, and a few bacteria are in this group. Short rotations, disease-free seed, and destruction of alternate hosts are means of control. The facultative saprophyte can grow on dead tissue but grows more commonly on living tissue like the obligate parasite. The facultative parasite has the ability to become a parasite but is more like a saprophyte than is a facultative saprophyte. These organisms are carried over in the soil on crop residue. The closer they are to being saprophytes, the longer is the time period between susceptible crops required to help control them. They may also survive on other living hosts, both weeds and crop plants. Many bacteria and fungi are in this category. The saprophyte lives on dead tissue and is usually a secondary problem after other damage has been done by insects or diseases that attack living organisms. Toxins formed by saprophytes growing on rotting plants can affect the plant. Some fungi and bacteria are represented in this group.

In summary, the plant breeder has more difficulty breeding resistance to a disease which exhibits a lower order of parasitism (i.e. one that is more saprophytic). The more obligate the parasite, the more readily the plant can be changed to a genetic and chemical composition that resists the disease.

Varietal resistance to disease is referred to as genetic resistance but may be separated more specifically into two groups: mechanical and physiologic. Mechanical resistance may be described as follows. The disease organism may be dependent on a natural or artificial opening for entry. For example, the fungal hyphae may be too large to enter the natural stomatal opening. This type of resistance is common in seedling plants. Potato varieties with heavy leaves allow less late blight infection than varieties with smooth leaves. Wheat seeds that are free of hairs do not lodge as many smut-producing organisms. Barley does not open its flower during pollination and less loose smut infection results. Physiologic resistance is more related to the disease organism's growth and development once it has entered the plant. For example, the plant tissue may exhibit hypersensitivity; in a resistant plant, the cells when attacked collapse and do not support the further growth of the parasite. Or the cells may repel the organism and chemically react to it.

SPECIFIC DISEASE PROBLEMS

Small Grains. The small grains—wheat, oats, and barley—share certain disease groups in common. Rusts are among the most serious diseases of wheat and oats. The rust organism is an obligate parasite which has made possible the development of many resistant varieties. Conversely, the alternate host phenomenon has allowed the rusts to hybridize readily and form new races, which then require the development of a new resistant variety. Stem, leaf, and stripe rust of wheat and stem and crown rust of oats may have their spores spread widely by the wind. In the United States stem rust is most severe in the hard red spring wheat region with yield losses as high as 85–90%.

Stem rust of wheat is used as an example of this important group. It is characterized by elongated reddish-brown pustules (uredospore stage) that occur on the stems, leaves, sheaths, floral bracts, and sometimes on the awns of the wheat plant. These pustules, which often break through the epidermis, contain brick-red uredospores (Fig. 14.4). As the season advances, the pustules as well as the spores are replaced by black ones (teliospore stage). It is the red-spore stage which is responsible for the direct damage to plant growth by utilizing the water and nutrient materials needed for the development of the wheat kernel as well as causing the plant to transpire at a greatly accelerated rate. In the southern states and northern Mexico, the uredospores live throughout the year on susceptible wheat and other grasses. In the North, the black teliospores overwinter on stubble, then produce sporidia in the spring which infect the leaves of the common barberry. Pycniospores which develop on the barberry constitute the sexual reproductive cells of the fungus. The union of these cells produces tissues which give rise to cluster cups (aecia) in which aeciospores are borne. The aceciospores

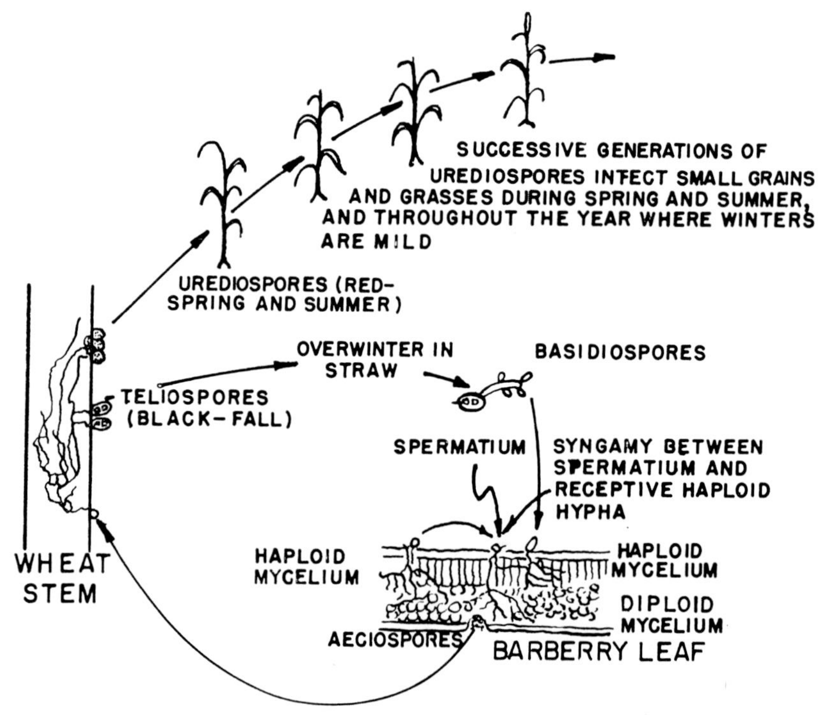

FIG. 14.4 ☙ Life cycle of the fungus *Puccinia graminis* which causes stem rust of wheat and other cereals (Pearson, 1967).

then are blown from the barberry bush to wheat or grass plants where uredospores are produced to complete the life cycle.

Crown rust has buckthorn as its alternate host and goes through the same general phases as described for stem rust. It has been the most destructive rust in oats and one for which many resistant varieties have been developed.

Smuts are obligate parasites which are problems in all three crops: bunt or stinking smut in wheat; covered and loose smut in oats and barley. These diseases cause the formation of spore masses in place of the grain and the kernel is destroyed. Fungicides give some control but resistant varieties are most useful.

Scab or *Fusarium* head blight is a particular problem with barley. Head blight causes a reduction in grain yield as well as a deterioration in grain quality. Scabby barley will cause vomiting in monogastric animals and also is unsuitable for malting.

Corn. Stalk rot is a disease of major importance and its incidence may be increased by hail, leaf blight, or a high plant population. Stalk rot is caused by a complex of organisms in the *Fusarium, Gibberella,* and *Diplodia* genera. Breeding resistance to such a broad complex is difficult. These organisms are of a lower degree of parasitism and their lack of specific host requirements adds to the difficulty of developing resistant varieties. Resistance can be obtained but is not well associated with high yield. First symptoms include a premature drying of the lower leaf sheaths and then the pith is destroyed. Numerous research reports indicate that stalk rot damage increased as the N:K ratios in the corn leaf increased. A balanced NPK fertilization program gives the best root development, and stalk lodging is commonly reduced by K fertilization. However, this may be related to the anion Cl, which is commonly present in conjunction with K. The Cl may compete with NO_3 and reduce the NO_3 uptake. This in turn may place the plant's mineral nutrition in better balance, leave more sugar in the stalk, and thus reduce attack by the stalk rot organisms. This possibility is still very much under discussion.

Helminthosporium leaf blight causes considerable yield reductions, especially in wet seasons. Northern corn leaf blight occurs first as small grayish-green, water-soaked elliptical lesions on lower leaves. They progress in size and move up the plant, resulting in an "early-frosted" appearance. Resistant varieties are available and the best means of control. Southern corn leaf blight develops smaller, lighter-colored lesions. The blights overwinter on old infected leaves. Their yield reductions are the result of premature death of the plant.

Both smut and rust can be problems in corn. However, genetic resistance to smut is readily available and is present in many hybrids. Rust is much less of a problem than it is in the small grains. Both diseases are obligate parasites in corn as in the small grains.

A new disease confronting the corn breeder is maize dwarf mosaic. This obligate parasite may be transferred from plant to plant mechanically or by aphids, while a disease with similar symptoms, corn stunt virus, is spread only by an insect vector. These diseases stunt the plant and may cause reddish leaves to form and yield to be greatly reduced. It appears as if resistant varieties can be readily developed, and after initial concern, discussion of this problem has greatly diminished.

Seed rots and seedling blights could be a problem in corn, but they are very effectively controlled with seed-applied fungicides.

Soybeans. Soybeans are beginning to develop significant disease problems as their acreage increases. Phytophthora root rot is a disease associated with wetter soils and first occurred in the eastern and southern soybean production areas. Considerable breeding work has been done and resistance to this disease is available in many varieties. Brown stem rot is increasingly a problem in the north central soybean production area. This organism is a soil-inhabiting fungus and a facultative parasite. In addition, it is not strongly host-specific but will attack red clover and alfalfa as well as soybeans. Brown stem rot may build up to very high levels on fields continuously planted to soybeans. In mid-July a brown discoloration occurs in the vascular tissue of the primary root and the lower stem symptomatic of this disease. A rotation of crops in which the soybean is not grown on a field for three years after an epiphytotic is presently the only means of control. Genetic material resistant to the disease now appears available after a search of several years but it is not yet incorporated into varieties.

In the southeastern and south central areas of the United States the cyst nematode is a major problem. This obligate parasite attacks the vascular tissue and causes increased sensitivity to periods of dry weather, especially on sandy soil. Practices to offset the disease relate primarily to creating the best growth conditions possible.

Leaf diseases may cause loss of photosynthetic area and a reduction in yield. Bacterial blight is seed-borne and attacks the lower leaves, causing them to disintegrate and fall prematurely. Brown spot is a fungus-caused disease which destroys large sections of leaf tissue. Bacterial pustule is characterized by small yellowish-green spots and gives small yield reductions. Two virus diseases, soybean mosaic and bud blight, may cause serious damage occasionally.

Forage Crops. Bacterial wilt and brown leaf spot are diseases of prime importance in alfalfa production. Bacterial wilt infects a susceptible plant the second year, plugs the xylem the third year (giving a brown discoloration), and the plants will wilt and die under mild water stress in the third or fourth year. However, this alfalfa disease is an obligate parasite for which many resistant varieties have been developed. Brown leaf spot tends to be a seasonal problem. It may cause a rather heavy drop of lower leaves in the first cutting of alfalfa and then be of little consequence the remainder of the season.

The major diseases in red clover are root and crown rot and anthracnose. Root rot is a prime factor in the poor longevity of red clover stands. It, in conjunction with the clover root borer, have made this perennial a biennial under field conditions. Northern and southern anthracnose have also been responsible for yield reductions, and considerable breeding work has been done to develop resistant lines.

Disease problems have not been extensive in the forage grasses although rusts have sometimes required breeding for resistance. ❧

LITERATURE CITED

American Society of Agronomy. 1966. *Pesticides and Their Effects on Soils and Water.* ASA Spec. Publ. 8. Soil Sci. Soc. Am.

Barker, R. J., C. F. Cohen, and A. Mayer. 1964. Photoflashes: A potential new tool for control of insect populations. *Science* 145:1195–96.

Doutt, R. L. 1958. "Vice, virtue and the vedelia." *Ent. Soc. Am. Bull.* 4 (4):119–23.

Edwards, C. A., and G. W. Heath. 1964. *The Principles of Agricultural Entomology.* Chapman and Hall, London.

Harris, M. R. 1964. Diseases of cereal crops in Washington. *Wash. State Univ. Ext. Bull.* 559.

Huber, D. M. 1966. How nitrogen affects soil borne diseases. *Crops and Soils* 18:10–11.

Jones, F. G. W., and M. G. Jones. 1964. *Pests of Field Crops.* Edward Arnold, Ltd., London.

Martin, J. H., and W. H. Leonard. 1967. *Principles of Field Crop Production,* 2nd ed. Macmillan Company, New York.

Metcalf, C. L., W. P. Flint, and R. L. Metcalf. 1951. *Destructive and Useful Insects,* 3rd ed. McGraw-Hill Book Co., Inc., New York.

Pearson, L. C. 1967. *Principles of Agronomy.* Reinhold Publishing Corporation, New York.

Pederson, V. 1967. New chemical control for loose smut of barley. *S. Dakota Farm and Home Res.* 18:14–17.

Pfadt, Robert E. 1962. *Fundamentals of Applied Entomology.* Macmillan Co., New York.

Richter, Paul O. 1966. Biological control of insects and weeds in Oregon. *Oregon State Agr. Exp. Sta., Tech. Bull* 90.

Rolston, L. H., and C. E. McCoy. 1966. *Introduction to Applied Entomology.* Ronald Press, New York.

Stakman, E. C., and J. G. Harrar. 1957. *Principles of Plant Pathology.* Ronald Press, New York.

USDA. 1952. *Insects. The Yearbook of Agriculture.*

USDA. 1953. *Plant Diseases. The Yearbook of Agriculture.*

Viglierchio, D. R., and P. K. Yu. 1965. Plant parasitic nematodes: A new mechanism for injury of hosts. *Science* 147:1301–3.

CHAPTER FIFTEEN HARVEST AND STORAGE

THE HARVEST AND STORAGE of field crops often encompass many physiological principles. For example, moisture management at the time of harvest is a primary factor in obtaining a quality product for feeding or processing. This moisture content may be controlled by the physiologic stage of growth of the species or variety, in which case knowledgeable management can manipulate the moisture content without costly drying or other mechanical procedures. In other instances it may not be possible to lower the moisture content to the desired level except by mechanical procedures. However, in these cases it may be important to predict when the crop is nearing maximum dry matter accumulation or has reached the stage of growth which provides a high protein yield or has enough stored reserves for a satisfactory longevity of the stand. These latter factors may be important when determining the time to harvest. Finally, the respiring plant tissue has significant effects on storage procedures.

The field crop may be harvested by grazing (thus entailing no storage procedure), via a harvesting-storing combination that involves drying, or by ensiling as part of the storage process. These three methods encompass most crop processing systems. Overall objectives of crop harvest may include one or more of the following:

1. Selection of a harvest date when maximum yield of protein and/or energy may be obtained.
2. Retention of a high percentage of this yield either by careful grazing or processing.
3. For hay, grain, and silage, a product that can be stored for a long period without deterioration.

DRYING

GRAIN CROPS

Dry grain that contains 11–12% moisture can be stored in weatherproof bins for many years in most climates without appreciable deterioration, provided it is protected from insects, rodents, external moisture, and high humidity. The composition of dry grain remains almost unchanged except for some increase in fatty acids and a slight loss of energy from respiration (1% over 20 years). The only major change is the rather marked loss of carotenoids in long-term storage (Dua et al., 1965).

Insect Damage. Stored grain suffers damage when its moisture content and temperature are sufficiently high to permit organisms (both insects and diseases) to thrive. Grain-infesting insects (granary weevil, rice or black weevil, lesser grain borer or Australian wheat weevil, and Angoumois grain moth) usually are inactive when the moisture content is below 9% and the temperature is 40°F or less (USDA, 1962). At 13% moisture and greater than 70°F insects are very active. The dry grain they consume is converted to growth energy and respiration, releasing carbon dioxide and water. The released water raises the moisture content of the mass of grain; thus spoilage by heating, molding, and decay occur in addition to direct damage by the insects.

Disease Damage. The chief damage to stored grain protected from insects and rodents results from fungal activity. The major factors that determine when stored grain will be damaged by storage fungi are temperature, moisture content, physical damage, whether storage fungi have already begun to invade, and length of time the grain is to be stored. Storage fungi grow slowly at 40–50°F and rapidly at 80–90°. However, fungi may grow at temperatures below freezing. Some fungi grow at 13.5–15.5% moisture and others at 16–23%. Storage fungi cause loss of germination, dark germs, mustiness, and heating. Therefore, the type of spoilage is largely determined by the moisture content of the cereal grain in the sequence described below:

1. Grain that contains somewhat more than 20% moisture usually sours due to the fermentation of soluble carbohydrates. Alcohol and organic acids form and heat is generated. Souring is followed by rotting if oxygen is present. However, wet grain in an airtight silo exhausts the oxygen, the fungi and enzymes present are destroyed by heat, and good livestock feed results.

2. Grain at 16–20% moisture heats because of fungal growth and becomes moldy. The heating causes a "bin burnt" brown color.

3. At 14–15.5% the grain develops a musty odor from limited fungal growth.

4. At 14–16% moisture and under limited oxygen, dead germs (i.e. "sick" wheat) occur. The dead germs have a high content of fatty acids and give a rancid flavor to wheat flour.

Moisture Control in Stored Grain. Moisture movement in grain allows well-mixed wet and dry grain to reach a uniform moisture content within 24 to 48 hours. However, moisture and air movement through a mass of bulk grain is extremely slow, and consequently a small area of damp grain in a bin may spoil when it is surrounded by a large bulk of dry grain. Furthermore, the moisture content of grain at the top of a pile quickly comes into equilibrium with the water vapor pressure of the immediate atmosphere. Grain exposed to an

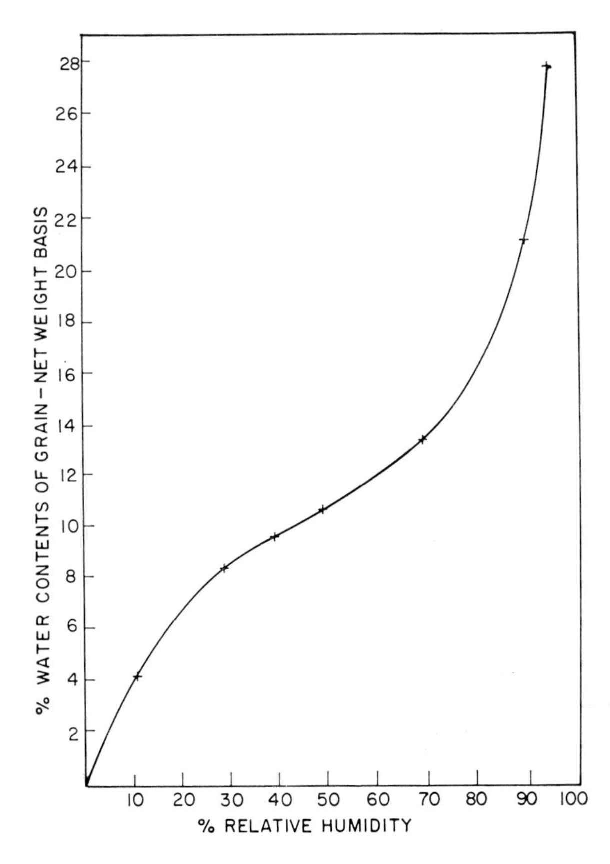

FIG. 15.1 ☙ Relative humidity-moisture relationships in stored grain (adapted from Oxley, 1948).

atmosphere of 65% relative humidity and 70°F temperature has a moisture content of about 14%, but the moisture content rises rapidly at higher humidities, especially above 80% (Fig. 15.1).

An established practice in elevators is to pull the air out of the bottom of the elevator at a rate of 0.05–0.10 cubic feet per minute per bushel of grain when the air temperature is at least 10°F below the grain temperature. This aeration process with cool air prevents moisture migration. Without aeration the moisture vapor moves to the center top of the bin and condenses there, causing grain spoilage in the center top zone. The circulating air reduces mold and insect activity and helps to cool hot spots. Aeration accomplishes very little drying but can cool the grain down to air temperature in 80–100 hours, and it is much more economical than the previous practice of "turning" the grain to cool it by moving it from bin to bin by elevator and conveyor. Aeration also causes much less of a problem in terms of cracking the grain or making it dustier.

Harvesting Considerations. The first point commonly evaluated in determining the readiness of an annual crop for harvest as grain is its physiologic maturity. This is defined as the point at which maximum dry matter accumulation has occurred in the seed. While this is not an absolute value (varying among species, varieties, and years), it is rapidly approached when the seed is below 40% in water content and usually is complete before the grain reaches 30% moisture. The moisture relationships in the grain might be summarized as outlined in Table 15.1. As the seed changes from the 85–90% moisture level at pollination down to 35% moisture, there is in effect a "dilution" of the water with carbohydrate. From 35 to 13%, reduction in the water content occurs by desiccation. Since grains never ripen uniformly (the range of date ripened may vary from 3 to 10 days among individual kernels), some increase in dry matter may occur until the average water content of a field sample has dropped to 26–30%. The pattern described here is a normal one in annuals. For perennials certain management practices to be discussed later are designed to simulate this progression.

Secondly, the characteristics of the crop and the means of harvest influence the moisture percentage chosen for harvest. Grain crops in

TABLE 15.1. Moisture relationships in grain

	Pollination	Physiologic Maturity	Harvest Maturity	Storage
Moisture %	90	30–40	16–28	13–14

various parts of the world are still bound into bundles and allowed to dry in small shocks in the field. These crops can be cut at 30% moisture and provide a quality product. If the crop is to be dried with forced heated air, it may also be harvested in this moisture range. Corn is commonly shelled or combined in the field at 26–28% moisture (or harvested on the ear for seed at moistures up to 32%) and then dried. Rice to be artificially dried may be combined at 26% moisture. It is important to emphasize that once the grain is removed from the inflorescence (ear, head, etc.) it must be dry enough to store (11–15%) or be dried artificially. Grain stored in cool weather will keep at 15% satisfactorily, but for warm summer conditions the moisture must be down in the 11–13% range. Waiting to harvest the crop at these low moistures presents problems in terms of lodging, shattering, sprouting, cold or otherwise inclement weather in which to harvest, and loss of grain quality generally; therefore, artificial drying is practiced increasingly.

In summary, storage of high-energy grains rich in carbohydrates and fats is most commonly done in a desiccated condition. This dryness inactivates the respiratory enzymes and minimizes the activity of insect and disease organisms that might damage the grain. Specific examples are given below for small grains, corn, sorghum, cotton, and castor beans. Although cotton is harvested primarily for the fiber on its seed coat rather than for the seed, the crop is included here because the harvesting practices are closely related to those for grains.

Small Grains. Small grain production is concentrated in semiarid and subhumid sections of the world. Therefore, weather can be a very beneficial factor in getting the grain dry enough for storage (Brooker and McQuigg, 1963). The rate of reduction of kernel moisture of wheat in Saskatchewan between psyhiologic maturity and harvest maturity was shown to be influenced by: (1) the vapor pressure deficit of the air, (2) the hours of sunshine, (3) evaporation, and (4) wind (Dodds and Pelton, 1967). Even under relatively dry conditions, there is a wide variation in drying rate from year to year, and once physiologic maturity is reached, artificial drying has a tremendous potential for smoothing out the work load.

HEATED AIR. If heated, forced air drying is used on wheat, great care must be taken to maintain grain quality. High heat (above 130–140°F) will damage the baking quality by causing an uneven response in the rising process. High temperatures will also reduce the completeness with which the starch and gluten of the kernel can be separated in centrifugation, one of the key steps in wheat and corn milling. Malting barley has such special germination response requirements that only natural drying appears satisfactory for the maximum malt production.

WINDROWING. To circumvent some of the problems above in achieving the desirable moisture for harvesting small grains, windrowing or swathing is commonly done. The head and part of the stem is severed from the stubble and allowed to dry in swaths in the field. The grain remains attached to other plant parts and dries by natural aeration. Koenig et al. (1965) report that swathing barley at moisture contents of from 15 to 42% did not have a significant effect on yield, but test weight decreased as the kernel moisture at the time of swathing decreased. (These results are not fully in agreement with desiccation studies described later which indicate that desiccation at high moisture levels reduced test weight.) Plumpness of the grain was increased by allowing maturation to 25–28%. Koenig et al. concluded swathing could be done between 18 and 25% without any appreciable loss in quality.

Another system used very frequently in the Great Plains states is to pile grain of 15–18% moisture on the ground. It will dry down to 14% in a three-week period if dry weather prevails. These piles are made long and shallow so that all material is within two feet of the surface and 70% is within one foot of the surface.

Corn. Physiologic maturity of corn varieties in the major section of the Corn Belt occurs about 55 days after 75% of the plants have visible silks; the moisture content may range from 28 to 38% at that time. Hillson and Penny (1965) observed that two-thirds of the hybrids in their study reached maximum dry weight 54–57 days after pollination. Gunn and Christensen (1965) studied a wider range of maturity classes and found this period to be from 45 to 70 days. They further suggest a relationship of this span to temperature, although most workers have suggested the period to be rather independent of environmental conditions (Hallauer and Russell, 1962; Hanway, 1963; Shaw and Thom, 1951). Figure 15.2 provides a diagrammatic representation of dry matter accumulation and moisture loss for several maturity classes of hybrids studied by Gunn and Christensen. They did not observe any significant differences in the rate of drying of the 49 hybrids they studied, although they point out that this does not necessarily mean differential rates of drying do not exist.

As expected, the average daily rates of kernel moisture reduction varied considerably from year to year within each of five phases in corn

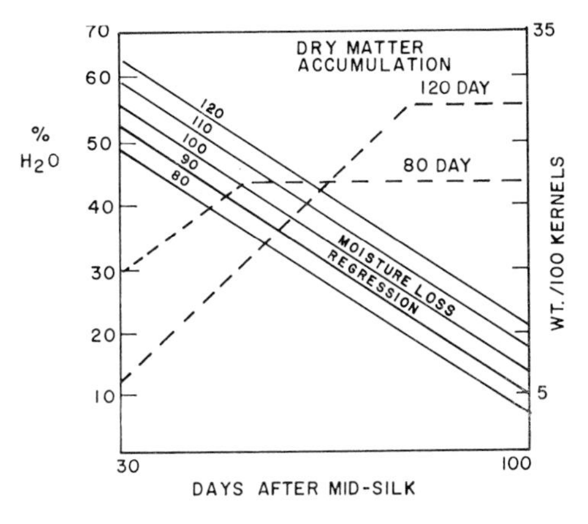

FIG. 15.2 ❧ Diagrammatic dry matter accumulation and moisture loss profiles of 80-, 90-, 100-, 110-, 120-day maturity reference hybrids (Gunn and Christensen, 1965).

seed maturation (Schmidt and Hallauer, 1966). Above 30% kernel moisture, the rate of moisture reduction was significantly related to air temperature; below 30% kernel moisture, relative humidity, wet bulb depression, and saturation deficit of the air were significantly related to moisture reduction rate. As noted in Figure 15.3, the range of moisture content around the average is greater as the kernel becomes drier.

COB AND KERNEL MOISTURE. At physiologic maturity the corn kernel moisture content may vary from 28 to 38%. The cob dries much more slowly than the kernel (Miles and Remmenga, 1953). When the kernel is 20–21% the cob is 36–40%; the average for the ear is 24–25%. The kernel and cob are equal in moisture content at 13%

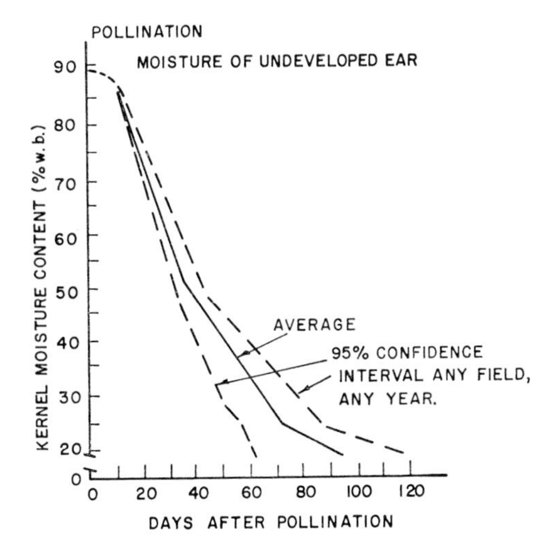

FIG. 15.3 ❧ Relation between kernel moisture content of corn in the field and days from pollination (Schmidt and Hallauer, 1966).

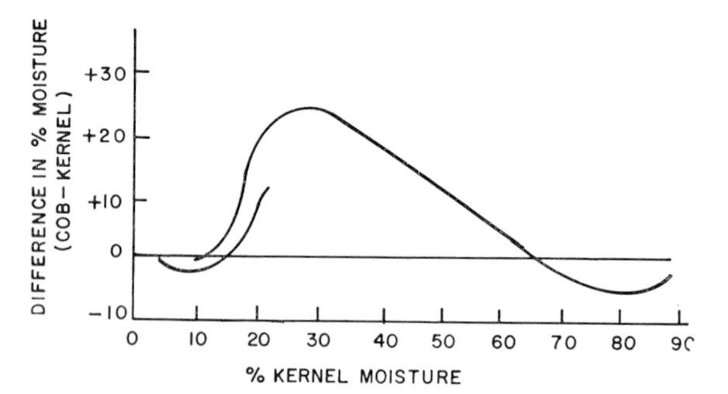

FIG. 15.4 ❧ Curves describing the difference between cob and kernel moisture. Percentages are on a wet basis (Miles and Remmenga, 1953).

and at 64%; within this range the cob is high in moisture, above this range the kernel is higher (Fig. 15.4). The maximum difference is at 28% kernel moisture when the cob is 23.5% wetter than the kernel. Knowing this, the commercial corn producer does not dry the cob. Seed companies do so because they want to minimize the danger of damage to the embryo (germ) by frost. They also wish to maximize the use of expensive drying and processing equipment over the longest harvest season possible.

During filling, moisture reduction occurs at about 1.3% per day. Once the corn is down to physiologic maturity, the drying rate is approximately 0.5% per day. However, in some seasons it may exceed 1% per day or it may be less than 0.1% a day, depending on humidity and temperature. This suggests approximately a two-week span between physiologic maturity and harvesting maturity. It is important to note that there is little relationship in corn varieties between moisture content at physiologic maturity and rate of drying.

Harvesting maturity (26–28% moisture content) is defined as the point at which the grain retains satisfactory structural condition (minimal scarring or grinding) when it is removed from the cob. The corn combine and the picker-sheller seldom function satisfactorily above this moisture percentage, although some experimental models remove the kernel from the cob with minimal damage at 32% moisture. There is a distinct advantage for the farmer in doing corn harvest while the moisture content is in the 20% range (Fig. 15.5), because harvesting losses can be greatly reduced as compared to harvesting at lower moisture contents. Harvesting too early, of course, results in yield reduction both because of harvesting losses and, if very early, because maximum dry matter is not accumulted.

DRYING PROCEDURES. Corn has long been dried "naturally" on the ear in slatted cribs. With this storage system, harvest can be done when the kernel reaches 22% moisture. To facilitate having the crop at a

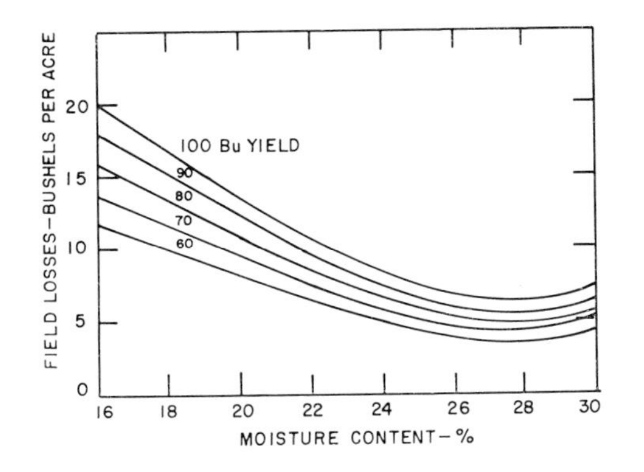

FIG. 15.5 Harvesting losses at different yield levels and moisture percent of the grain (Van Fossen and Stoneberg, 1962).

desirable moisture content when it will fit the work load, "calendarization" has been suggested. This means planting some early, some midseason, and some full-season hybrids, in that order, to spread the time of maturation. The producer might have 25% of his acreage in early maturing hybrids, 50% in midseason, and 25% in full-season hybrids. Thus he could harvest each maturity group while it was in the most favorable moisture range. This same principle could be made use of by practicing a cropping sequence including different crops, but current interest leads toward more specialization on one species. Therefore, a variation of maturity within that species and between varieties or lines is desirable.

A second method of managing the rate of drying in the field is the use of desiccants which by chemical action dry the leaf and sometimes the stem. Contact herbicides can be used for this purpose. In the north central states, use of desiccants on annual species is very marginal, balancing between early application, in which maximum drying benefits are sought but with the danger of killing the plant before physiologic maturity, and later application which does not cut the final yield but may give little benefit in drying.

Once corn is removed from the cob, it must be dried rapidly, usually with forced air, down to 13–15% moisture. Ear corn is dried by seed companies between 100 and 110°F in enormous batch driers. Drying is done over a 48–72-hour period and then the corn is shelled and stored. As the harvest season begins, the moisture must be lowered from 32 to 13%. As mentioned before, drying seed corn on the ear is expensive because of the high moisture in the cob; however, this expense is justifiable to obtain a seed of high germinability.

The commercial grain producer dries shelled corn with various systems including batch driers, continuous flow driers, and in-storage bin driers, all of which may be used alone or in combination. For example, combination systems in which a batch drier takes the moisture down to 20% and then completion of drying in an in-storage bin help to keep the quality high and hold drying costs down. The time required for drying corn down to storage moisture levels will range from 3 to 24 hours. This time is inversely proportional to the temperature used for drying, but the drying speed must be adjusted if stress crack formation is to be minimized. Thompson and Foster (1963) found that shelled corn dried with heated air (140–240°F) was two to three times more susceptible to breakage than the same corn dried with unheated air. Stress cracks (endosperm fissures), while practically nonexistent in crib-dried ear corn, were found in all samples of shelled corn dried artificially and accounted for much of the increased susceptibility to breakage. Shelling of dried ear corn causes some cracks but fewer than with drying shelled corn. Kernel pericarp breakage contributes to downgrading of corn and to its susceptibility to molds and insect damage. Drying speed was the most significant factor in stress-crack development. The number of stress cracks increased with increased drying temperatures, and air-flow rate and rapid cooling added to drying stress. If dried below 13.5%, much breakage occurs. Overdrying can also be detrimental by reducing profits from unnecessary weight reduction as well as increasing breakage. For every 1% water removed, there is a 1.1% loss in weight. Therefore, a 1,000-bushel unit testing 15.5% and dried to 13.5% becomes 978 bushels.

If the corn is to be sold for wet milling the temperature of the grain should not be raised above 130–140°F. If dried at higher temperatures it is more difficult or even impossible to separate the starch from the gluten by the standard centrifugation method. One practical test a corn processor can use to determine if corn has been overheated is a germination test, because germination drops rapidly in corn heated above 130°F for any length of time. A hot air temperature of 180° F in cool weather will hold grain temperature between 130 and 140° F in most commercial dryers.

For feed the corn can be dried at 190° F apparently with no risk of reducing the energy content. Some studies suggest that protein digestibility may be reduced if the corn is dried above 170° F. However, Purdue animal scientists reported no differences in rate of gain with pigs fed the following: crib corn; 24% moisture corn; 13% moisture corn dried at 190° F, 240° F, or 290° F; corn dried to 10.6% moisture at 190° F (Taylor et al., 1964).

Sorghum. The morphology of the sorghum plant markedly affects the harvest practices needed for this crop. The head or inflorescence is supported in an upright position by a peduncle that is especially susceptible to disease. When the peduncle is diseased and then becomes dry during maturation it is very prone to breakage. Therefore, the grain may need to be harvested before it has dried naturally to a storage moisture content, especially in more humid areas. This means that in these humid areas drying by heated forced air is a recommended practice.

A second means of drying the plant down more rapidly in the field is desiccation. Bovey and McCarty (1965) point out how two varieties of sorghum respond quite differently to desiccation. For the variety Martin, the use of desiccants allowed harvest of the grain at less than 18% moisture 10 days earlier than the untreated. Both varieties gave much less response in rate of drying when the desiccant was applied at less than 35% moisture. Severing the sorghum plant stem either below the seed head or at ground level usually resulted in the most rapid grain-moisture loss of all treatments studied. Magnesium chlorate, DNBP, and diquat were all effective as chemical desiccants (Table 15.2).

Desiccation or severing of the plant or plant part at moisture levels above 30% reduced both the seed weight and test weight. These observations emphasize that the plant had not reached physiologic maturity at the time of treatment and point again to the difficulties in trying to use a desiccant with a grain crop. The balance between keeping the leaves alive long enough to give a maximum photosynthetic product and yet speeding the rate of drying is delicate and has not been solved for many crops.

Cotton. For cotton the most critical factor in harvesting is to get it dry and as leaf-free as possible for mechanical cotton pickers to work most effectively.

Before the use of defoliants or desiccants, cotton growers in some

TABLE 15.2. Days required after desiccation or cutting treatment for grain to reach 18% moisture or below (Bovey and McCarty, 1965)

Variety	Grain Moisture at Time of Treatment	Treatments				Cut at Soil	Cut at Head
		Untreated	Mg ClO_3	DNBP	Defoliated		
Martin	Aug. 28, 57%	45	38	28	23	23	17
	Sept. 5, 39%	29	19	14	19	14	14
	Sept. 17, 26%	17	7	17	17	17	7
Combine Kafir-60	Aug. 28, 54%	45	38	38	38	23	23
	Sept. 5, 43%	36	29	19	29	19	19
	Sept. 17, 30%	24	24	17	17	17	17

areas had to wait for frost to defoliate their cotton. In some years this meant a late harvest—sometimes dangerously close to the time when heavy fall rains were imminent. The fact that cotton is inherently perennial means that it can often complete one cycle of growth during a season and start another. The second growth can be a very difficult problem. Therefore the use of harvest-aid chemicals has become a very useful practice, and over half the cotton acreage is conditioned each year. Pentachlorophenol is one desiccant recommended to dry cotton leaves. Adding 2,4-D or 2,4,5-T to the mixture almost guarantees an efficient kill and exclusion of second growth. Calcium cyanamide, sodium-chlorate-borate mixtures, magnesium chlorate, aminotriazole, tributylphosphorotrithioite, and pentachlorophenol have all been used as defoliants or desiccants (USDA, 1960).

Defoliation is preferred over desiccation but is more difficult to achieve. The objective is to cause the leaves to drop. This allows the remainder of the plant to dry more rapidly under natural conditions. Mature cotton picked from either spontaneously or artificially defoliated plants is usually higher in quality than desiccated or untreated cotton for two reasons: (1) few green leaves to stain the cotton and (2) fewer leaf stalks, crushed leaves, and other debris. However, cotton defoliated prematurely is very undesirable because premature defoliation may cause: (1) lowered oil content of the cottonseed, (2) increased fiber fineness, (3) lowered seed viability, and (4) reduced gain appearance and increased loss during processing.

Desiccants are used to dry leaves and other plant parts quickly; generally chemical desiccation causes little or no rapid leaf fall. A California study by McMeans et al. (1966) indicated that desiccants were more practical and predictable than defoliants. The desiccants can be applied 10 to 14 days later and still obtain useful results both in the rate of drying and in allowing more time for the bolls to open. For this reason desiccants are used more, even though defoliants have more theoretically desirable characteristics.

Castor Beans. The castor bean capsule may dry out either naturally or because of disease before the remainder of the plant is dry enough for the harvesting operation to be performed. This is especially true when the crop is grown in marginal areas of adaptation. Therefore, chemical desiccation is of interest. Culp (1964) suggests that chemical desiccation and early harvesting of castor beans may allow the production of a profitable crop in areas where otherwise diseases destroy the capsules.

HAY CROPS

Storage of forage crops in a dry condition has been a standard agricultural practice for centuries, and hay continues to be one of the major crops in nearly every state in the United States. Emphasis has often been placed on the relative cheapness of hay for livestock feed and its value as a source of protein, vitamins, and minerals. However, the fact must be kept in mind that energy (TDN or digestible energy) is the most expensive part of a livestock ration. For this reason any evaluation of hay must be heavily based on its energy value. Furthermore, if hay is to be a low-cost feed, it must be handled efficiently and harvested at the most desirable stage of growth.

Quality. The quality of hay is a product of its palatability, digestibility, and nutritive value. To evaluate forage quality three primary methods have been used: (1) visual inspection, (2) chemical analysis, and (3) animal feeding trials.

VISUAL INSPECTION. Visual evaluation is based on correlation of visible characteristics and feeding value. Hay grades developed by the USDA provide a common language of hay quality; the grades are based on five factors noted below.

1. Leafiness. Two-thirds of the protein and nine-tenths of the minerals in hay are found in the leaves. For first-cutting hay there is a high correlation between the quantity of leaves present and the digestible dry matter, but this correlation is less distinct for later cuttings.
2. Color. Green color is used to evaluate the conditions that occur during curing and the length of exposure to sunlight. Retention of chlorophyll and retention of carotene (the vitamin A precursor) are closely correlated under most natural drying conditions. However, carotene is lost in storage if oxygen is present; chlorophyll is not. Therefore, the color correlation is good only for field-cured hay that has not been stored very long.
3. Presence of foreign material such as burrs or sticks.
4. Condition. Does it look moldy or smell musty?
5. Texture, most commonly measured by stem diameter.

CHEMICAL ANALYSIS. The most widely used chemical analyses in hay testing programs are crude fiber and crude protein. These factors may be used to predict TDN. While sampling and interpretation difficulties are significant, many states feel that these tests are a useful guide in forage evaluation and provide forage testing laboratories.

Because energy value is of primary importance, several tests have been developed which either directly measure or correlate well with energy. One key group of energy-supplying compounds—the carbohydrates—is divided into two main classes: crude fiber and nitrogen-free extract (NFE). The crude fiber includes the relatively insoluble carbohydrates like cellulose, which may be only 35–75% digestible. The nitrogen-free extract includes the soluble portions of the carbohydrates (starches, sugars, and pentosans) as well as the organic acids (lactic and acetic). This fraction is highly digestible. Forage crops commonly contain 60–85% carbohydrate with the content being influenced primarily by the stage of maturity and the species composition of the herbage. A chemical analysis high in NFE and low in crude fiber is desired.

The maturation process affects the feeding value of forages more consistently than any other single factor. Immature, actively growing herbage has high feeding value. During maturation, increased concentrations of lignified fiber accumulate in the structural framework of forage plants, and forages may contain 3–20% lignin, depending upon the stage of maturity. Many researchers feel that the lignin content of a forage is the most reliable method of predicting forage digestibility. The inverse relationship between lignin content and forage quality is based on the assumption that lignin is indigestible. Lignin may limit digestibility more by its encrusting action around cell walls than by total content. Because of the difficulties of lignin analysis this method is not widely used in hay testing.

ANIMAL FEEDING TRIALS. While earlier research laid heavy emphasis on nutritive content, more recent work has shown that the amount of forage an animal will eat is an even more significant measure of forage quality. Therefore, feeding trials must be performed to give guidance on expected intake of various kinds of forage. Artificial rumen studies in which a laboratory culture of rumen microorganisms is used to simulate natural rumen have been useful in screening feed value but are not as yet a satisfactory laboratory technique for estimating intake. Feeding value is also significantly affected by fertilization practices and by the stage of plant maturity.

Harvest Date. The stage of plant development has been emphasized above as a major factor in influencing the feeding value of hay. The three primary factors to consider in deciding when to harvest hay are: (1) yield, both of total protein and total dry matter (Fig. 15.6), (2) quality as represented by a high percentage of protein and a low percentage of fiber, and (3) level of accumulated or stored carbohydrates in the storage organ (e.g. taproot, rhizome), because if these are overdepleted for perennial forage plants, stand longevity will be decreased.

BLOOM PERCENTAGE. Many harvest date recommendations have been based on phenological phenomena such as head or floral development. For example, legume harvest date has been recommended on the basis of percentage of bloom. This is generally accurate for the first harvest and probably remains the best guide available, but the blasting (flower destruction by insects or environmental conditions) of legume flowers, particularly in alfalfa, does cause some inaccuracy in this method. Harvest date for grasses has been based on the stage of head emergence or flowering. This evaluation is useful for the first harvest but may not be used for later harvests since the grasses most commonly head only once per season.

CALENDAR DATE. Calendar date has been proposed by J. T. Reid at Cornell University as a basis for determining when to cut hay. He concluded that the energy value reaches a peak at a given number of days after the crop starts growing in the spring. This value has been suggested to be rather constant at a given latitude. Using April 20 as

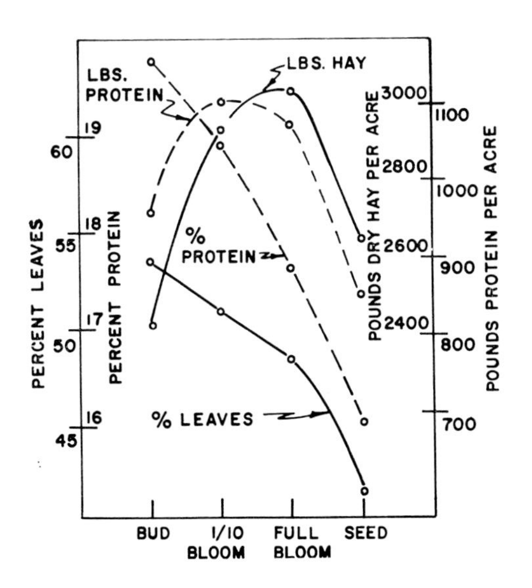

FIG. 15.6 ☙ Effect of stage of maturity on pounds of protein and hay and percent of protein and of leaves in alfalfa hay (after Hughes et al., 1962).

a base point, the first harvest is suggested for 55–60 days later or sometime in early June. Reid's equation and analysis include a wide array of forages. Some questions have been raised as to whether this idea is applicable to evaluating a particular species, but it does apply broadly to forages as a group. It may be that the straight-line regression developed by Reid is not as useful in working with pure stands, because alfalfa retains its digestible energy longer and grass in contrast drops away more rapidly.

STAND LONGEVITY. The harvest date must be selected to allow sufficient carbohydrate to accumulate for adequate stand longevity. The sensitivity of alfalfa to grazing or cutting at certain seasons is documented by decades of research. While new varieties have different recovery requirements, fall replenishment of reserve is still necessary for satisfactory winter survival (see Chapter 12). Early research results supported the recommendation that alfalfa should reach full bloom before harvesting. While this gave maximum storage of reserve carbohydrates, stemmy forage of low nutritive value resulted. Modern management permits harvest earlier in the flowering cycle without injury. Improved varieties and liberal liming and fertilization allow harvesting at full bud stage in the spring, followed by aftermath harvests at about 1/10 bloom. This full bud-early bloom harvest schedule offers a satisfactory compromise between forage quality and long-term persistence and yield. Because of the vagaries of flower development, however, the alfalfa producer may find that a regular harvest at 4–5-week intervals during the season is an equally useful plan. Periods of drouth or other environmental stresses might modify this pattern, but it has merit in planning. An additional indicator which has found particular favor in the western states is to cut when new growth appears from the crown buds. This method has been generally unsatisfactory in the Midwest.

Alfalfa appears to withstand intensive defoliation very well in midseason. In contrast bromegrass and timothy, major hay grasses of the North considered in earlier years to be quite insensitive to harvest management, may be seriously damaged by defoliation in the preheading stage. The injury results from inadequate carbohydrate storage and poor tiller bud development at certain stages of growth. The first harvest of stands in which these grass species comprise a significant percentage is best done at early- to mid-head stage. Forage intake is reduced, but digestibility is still adequate at this time (65%). Second crop harvest should be delayed until vigorous grass tiller buds are noted near the soil surface. The ideal harvest period is less than two

TABLE 15.3. Effect of cutting date on yield of dry matter, protein, and TDN (*Crops and Soils,* 1965)

Treatment (cuttings)	Cutting Dates	Dry Matter (tons/acre)	Protein (lb/acre)	TDN (lb/acre)
3 early	June 1, July 15, Aug. 31	3.4	1,295	4,435
3 medium	June 14, July 26, Aug. 31	3.8	1,382	4,770
2 late	June 23, Aug. 15	3.6	1,105	4,125
2 very late	July 1, Aug. 31	3.8	1,059	4,160

weeks in length for a given forage species; therefore, different mixtures are probably needed to spread the harvest schedule.

Table 15.3 summarizes Minnesota data on the effect of cutting date on yield of nutrients and the feeding value of alfalfa. The three medium dates of harvest provide the largest yield of TDN, of protein, and of dry matter per acre. Two very late cuttings provide an equal dry matter yield, but both protein and TDN are strikingly diminished.

Harvest and Storage Methods. Once the proper stage of plant development has been selected for hay harvest, the major objective is to get the forage stored without damaging effects from the environment. Development of haymaking methods has laid major emphasis on reducing the time in the field after cutting and reducing the labor required. Figure 15.7 describes the moisture relationships for forage as it might be stored by various procedures. The values in this figure emphasize the higher moisture content of silage and haylage. This allows a shorter time in the field after cutting than for any of the hay storage methods. However, hay may still be of value in a particular feeding system, and the following considerations are pertinent in thinking about hay harvest.

DRYING CONSIDERATIONS. Hay is most economically dried by maximizing the use of environmental factors, particularly radiant energy and to some extent the wind and general aeration of the crop after cutting.

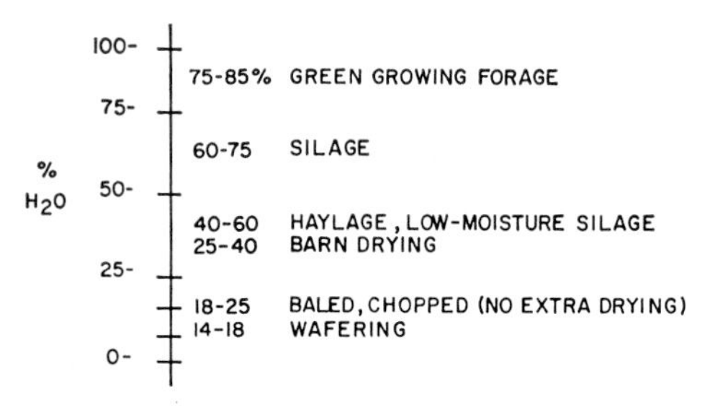

FIG. 15.7 Moisture percentage scale for harvest and storage of several forage handling systems.

Several procedures are helpful to achieve maximum utilization of the environment to speed the rate of drying. First, weather forecasts should be utilized as fully as possible. The recent passing of a cold front and a rising barometer are indicative of good drying weather. Once cut, the hay should be handled in ways to minimize time in the field. Probably the best single method of doing this at low cost is to use a stem conditioner. The stem crusher, crimper, and flail harvester bruise the stem and develop splits that speed the rate of moisture loss from the stem and also cause the stem to dry out at about the same time as the leaf. Such treatment cuts field drying time in half. Because radiant energy dries the foliage, maximum exposure in the full swath favors less field exposure time. The hay should be raked (if alfalfa is in the mixture) when the upper leaves of the alfalfa begin to shatter. Once windrowed, damp forage may be speeded in its drying by raking the windrow one half turn.

Where very high quality hay is sought or the drying climate is particularly undesirable, barn driers using forced heated or unheated air have been installed. These systems can satisfactorily handle hay after it is down to 40% moisture. Such heated or unheated forced air systems can reduce time in the field by half but do incur considerable expense and require the handling of very heavy, water-laden forage as it is initially brought from the field.

STORAGE REQUIREMENTS. The amount of space necessary to store hay handled in various forms is described in Table 15.4. The greater density material economizes on storage space but requires more expense to prepare for storage.

The harvest method has a distinct influence on the amount of nutrients preserved for feeding (Table 15.5) as it influences the weather damage incurred, but there is no appreciable difference in the feeding value between field-cured or barn-dried hay and hay crop silage when they are harvested at the same maturity with no weather damage.

CHEMICAL ADDITIVES. To date, no chemical additives have been found that preserve hay when the moisture content is too high for normal

TABLE 15.4. Space requirement for storing forage (after Hughes et al., 1962)

Storage Form	Lb/Cu Ft	Cu Ft/Ton
Loose-shallow mow	4	512
Baled loose	10	200
Chopped, short cut	12	167
Wafer	25–30	60–80

TABLE 15.5. Comparison of field and storage losses and % forage preserved for feeding with various harvest methods (after Hughes et al., 1962)

Methods of Harvest	Field Loss	Storage Loss	Preserved
	(percent)	*(percent)*	*(percent)*
Field cured—rain damage	32.6	4.0	63.4
Field cured—no rain	17.4	3.6	79.0
Barn dried—no heat	12.6	6.4	81.0
Barn dried—heat	13.4	1.8	84.8
Wilted silage	5.8	11.0	83.2
Dehydrated	5.6	4.1	90.3

storage. Salt and other materials have been used as "fire-conditioners" and desiccants, but none has been shown to be beneficial in preventing fire or maintaining hay quality.

WAFERING. Wafering machines have been of interest in reducing the labor necessary to handle hay, and the procedure may have merit in supplying a market (e.g. dairy) which is removed from the area of forage production. One major advantage to the livestock feeder is that wafers improve feeding efficiency by increasing the TDN intake per animal. Cubes approximately 2 x 2 x 2 inches are developed under pressure and can be handled mechanically. They also require much less storage space. One major drawback of the wafering process is that the forage must be very dry (14–18% moisture) before it can be handled by the wafering machine. To get the material this dry calls for an undesirable length of field exposure time. In addition, the forage wafers must be redried for satisfactory storage after they have been through the machine because water is added during the wafering process.

Dehydration. Forage crops, particularly alfalfa and Bermudagrass, are dehydrated to obtain a feed supplement high in protein and carotene (vitamin A precursor). The forage is cut directly and contains 75% water. It is dried in large revolving drums at rather high temperatures. By careful planning, the average length of time between cutting the hay and feeding it into the dehydrator is less than one hour, thus minimizing carotene loss.

The operator of an alfalfa dehydration unit seeks to develop a product containing at least 17% protein and high in carotene. In addition dehydrated alfalfa is an important source of vitamins and minerals.

Dehydrated alfalfa meal will suffer a loss of carotene when stored in air, but storage in large steel tanks under inert gases (most commonly N_2 and CO_2) has been used to counteract this undesirable carotene oxidation. With the inert gas system, very little loss of carotene takes place even under prolonged storage if the moisture is held below 7%. Chemical reagents (e.g. diphenylamine, quinolines) have been successfully used as antioxidants in bagged alfalfa meal. The antioxidant is dissolved or suspended in a suitable vegetable oil and added to the dehydrated material as it enters the hammer mill for grinding. The vegetable oil has also been valuable in reducing the dustiness of the dehydrated meal. Pelleting is now commonly done for ease of handling, and it also contributes to a reduction of dustiness in the meal.

When alfalfa is to be used as a dehydration crop in the north central states, the variety selected should have somewhat greater winterhardiness than the average for other forage uses, because the dehydrator operator imposes a rigorous cutting schedule on the field in the effort to obtain high protein content.

ENSILING

Preservation of plant material as ensilage is accomplished by an acid storage environment. The acid may be added directly as inorganic acid, formed indirectly from inorganic substrates, or formed from organic substrates by fermentation. Each of these reactions has merit under certain agricultural systems. Fermentation is used most in the United States.

INORGANIC ACID

The work of A. I. Virtanen in the 1920s laid heavy emphasis on low pH which he accomplished by the direct addition of inorganic acid. His work was highly regarded and merited a Nobel prize in 1924. The process consisted of adding 6 liters of H_2SO_4 or some other single or mixed strong inorganic acids to a metric ton of fresh material. Such a drastic procedure was necessary because the material to be ensiled was long (unchopped plant material) and stored in shallow silos. This meant that compaction and airtight conditions achieved by fine chopping and tall silos did not occur. For the northern sector of Europe where hay drying days were few in number and dairying was an important enterprise, this procedure made a great contribution. The AIV process is not useful in the United States because the acid

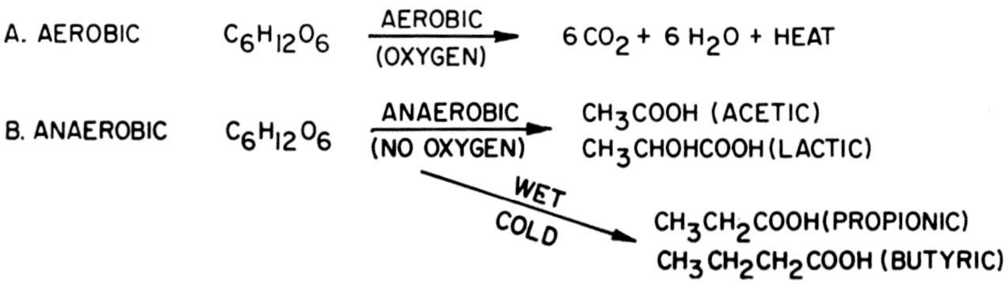

A. AEROBIC $C_6H_{12}O_6 \xrightarrow[\text{(OXYGEN)}]{\text{AEROBIC}} 6\,CO_2 + 6\,H_2O + \text{HEAT}$

B. ANAEROBIC $C_6H_{12}O_6 \xrightarrow[\text{(NO OXYGEN)}]{\text{ANAEROBIC}}$ CH_3COOH (ACETIC), $CH_3CHOHCOOH$ (LACTIC)

WET, COLD → CH_3CH_2COOH (PROPIONIC), $CH_3CH_2CH_2COOH$ (BUTYRIC)

FIG. 15.8 Products of aerobic and anaerobic processes which occur during ensiling.

used is too corrosive on the handling equipment. On occasion there has been interest in using organic acids, but for the most part they are too expensive.

FERMENTATION

The most useful procedure for silage making is fermentation of organic materials to acid. The following reaction set provides a framework in which to consider the several steps possible during fermentation (Fig. 15.8).

Sugar ($C_6H_{12}O_6$), when it is respired under conditions where oxygen is present, will be converted to energy and heat with carbon dioxide and water being given off. This in fact is an early step in the ensiling process, as will be described in detail later. However, if the forage to be ensiled becomes oxygen-free (anaerobic) and certain other requirements are met, acetic and lactic acid will be formed. A lactic fermentation is considered a very desirable one. Under wet, cold, partially aerobic conditions, the undesirable propionic and butyric acids are formed. A quality silage cannot be exactly described, but very often it would have: (1) pH less than 4.5, (2) a low ammonia content, (3) little or no butyric acid, and (4) a lactic acid content of 3–13%. To favor the formation of the desirable acids four conditions are needed: (1) an adequate sugar supply, (2) anaerobic conditions, (3) a water content of 40–75% depending on several factors, and (4) temperatures between 80 and 100° F.

Because bacteria are heavily involved in acid fermentation, there has commonly been interest in providing additional bacteria to forage going into the silo. This is not necessary and no benefit has been obtained from such inoculations.

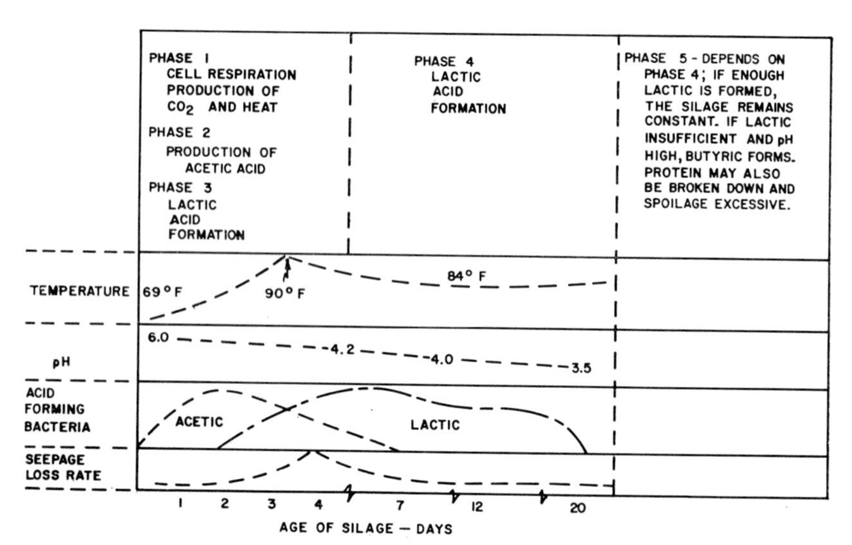

FIG. 15.9 ☙ Phases of silage fermentation.

During fermentation several phases may be identified, as described in Figure 15.9 and in the descriptions below:

PHASE 1: The living cells continue to respire, using oxygen and some of the simple carbohydrates. This activity consumes the residual oxygen captured in the mass and allows the silage to become anaerobic in four to five hours. Water is released both by respiration and compaction or settling. It is lost as runoff (effluent) and will carry from 0 to 20% of the dry matter with it. Phase 1 may go on for a number of hours. Heat released in these aerobic processes can assist in raising the temperature of the mass. While some workers suggest this is desirable, others feel it is only an indicator of other changes occurring.

PHASE 2: Acetic acid is formed from coliform bacteria. The plant cells die and release their water more rapidly as cell membranes disintegrate. From this cell breakdown additional carbohydrate is also released to the bacteria. This phase may continue for three or four days and then the coliform bacteria diminish in activity.

PHASE 3: From the 3rd to 21st day lactic acid formation is the dominant activity. Lactic acid is formed by lactobacilli and streptococci organisms if adequate carbohydrate is present. The lactic formers are quite tolerant to acidity and continue to function until the pH drops into the 4–5 range.

PHASE 4: The lactic formation passes a peak after this eighteen-day period. If the pH has dropped low enough and other conditions are favorable the silage will stabilize for many months. However, if the pH is not low enough and the silage is cold, wet, and possibly somewhat aerobic, phase 5 may occur. In good silage there would be no phase 5.

PHASE 5: Butyric formers begin to function; they can use the lactic acid and other carbohydrates as energy sources. Not only do they use up more energy but they also create a very undesirable odor to the silage. In addition, proteolytic bacteria (which release amino acids) and putrefactive bacteria (which deaminate amino acids and release ammonia) are favored by the same conditions as are the butyric formers. Very frequently the ammonia thus released is a factor in the reduced palatability of this silage.

MANAGEMENT TO FAVOR DESIRABLE ACIDS

Cereal Crops. In the cereal crops the conditions necessary for a quality fermentation occur rather naturally. As the grain increases in dry matter in an annual, the rest of the plant begins to diminish in moisture. By the time the grain is in the dough stage (approaching physiologic maturity), a significant quantity of carbohydrate has accumulated and the desirable water content of 65–70% is approached. Thus the proper harvest condition is achieved automatically in these crops; they have the right moisture content, a high carbohydrate content, and excellent density. Nevertheless, the cereal crops can be harvested too early, and if so, will not provide the maximum feeding value (Table 15.6). Grain makes up a greater portion of total silage yield in the later harvest. This silage is lower in protein but has slightly more TDN and is more palatable.

Forage Crops. Included here is a range of species that historically have been used for hay. They include legumes and forage grasses and as a group have the common characteristic of not converting a major portion of their dry matter into seed. These species are very different in their initial characteristics when compared to the cereal crops and

TABLE 15.6. Effect of maturity of corn silage on chemical composition, dry matter, intake, ar apparent digestibility (Bryant et al., 1966)

Stage of Maturity	% Protein	% Nitrogen-Free Extract	% Fiber	% Dry Matter	Intake (lb/day)	TDI
Immature (milk)	9.1(57.4)*	52.5(68.4)	32.2(70.8)	21.8	0.2	67.
Mature	8.1(53.9)	62.6(75.0)	22.9(63.2)	32.0	2.0	70.

* Parenthesized figures indicate digestion coefficients of the protein, nitrogen-free extrac and fiber.

therefore require an entirely different management scheme if they are to be used as silage. At the time their feeding value is highest, they are high in water content, rather low in carbohydrate, and high in protein. To bring the moisture content into the desirable range it is necessary to wilt these crops in the field. When cut they may contain as much as 80% moisture, and effluent losses will commonly be 15–20% if this material is placed in the silo directly, but wilting to 68–70% moisture can reduce effluent losses to nearly zero. The wilting process may require from one to several hours. Then the forage should be cut fine (¼–⅜ inch) and put into the silo rapidly to favor compaction. If the silo walls are tight, these steps will provide a suitable fermentation without additives or special storage structures.

SILAGE ADDITIVES

There has been interest through the years in adding a number of materials to silage to improve the fermentation process, to save more nutrients, or to improve the nutrient value or palatability of the silage. These materials have been referred to as additives which are discussed here under two subgroupings, conditioners and preservatives.

Conditioners may be defined as materials which absorb water and/or provide carbohydrate. Two examples of conditioners are molasses and ground corn. Molasses can provide an additional sugar source and a pleasing scent. If the silage is dry enough to minimize seepage, most of the molasses feeding value is retained, but if the silage is wet, considerable feeding value is lost. Ground corn or other grains will absorb moisture and provide more carbohydrate for fermentation, but if the silage is wet, the amount of dry matter saved in the silage will not be much more than that lost from the grain. In other words, there will be no net gain in dry matter from the use of such an additive.

Preservatives serve to form mild inorganic acids. For example sulfur dioxide may be added in a gaseous form which combines with water to provide sulfurous acid (H_2SO_3). Because SO_2 must be handled as a gas under pressure it is inconvenient to use. Sodium metabisulfite ($NA_2S_2O_5$) is a dry granule which releases SO_2 in the presence of water and plant enzymes. It is easier to handle than gaseous SO_2 and has been used frequently as a means of improving silage odor.

Limestone has been added to neutralize some of the acids as they are formed and thus allow the lactic acid bacteria to work longer and to produce more acids. There is no research evidence to indicate that this additive increases feeding value.

Mold and bacteria inhibitors (such as salt, bacitracin, and sodium propionate) have been suggested as additives, but none of these has shown a beneficial result in reducing undesirable microorganism activity. In addition, supplying bacterial or yeast cultures and enzymes has been tried with no improvement in the fermentation process.

To summarize, none of the many conditioners or additives proposed has been shown to enhance the amount of nutrients saved or the feeding value of the material. Properly wilted, finely cut material will ferment satisfactorily with none of these materials added (Olson et al., 1966).

TYPES OF SILOS

The type of silo to be used for storage may range from an inexpensive horizontal structure of the bunker or trench type to an upright conventional concrete stave or gastight type. The horizontal structures may be built for short-term use and loaded and unloaded with mobile farm equipment, including wagons and front-end tractor-mounted loaders. The upright types are much more permanent and require blowers and, desirably, silo unloaders in their use. They are more expensive and a long-term investment, but they do provide optimal conditions for compaction and fermentation and expose a minimal surface to air, thus reducing losses.

Table 15.7 compares the several types of losses for different silo structures. The variable moisture contents represent the moisture at which experience and experimentation have shown it is best to store forage in the particular structure. It is important to note that over a wide range of moistures the amount of dry matter per unit volume of silo changes very little. The manufacturers of gastight silos commonly have recommended the storage of lower-moisture material and indicate this increases animal intake. Experiment station feeding trials do not support this idea. To consider the question and compare the several

TABLE 15.7. Estimate of minimum dry matter losses (in percent) in forage stored as silage at different moisture levels (Hughes et al., 1962, Table 51.4)

Kind of Silo & H_2O% of Forage as Stored	Surface Spoilage	Loss Gaseous	Seepage	Total Silo Losses	Field Losses	From Cutting to Feeding
Conventional tower						
80	3	9	7	19	2	21
70	4	7	1	12	2	14
65	4	8	0	12	4	16
Gastight tower						
70	0	7	1	8	2	10
60	0	5	0	5	6	11
50	0	4	0	4	10	14
40	0	4	0	4	13	17
Trench or bunker						
80	6	10	7	23	2	25
70	10	10	1	21	2	23
Stack						
80	12	11	7	30	2	32
70	20	12	1	33	2	35

structures, the moisture contents most commonly suggested for each one have been used in the discussion which follows.

The horizontal structures have the greatest surface spoilage, although this can be reduced by plastic covers. Gaseous loss represents a prolonged aerobic period and more fermentation to undesirable products like butyric acid. Seepage is strongly related to the moisture content at which the material is placed in the silo and therefore indicates why it is so important to wilt forage material before placing it in the silo. The column of "Total Silo Losses" in Table 15.7 compares the sum of the above three losses. However, to get the material to the suggested storage moisture percentages indicated in the left column of the table, it must be dried in the field. This means field losses will increase. Therefore, the most important column for comparison is the one entitled "From Cutting to Feeding." The underlined values for the conventional tower and the gastight tower at their suggested moisture contents for harvest show no difference in the total losses of dry matter from cutting to feeding. Therefore, no apparent agronomic advantages occur between total losses from harvest and storage in these two structures, and a decision to select one over the other must be based on frequency of filling, unloading convenience, or other factors rather than a reduction in dry matter losses.

One of the difficulties involved with ensiling crops is related to gases which form during fermentation. Silo gases may be formed from plant material containing high levels of nitrate and have been report-

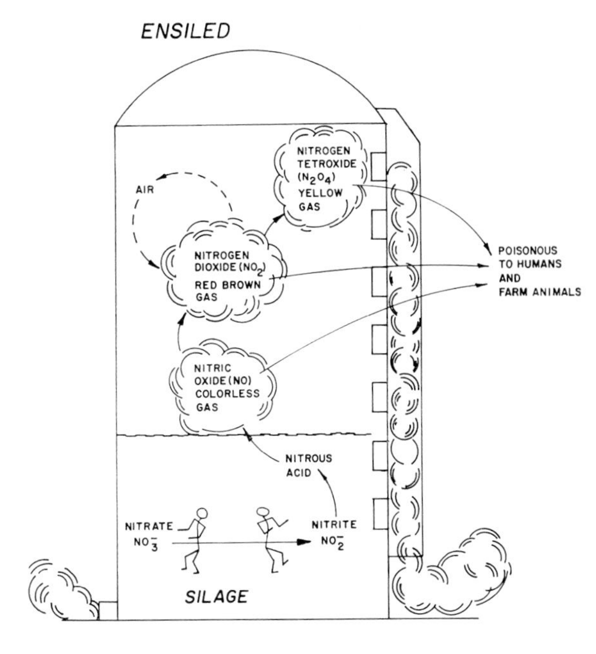

FIG. 15.10 ☙ Silo gases. Nitrogen oxide gases poisonous to humans and animals are formed when nitrate is present in the ensiled material (Crawford and Kennedy, 1960).

ed frequently both on the farm and in experiment stations. The formation of the oxides is described in the sequence of reactions shown in Figure 15.10. These gases may be formed after a few hours in the silo and, because they are heavier than air, may remain in the silo and collect at the surface of the silage until removed by forced ventilation or prolonged aeration.

All of the oxides (nitrogen oxide, dioxide, and tetraoxide) are poisonous, and precautions in removing this gas from the silo are important both during filling and as feeding is begun. It is important to stay out of the silo and its vicinity the two weeks following filling. If the silo must be entered, it should be ventilated with forced air first. The silo should be ventilated again for a two-week period before feeding. Since it is gaseous in nature, the material is quickly removed by these ventilation precautions and does not cause a residual problem.

GRAZING

PASTURE MANAGEMENT

To discuss the third harvesting system—direct harvest by the grazing animal—it is useful to consider the types of pastures to be harvested and the general goals for a pasture. Three general classifications of pastures are commonly used: (1) temporary or emergency, (2) rotation, and (3) permanent.

Temporary Pasture. The temporary pasture is used for a short time, usually for one grazing season or less. Cool-season temporary needs are commonly met with oats or rye. Sudangrass has been the outstanding warm-season temporary pasture species for many years and continues to maintain that position. Sorghum-sudangrass hybrids (Chapter 10) may outyield sudan under a green chop system but not under grazing.

Two primary characteristics of these temporary pasture species determine that rotational grazing (see below) is the best means of harvest. First, they are all upright in growth, and continued growth will come only from remaining leaf surface. This means they should be grazed off in a relatively short period and then the livestock removed to allow recovery. The erect growth does not provide sufficient leaves below the grazing level for such species to recover while defoliation continues. Secondly, the young cereals may contain high levels of nitrate and the young sudangrass shoots are high in prussic acid. This means it is important for this tissue to become older and thereby dilute the concentration of nitrate or prussic acid with carbohydrate or other nontoxic plant constitutents. This requirement has most commonly been met by recommending that annual species not be grazed until they are at least 1½–2 feet tall. For the sorghum-sudan hybrids, which are higher in prussic acid than sudangrass, it is recommended that grazing be delayed until the plants are 2½ feet tall.

Rotation Pasture. Rotation pasture is seeded for use over several years but is rotated with row crops on high-productivity land. The rotation pasture species are selected for their high productivity potential under optimal levels of management. This usually means the use of tall-growing grasses and legumes such as smooth bromegrass and alfalfa. These species are well suited to a system of management analogous to hay production, that is, a period of undisturbed growth for several weeks and then defoliation over a short period. Following this defoliation another period of undisturbed growth is desirable for recovery. The rotation or strip grazing scheme described in the next section is useful for this purpose.

Permanent Pasture. Permanent pasture species can withstand continuous close grazing as contrasted to the rotation pasture species which require periods of recovery. While the permanent pasture species do have some accumulated carbohydrates, their tolerance to close grazing is primarily because they maintain some leaf surface below the bite level of the grazing animal. In a mixed stand of tall and short species most of the permanent pasture species are actually favored by regular defoliation. If harvest is not frequent, the taller plants will shade out the permanent pasture type plants.

GRAZING SYSTEMS

Continuous Grazing. Continuous grazing is essentially one of leaving the livestock on the pasture throughout the growing season. It has commonly been shown to require the least labor and the least dollar investment per acre. In addition, animal performance is as good under this system, if the pasture is not overstocked, as for any of the systems discussed. However, there are distinct disadvantages to continuous grazing. Since the livestock have ranged the entire area, there is not an opportunity to harvest the surplus that commonly occurs, especially early in the growing season. In addition, there is very often both overgrazing and undergrazing in the same field. The livestock tend to return to the new growth that is succulent and palatable and leave more mature material. Additionally there is a high waste level because of tramping and fouling of the herbage. As a general rule, one-third of the yield is lost because of these problems with continuous uncontrolled field grazing. Thus production per animal may be the highest of any of the systems, but production per land unit is low.

Rotation Grazing. The rotational or paddock system is one in which the field is divided into several subunits. This may be a simple subdivision by electric fencing into as few as three or four units or may be extended to elaborate systems in which a large number of paddocks are permanently fenced. For upright growing species this system increases carrying capacity 10–25% over continuous grazing, and there is an opportunity to harvest any surplus in ungrazed paddocks during the seasons of flush growth. While this system does provide more control of animal intake, it does not improve animal performance over continuous grazing. The rotation program requires more investment in terms of labor and fencing. In addition, there is the challenge to the manager to regulate the number of livestock units and the length of the grazing period on a particular field. This can be learned but requires even greater care and experience than the continuous grazing system.

Ration Grazing. Ration grazing, also known as strip grazing or close folding, represents the most intensive grazing system. The idea is to provide one day's ration for the herd and then move them to a fresh supply of forage the next day. This system may provide a 15–40% yield increase over the rotational system and provides an optimal condition for the upright growing pasture plants in terms of recovery after defoliation. However, judgment on the quantity of forage needed is a challenge and calls for considerable experience. More labor and capital are involved, and very often one problem in developing the system is to provide a readily available water supply.

Soilage and Stored Feeding. Although soilage and stored feeding are not grazing systems, they are considered as alternative ways of harvesting forage. Soilage, green chop, or zero grazing is a mechanical harvest process. Feed may be cut once or twice a day and hauled to the herd. The system results in almost no bloat because the animal must consume a mixture of stem and leaf simultaneously. Animal performance is not improved, but carrying capacity per acre is increased over any of the direct grazing systems. Stored feeding, i.e. feeding animals silage or hay in dry lot, provides still another increase per acre in carrying capacity. It is favored by some livestock producers because they feel the more uniform supply of feed of comparable feeding value justifies the increased expense.

Hoglund (1960) suggested the following relationships of carrying capacity under these systems.

Rotational–100 acres.
Ration–84 acres.
Soilage–67 acres.
Stored feed–64 acres.

The major questions in the more intensive systems relate to the extra labor for machine operations and storage structures. Harshbarger et al. (1965) make some useful comparisons of these systems, and they conclude grazing is still the best harvesting method during the growing season unless land is limited and capital is plentiful.

PASTURE CALENDAR

The manager of any grazing herd seeks to maximize the months of the year during which his livestock may be on pasture. Because the dairy farmer seeks high output per cow, the forage available to his livestock must be of high quality and available in ample supply. This means he must fit together the preceding alternatives to provide the steadiest possible production and when these alternatives decline, to have stored feed available.

For the beef cow herd, and in some cases for fattening cattle during a portion of their feeding cycle, a somewhat wider range of alternatives is available. In addition to the standard pasture possibilities, other fields may be allowed to grow up and be left undisturbed for harvest during a midsummer period or during the late fall-early winter period. This practice, stockpiling or deferred grazing, provides an excellent opportunity to extend the pasture calendar and minimize the months when stored feed must be used. Stockpiling has been most successful with tall-growing species like sudangrass, reed canarygrass, or tall fescue. These species may be grazed early in the season, then allowed to recover and not be used again until other growth has stopped, perhaps in late October, November, and December.

BLOAT

With all the systems of forage harvesting by grazing livestock, the problem of bloat may occur. Bloat is induced by most legumes. However, two legumes (birdsfoot trefoil and crown vetch) and all the grasses do not cause bloat. Bloat results in a foamy, gaseous formation in the animal's rumen. The gas may put so much pressure on the animal's respiratory and circulatory system that death results. Bloat appears to occur most frequently when the animal has free choice and selects the youngest, most succulent tissue. The more intensive grazing systems reduce this tendency, and the mechanical harvesting systems reduce it even more by mixing old and young leaf and stem tissue together. The animal gets a less succulent diet and less bloat results. ☙

LITERATURE CITED

Addicott, F. T., and R. S. Lynch. 1957. Defoliation and desiccation: Harvest aid practices. *Advan. Agron.* 9:68–93.

Bovey, R. W., and M. K. McCarty. 1965. Effect of preharvest desiccation on grain sorghum. *Crop Sci.* 5:523–26.

Brooker, D. B., and J. D. McQuigg. 1963. Weather analysis for crop drying. *Mo. Agr. Exp. Sta., Res. Bull.* 837.

Bruinsma, J. 1962. Chemical control of crop growth and development. *Neth. J. Agr. Sci.* 10:409–26.

Bryant, H. T., R. E. Blaser, R. C. Hammes, Jr., and J. T. Huber. 1966. Evaluation of corn silage harvested at two stages of maturity. *Agron. J.* 58:253–55.

Burrowbridge, D. R., and P. H. Hoepner. 1965. An economic analysis of corn harvesting and handling systems. *Va. Agr. Exp. Sta., Bull.* 561.

Christensen, C. M., and H. H. Kaufmann. 1968. Maintenance of quality in stored grains and seeds. *Univ. of Minn. Ext. Folder* 226.

Crawford, R. F., and W. K. Kennedy. 1960. Nitrates in forage crops and silage: Benefits, hazards, precautions. *Cornell Misc. Pub.* 37. Ithaca, N.Y.

Crops and Soils. 1965. Forage quality—What you need to know. 17(5):7–17.

Culp, Thomas W. 1964. Chemical desiccation of castorbeans in the Southeast. *Agron. J.* 56(2):226–28.

Dodds, M. E., and W. L. Pelton. 1967. Effect of weather on the kernel moisture of a standing crop of wheat. *Agron. J.* 59:181–84.

Dotzenko, A. D., N. E. Humburg, G. O. Hize, and W. H. Leonard. 1965. Effects of state of maturity on the composition of various sorghum silages. *Colo. Agr. Exp. Sta., Tech. Bull.* 87.

Dua, P. N., E. J. Day, and C. O. Grogan. 1965. Loss of carotenoids in stored commercial and high carotenoid yellow corn, *Zea mays* L. *Agron. J.* 57:501–2.

Duncan, E. R., and H. E. Thompson. 1962. Calendarized row-crop production. *Iowa Farm Science* 16:175–77.

Gunn, R. B., and R. Christensen. 1965. Maturity relationships among early to late hybrids of corn (*Zea mays* L.). *Crop Sci.* 5:299–302.

Hallauer, A. R., and W. A. Russell. 1962. Estimates of maturity and its inheritance in maize. *Crop Sci.* 2:289–94.

Hammes, R. C., Jr., R. E. Blaser, H. T. Bryant, J. P. Fontenot, and R. W. Engel. 1966. The value of alfalfa-orchardgrass silage and hay cut at different maturities. *Va. Agr. Sta., Bull.* 567.

Hanway, J. J. 1963. Growth stages of corn (*Zea mays* L.). *Agron. J.* 55:487–92.

Harshbarger, K. E., E. E. Ormiston, J. R. Staubus, and R. V. Johnson. 1965. A nutritional assessment of methods of harvesting summer forage for dairy cows. *Ill. Agr. Exp. Sta., Bull.* 709.

Hillson, M. T., and L. H. Penny. 1965. Dry matter accumulation and moisture loss during maturation of corn grain. *Agron. J.* 57:150–53.

Hoglund, C. R. 1965. Some economic considerations in selecting storage systems for haylage and silage for dairy farms. *Mich. State Agr. Econ. Rept.* 14.

Hughes, H. D., M. E. Heath, and D. S. Metcalfe. 1962. *Forages* (Chaps. 49, 50, 52). Iowa State University Press, Ames.

Koenig, R. F., D. W. Robertson, and A. D. Dickson. 1965. Effect of time of swathing on malting barley (*Hordeum distichum* L.) quality. *Crop Sci.* 5:159–61.

McMeans, J. L., V. T. Walhood, and L. M. Carter. 1966. Effects of green-pick, defoliation, and desiccation practices on quality and yield of cotton, *Gossypium hirsutum*. *Agron. J.* 58:91–94.

Miles, S. R., and E. E. Remmenga. 1953. Relations of kernel, cob and ear moisture in dent corn. *Purdue Agr. Exp. Sta., Bull.* 599.

Miller, T. B. 1961. Recent advances in studies of the chemical composition and digestibility of herbage. Part I. *Herbage Abstr.* 31:81–85. Part II. *Herbage Abstr.* 31:163–67.

Olson, O. E., L. B. Embry, and H. H. Voelker. 1966. Silage additives. *S. Dakota Agr. Exp. Sta., Circ.* 178.
Oxley, T. A. 1948. *The Scientific Principles of Grain Storage.* Northern Publishing Co., Ltd., Liverpool.
Phillips, W. M. 1957. Response of several grain sorghum hybrids and varieties to chemical desiccants. *Res. Rept. NCWCC* 14:153–54.
Pratt, A. 1965. High dry matter silage. *Ohio Report* Nov.-Dec., pp. 83, 87.
Quackenbush, F. W. 1963. Corn carotenoids: Effects of temperature and moisture on losses during storage. *Cereal Chem.* 40:266–69.
Reid, J. T., W. K. Kennedy, K. L. Turk, S. T. Slack, G. W. Trimberger, and R. P. Murphy. 1959. Effect of growth stage, chemical composition and physical properties upon the nutritive value of forages. *J. Dairy Sci.* 42:567–71.
Schmidt, J. L., and A. R. Hallauer. 1966. Estimating harvest date of corn in the field. *Crop Sci.* 6:227–31.
Shaw, R. H., and H. C. S. Thom. 1951. On the phenology of field corn, silking to maturity. *Agron. J.* 43:541–46.
Smith, Dale. 1962. *Forage Management in the North.* W. C. Brown Co., Dubuque, Iowa.
Smith, Dale. 1964. Chemical composition of herbage with advance in maturity of alfalfa, medium red clover, ladino clover, and birdsfoot trefoil. *Wis. Agr. Sta., Res. Rept.* 16.
Sprague, M. A., and B. B. Taylor. 1966. Preservation of green chopped alfalfa at near freezing temperatures. *Agron. J.* 58:179–81.
Sprague, M. A., and L. Leparulo. 1965. Losses during storage and digestibility of different crops as silage. *Agron. J.* 57:425–27.
Taylor, M., R. A. Pickett, G. W. Isaacs, and G. H. Foster. 1964. Effects of drying corn on its nutritive value for growing-finishing swine. *Purdue Agr. Exp. Sta., Res. Prog. Rept.* 148.
Thompson, R. A., and G. H. Foster. 1963. Stress cracks and breakage in artificially dried corn. *USDA Marketing Res. Rept.* 631.
USDA. 1960. Harvest-aid chemicals for cotton: Defoliants, desiccants and second-growth inhibitors. ARS 22-58.
USDA. 1962. Stored grain pests. *Farmer's Bull.* 1260.
Van Fossen, L., and E. G. Stoneberg. 1962. What costs for harvesting, drying and storing corn. *Iowa Farm Science* 17(2):3–5.

INDEX